Recent Titles in This Series

155 **O. A. Ladyzhenskaya and A. M. Vershik, Editors,** Proceedings of the St. Petersburg Mathematical Society, Volume I

154 **V. A. Artamonov, et al.,** Selected Papers in K-Theory

153 **S. G. Gindikin, Editor,** Singularity Theory and Some Problems of Functional Analysis

152 **H. Draškovičová, et al.,** Ordered Sets and Lattices II

151 **I. A. Aleksandrov, L. A. Bokut', and Yu. G. Reshetnyak, Editors,** Second Siberian Winter School "Algebra and Analysis"

150 **S. G. Gindikin, Editor,** Spectral Theory of Operators

149 **V. S. Afraĭmovich, et al.,** Thirteen Papers in Algebra, Functional Analysis, Topology, and Probability, Translated from the Russian

148 **A. D. Aleksandrov, O. V. Belegradek, L. A. Bokut', and Yu. L. Ershov, Editors,** First Siberian Winter School in Algebra and Analysis

147 **I. G. Bashmakova, et al.,** Nine Papers from the International Congress of Mathematicians 1986

146 **L. A. Aĭzenberg, et al.,** Fifteen Papers in Complex Analysis

145 **S. G. Dalalyan, et al.,** Eight Papers Translated from the Russian

144 **S. D. Berman, et al.,** Thirteen Papers Translated from the Russian

143 **V. A. Belonogov, et al.,** Eight Papers Translated from the Russian

142 **M. B. Abalovich, et al.,** Ten Papers Translated from the Russian

141 **Kh. Drashkovicheva, et al.,** Ordered Sets and Lattices

140 **V. I. Bernik, et al.,** Eleven Papers Translated from the Russian

139 **A. Ya. Aĭzenshtat, et al.,** Nineteen Papers on Algebraic Semigroups

138 **I. V. Kovalishina and V. P. Potapov,** Seven Papers Translated from the Russian

137 **V. I. Arnol'd, et al.,** Fourteen Papers Translated from the Russian

136 **L. A. Aksent'ev, et al.,** Fourteen Papers Translated from the Russian

135 **S. N. Artemov, et al.,** Six Papers in Logic

134 **A. Ya. Aĭzenshtat, et al.,** Fourteen Papers Translated from the Russian

133 **R. R. Suncheleev, et al.,** Thirteen Papers in Analysis

132 **I. G. Dmitriev, et al.,** Thirteen Papers in Algebra

131 **V. A. Zmorovich, et al.,** Ten Papers in Analysis

130 **M. M. Lavrent'ev, et al.,** One-dimensional Inverse Problems of Mathematical Physics

129 **S. Ya. Khavinson; translated by D. Khavinson,** Two Papers on Extremal Problems in Complex Analysis

128 **I. K. Zhuk, et al.,** Thirteen Papers in Algebra and Number Theory

127 **P. L. Shabalin, et al.,** Eleven Papers in Analysis

126 **S. A. Akhmedov, et al.,** Eleven Papers on Differential Equations

125 **D. V. Anosov, et al.,** Seven Papers in Applied Mathematics

124 **B. P. Allakhverdiev, et al.,** Fifteen Papers on Functional Analysis

123 **V. G. Maz'ya, et al.,** Elliptic Boundary Value Problems

122 **N. U. Arakelyan, et al.,** Ten Papers on Complex Analysis

121 **D. L. Johnson,** The Kourovka Notebook: Unsolved Problems in Group Theory

120 **M. G. Kreĭn and V. A. Jakubovič,** Four Papers on Ordinary Differential Equations

119 **V. A. Dem'janenko, et al.,** Twelve Papers in Algebra

118 **Ju. V. Egorov, et al.,** Sixteen Papers on Differential Equations

117 **S. V. Bočkarev, et al.,** Eight Lectures Delivered at the International Congress of Mathematicians in Helsinki, 1978

(Continued in the back of this publication)

Proceedings of the St. Petersburg Mathematical Society

Volume I

American Mathematical Society

TRANSLATIONS

Series 2 • Volume 155

Proceedings of the St. Petersburg Mathematical Society

Volume I

O. A. Ladyzhenskaya
A. M. Vershik
Editors

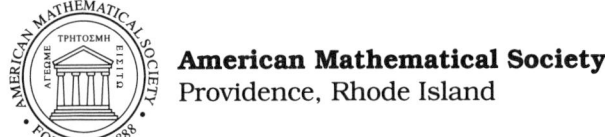

American Mathematical Society
Providence, Rhode Island

Translation edited by DAVID LOUVISH and SIMEON IVANOV

The translation, editing, and keyboarding of the material for this book was done in the framework of the joint project between the AMS and Tel-Aviv University, Israel.

1991 *Mathematics Subject Classification.* Primary 05A05, 22E40, 31A10, 31A35, 34A47, 41A10, 47A58, 47C15, 49J15, 58F12, 65Y20; Secondary 20D06, 35J55, 58F25.

ISBN 0-8218-7505-1
ISSN 0065-9290

COPYING AND REPRINTING. Individual readers of this publication, and nonprofit libraries acting for them, are permitted to make fair use of the material, such as to copy an article for use in teaching or research. Permission is granted to quote brief passages from this publication in reviews, provided the customary acknowledgment of the source is given.

Republication, systematic copying, or multiple reproduction of any material in this publication (including abstracts) is permitted only under license from the American Mathematical Society. Requests for such permission should be addressed to the Manager of Editorial Services, American Mathematical Society, P.O. Box 6248, Providence, Rhode Island 02940-6248.

The appearance of the code on the first page of an article in this book indicates the copyright owner's consent for copying beyond that permitted by Sections 107 or 108 of the U.S. Copyright Law, provided that the fee of $1.00 plus $.25 per page for each copy be paid directly to the Copyright Clearance Center, Inc., 27 Congress Street, Salem, Massachusetts 01970. This consent does not extend to other kinds of copying, such as copying for general distribution, for advertising or promotional purposes, for creating new collective works, or for resale.

Copyright ©1993 by the American Mathematical Society. All rights reserved.
Printed in the United States of America.
The American Mathematical Society retains all rights
except those granted to the United States Government.
The paper used in this book is acid-free and falls within the guidelines
established to ensure permanence and durability. ∞
This publication was typeset using $\mathcal{A}_{\mathcal{M}}\mathcal{S}$-TEX,
the American Mathematical Society's TEX macro system.
10 9 8 7 6 5 4 3 2 1   97 96 95 94 93

Contents

A. M. **Vershik**, The Leningrad Mathematical Society — ix

A. I. **Barvinok** and A. M. **Vershik**, Polynomial-time computable approximation of families of semialgebraic sets and combinatorial complexity — 1

Yu. N. **Bibikov**, Construction of invariant tori of systems of differential equations with a small parameter — 19

I. **Baesler** and I. K. **Daugavet**, Approximation of nonlinear operators by Volterra polynomials — 47

N. A. **Vavilov**, Subgroups of Chevalley groups containing a maximal torus — 59

Vladimir **Maz'ya** and Alexander **Solov'ev**, On the boundary integral equation of the Neumann problem in a domain with a peak — 101

A. V. **Megretskiĭ** and V. A. **Yakubovich**, Singular stationary nonhomogeneous linear-quadratic optimal control — 129

S. A. **Nazarov** and B. A. **Plamenevskiĭ**, The Neumann problem for selfadjoint elliptic systems in a domain with piecewise-smooth boundary — 169

S. Yu. **Pilyugin**, Limit sets of domains in flows — 207

On October 20, 1989, passed away the corresponding member of the Academy of Sciences of the USSR, president of the Leningrad Mathematical Society, senior professor of the Leningrad University, outstanding mathematician and teacher

Dmitriĭ Konstantinovich FADEEV.

As a member of the Board and later as president of the Society, Dmitriĭ Konstantinovich devoted much effort to the organization of mathematical journals in Leningrad, in particular an LMS journal. Unfortunately, he did not last to see the publication of the first volume of "Trudy of the LMS".

We hope to publish, in subsequent issues of "Trudy", materials on his notable contribution to mathematics and on the role of his work.

The Leningrad Mathematical Society

A. M. VERSHIK

The St. Petersburg Mathematical Society was founded in 1890. It was the third such in Russia—after the Moscow (1867) and Kharkov (1879) societies. Thus this year we can mark the first century of the Society. However, the presently acting Leningrad Mathematical Society, reconstituted thirty years ago (in September 1959), is the third in a row in our city; its on and off history is not accidental.

The constitution of the St. Petersburg Society was affirmed only in 1893, and by that time it had 98 members. It looks strange that in St. Petersburg, which was undoubtedly the main mathematical center of Russia in the nineteenth and the beginning of the twentieth centuries, a mathematical society was created that late and played, apparently, a nonessential role, being less known than its counterpart in Moscow, for example. However, one should keep in mind that the Academy of Sciences, and most of its members, resided in the then capital, and essentially, performed the functions of a mathematical society, such as conducting scientific meetings, evaluation of papers, awarding prizes, etc. Apparently, this explains why the Moscow Mathematical Society played a bigger role in the scientific life in Moscow than its counterpart in St. Petersburg.

The social tremors of 1917 and subsequent years destroyed many existing scientific institutions and structures. Although we do not have precise data, it seems that the Mathematical Society ceased to exist as well at that time. In 1919 Vladimir Andreevich Steklov was elected vice-president of the Russian Academy of Sciences and started his work on organizing the Society anew. He succeeded in 1921, when the Petrograd Physical-Mathematical Society was recreated. Nikolaĭ Maksimovich Gyunter was elected President of the Board.

Ya. V. Uspenskiĭ (who later emigrated to the USA), V. I. Smirnov, B. N. Delone, G. M. Fikhtengol′ts, and others played an active role in the work of the Society, V. A. Steklov, D. Hilbert, F. Klein, A. N. Krylov, Yu. V.

Sokhotskiĭ, O. D. Khvol′son, and others were among the honorary members. Regular sessions started being conducted in 1922 (a list of presentations can be found in volume 2 of the "Journal of the Leningrad Mathematical Society"; it is truly impressive). Lectures were given by V. A. Steklov, A. A. Fridman (Friedmann), V. A. Fok, Ya. V. Uspenskiĭ, A. S. Besikovich, S. N. Bernstein, V. I. Smirnov, Ya. D. Tamarkin, B. N. Delone, N. M. Gyunter, R. O. Kuz′min, N. I. Muskhelishvili, L. G. Loĭtsyanskiĭ, Yu. A. Krutkov, B. G. Galerkin, and many others. For the duration of the Society about 150 sessions took place. On April 1st, 1927 it had 102 members.

In 1926 V. A. Steklov finally succeeded in founding the "Journal of the Leningrad Physical-Mathematical Society", about which he had dreamt since 1910. Unfortunately, the scientist did not last to see the publication of the first volume,* which he personally edited and prepared for print.

The first issue of the "Journal" opens with the obituaries of Steklov and of the well-known mathematician and physicist A. A. Fridman, who died a little earlier. From 1926 to 1929 only four issues of the "Journal" were published (two volumes with two issues in each, once a year). They contained articles on mathematical physics, mechanics, number theory, topology, analysis, and geometry. Among the authors were N. M. Gyunter, B. N. Delone, I. M. Vinogradov, E. L. Nikolai, V. A. Fok, V. I. Smirnov, N. S. Koshlyakov, I. A. Lappo-Danilevskiĭ, A. A. Markov (Jr.), L. V. Kantorovich, and others. The editorial board was comprised of Ya. V. Uspenskiĭ (editor after the death of Steklov), N. M. Gyunter, B. N. Delone, G. M. Fikhtengol′ts, I. M. Vinogradov, V. I. Smirnov, A. F. Gavrilov, and K. V. Melikov. The "Journal" started its publication at a time when there were almost no mathematical journals in the country. The journal became widely known, although for a brief period, due to the high scientific level and the strong array of authors. At present it is all but forgotten.

Toward the end of the twenties the situation in the country was becoming more and more difficult for independently thinking people, scientists in particular; an open hunt was beginning on different thinking, real or invented. It did not pass by even such a science outside politics, one would think, as mathematics. Various societies fell from grace. The history of persecution, which was joined by semiliterate activists as well as by some serious mathematicians, needs special scrutiny. Objects of these persecutions in Moscow were, for example, the remarkable mathematician D. F. Egorov (president of the Moscow Mathematical Society, 1922–1931) who was sent to Kazan, where he died in a special hospital, and somewhat later his student N. N. Luzin (the prominent mathematician and founder of the modern Moscow school), and others. In Leningrad, very precarious was the position of the president of the Board N. M. Gyunter, whose independent character and courage had for a long time irritated those who "kept an eye" on the state of

---

* Steklov died on May 30, 1926.

affairs in the sciences. In this circumstance at a session of the Society, on the initiative of V. I. Smirnov (Gyunter's deputy on the Board) a resolution was adopted for self-dissolution of the Leningrad Mathematical-Physical Society. This unusual resolution preempted, possibly, the course of events usual for those times. In any case Gyunter continued his work and teaching. He died on May 4, 1941, in Leningrad.

Thus, in 1930 there occurred a second break in the history of mathematical societies in our city.

Starting with the mid-thirties, in connection with the relocation of the Academy of Sciences and its institutes to Moscow, many mathematicians, founders of scientific directions, moved to the capital. This process continued; in the fifties and sixties many leading mathematicians moved to Novosibirsk. Although this was a drain on Leningrad mathematics, nevertheless both the thirties and the post-war years were years of its intensive development, and the mathematical life of the city was not weakening. In 1953, on the initiative of Vladimir Ivanovich Smirnov, the Leningrad general mathematical seminar was organized; its sessions were held twice a month during the academic year, in the club of scientists. Smirnov was presiding. The organization of the seminar was accepted as a virtual creation of a Leningrad Mathematical Society (about the work of the seminar see Uspekhi Mat. Nauk [translated into English as Russian Mathematical Surveys]).

In the six years of its existence, about 150 reports of both survey and original character were heard, as well as informational reports on publication of mathematical literature and jubilee reports. Among the lecturers were A. D. Aleksandrov, A. A. Markov, S. M. Lozinskiĭ, L. V. Kantorovich, O. A. Ladyzhenskaya, P. S. Novikov, and many others. Members of the presidium of the seminar, in addition to Smirnov, were Yu. V. Linnik, B. A. Venkov, S. M. Lozinskiĭ, A. A. Markov, A. A. Ivanov, and M. F. Shirokhov.

During all that time attempts to organize the Leningrad Mathematical Society (LMS) never ceased, and, finally, after six years(!) they were crowned with success: on April 13, 1959, the Ministry of higher education of the USSR affirmed the constitution of the Society,[*] and on September 29, 1959, the constituent assembly of the LMS took place. A. D. Aleksandrov, the then rector of the Leningrad State University (LSU), rendered much help to Smirnov and others in the re-creation of the Society. The Leningrad Mathematical Society was created at the University rather than at the Academy of Sciences of the USSR and not as an independent society. Such a solution of the problem simplified the procedure of constituting the Society, and, maybe, there was no other possibility of creating it at that time. By its constitution the LMS is a voluntary scientific society whose purpose is to assist the development of the mathematical sciences. It may be considered the direct

---

[*] Presently the constitution of the Society is under review. The new wording will be published in the next volume of Trudy of the LMS.

successor of the Leningrad Mathematical-Physical Society (1921–1930) and thus of the St. Petersburg Mathematical Society.

At its founding the Society had 49 members—mainly all doctors in mathematics and in mechanics in Leningrad. V. I. Smirnov was elected honorary president, and Yu. V. Linnik was elected president. By 1963 LMS already had 92 members. The members were, as a rule, local mathematicians. However, there were also exceptions; for example, M. G. Kreĭn (1907–1989). By the Constitution, the members of the Society were required to have a scientific degree, but here also there were exceptions. The Constitution envisaged instituting honorary members. In the second half of the sixties the activity of the Society dropped sharply. This was due to various causes, one of which being that there was no intensive co-opting of new members; from 1963 to 1970 only six members were accepted. In May, 1970, the composition of the Board was renewed, and it was decided to organize a cycle of survey reports on contemporary problems in mathematics and to widely engage mathematicians from other cities, as well as to accept as members a large number of young mathematicians. By May, 1973, LMS had 123 members, in 1978 150, in 1984 209, and in 1985 224 members. On January 1, 1990, the Society had 241 members.

In all the years of its existence (since 1959) the Society had about 350 people as members. Besides the ones who passed away, some moved to other cities or emigrated. The current membership of the Society encompasses the majority of the active mathematicians in the city.

From 1962 on, practically annually, a prize for outstanding work has been awarded to a young mathematician (up to age 30). Since 1987 LMS has organized a competition of students' scientific articles.

The main work of the Society, as before, has been holding sessions dedicated to scientific problems. In the thirty years of the existence of the Society, about 350 reports on a wide variety of themes have been heard: approximately two-thirds of them delivered by Leningrad mathematicians, the rest by Muscovites and mathematicians from other cities and from abroad. An account on all sessions of the Society, with the exception of the fall semester 1969, is published in "Uspekhi" (Russian Math. Surveys). An analysis of this material for the period up to 1984 was presented by the president of the LMS, S. M. Lozinskiĭ, at the session devoted to the twenty-fifth anniversary of the Society (see Uspekhi, 1986, vol. 1, issue 1).

We do not intend, in this first volume, to go into a detailed analysis of the themes covered, a survey of most interesting reports, discussions, memorial and jubilee sessions for all these years, or into a description of all the types of activity of the Society. Possibly, this will be done later. We restrict ourselves to considering only several facts.

In 1976 a Mathematical seminar for students was created at LMS, where special popular lectures for a general audience are read. A list of these can also be found in "Uspekhi".

In 1980, with the relocation of the department of mathematics and mechanics of the LSU to Petergof, the question arose of where to hold the seminars. Before that they always took place twice a month, on Tuesdays, at the department. A solution was found in organizing a mathematical section at the club of scientists of the Academy of Sciences of the USSR, where it was decided to hold one of two sessions. On occasion the all-city mathematical seminar held its sessions there. The president of the section from its inception in October 1981 till 1985 was S. M. Lozinskiĭ, and since May 1985 was A. M. Vershik. The transfer of the sessions and organization of the section gave the work of the Society a more "open" character, by allowing the sessions to be attended not only by mathematicians, but also by specialists from other scientific sections.

In 1988, the educational council of the LMS was created. Among its tasks are work with teachers, organization of mathematical education in schools, conducting olympiads, etc.

Finally, it should be mentioned that for a long time the leadership and the members of the LMS have endeavored to organize a Leningrad mathematical journal; these efforts have been fruitless for a very long time (almost 30 years!). Only now the reader holds in his hands the first volume of "Trudy of the LMS".

On February 6, 1990, O. A. Ladyzhenskaya was elected president of the Leningrad Mathematical Society.

Polynomial-time Computable Approximation of Families of Semialgebraic Sets and Combinatorial Complexity

A. I. BARVINOK, A. M. VERSHIK

In this paper we propose a new method to estimate the complexity of computational and combinatorial problems based on inductive families of semialgebraic sets and special approximations of them. A major role is played here by a broad class of families of algebraic origin that include almost all discrete and continuous combinatorial problems. This class enables us to describe a quite general scheme, defined in §1, according to which any problem can be considered as a problem about whether a point belongs to an inductive family of semialgebraic sets over $\mathbb{R}$, while complexity is estimated with the help of a real Turing machine as defined in [1]. The important notion of a polynomial-time computable family enables us, in accordance with a standard scheme, to define what we call equinomial families (their projections) and the category of such families with polynomial-time computable morphisms. Finally, we define universal families and formulate a real analogue of the $P = NP$ problem. In §3 we introduce the basic notion of polynomial-time computable approximation (PCA) and present some basic examples. A PCA always exists, although the construction of an effective PCA is a central problem, still far from a general solution. The classical algorithms of linear programming theory are simplex algorithms, and the Ellipsoid Method [2] defines such an approximation. The present authors in [3]–[5] have introduced and studied the π-assignment problem, which is connected with representations of particular groups, mainly symmetric groups. This problem is studied from the point of view of approximations in §§3,4. Our main results give descriptions of several π-assignment problems for which a polynomial-time computable approximation exists. A powerful method for constructing approximations is the method of statistical sums, in which one replaces computation of a

---

1991 *Mathematics Subject Classification.* Primary 05A05, 65Y20.

maximum by the computation of sums (compare the spin glass method [13]). This question is taken up in §4.

The general program of the algebraic geometry treatment of combinatorial and related problems using representation theory was first suggested by Vershik [6], [7]. The recently introduced concept of a real Turing machine [1] fits in very well with our approach, but the only thing we borrow from [1] is the definition of polynomiality—in all other respects our treatment is different from that of [1]. Other aspects of the discussed program will be considered elsewhere. The general goal of the program is to extend the use of "continuous" mathematics (in particular, singularity theory) and algebraic methods (representation theory) to the analysis of combinatorial and optimization problems. A relationship is thus created between combinatorial and real complexity and the complexity of singularities or representations of certain objects.

§1. Basic notions

1. Semialgebraic sets.

DEFINITION. An *elementary semialgebraic set* is the set of real solutions of a system of polynomial equalities and inequalities $p_i(x) = 0$, $i \in I$; $p_j(x) > 0$, $j \in J$, where $x = (x_1, \ldots, x_n) \in \mathbb{R}^n$; I, J are disjoint finite sets. A *semialgebraic set* is a union of finitely many elementary semialgebraic sets. A *semialgebraic map* is a map $\varphi : M \to N$ of semialgebraic sets whose graph is a semialgebraic set. A *semirational* map is a map P/Q, where P, Q are semialgebraic.

The Seidenberg-Tarski Theorem (see, e.g., [8]) states that if M is a semialgebraic set in $\mathbb{R}^n$ and φ is a map whose coordinate functions are polynomials, then $\varphi(M)$ is a semialgebraic set, or, in short, the projection of a semialgebraic set is semialgebraic. A detailed account of the theory of semialgebraic sets and a review of the literature can be found in [8]. In what follows we will denote finite-dimensional affine spaces over the field of real numbers $\mathbb{R}$ by $\mathbb{A}$, possibly with an index (which need not indicate the dimension).

2. The family of semialgebraic sets.

Let I be a partially ordered set; this will usually be a set $\mathbb{Z}_+ \oplus \cdots \oplus \mathbb{Z}_+ = \mathbb{Z}_+^k$ of ordered k-tuples of nonnegative integers. The following order is defined on I:

$$(i_1, i_2, \ldots, i_k) \leq (j_1, j_2, \ldots, j_k) \Leftrightarrow \forall s\ i_s \leq j_s.$$

We will consider a family of semialgebraic sets $M_i \subset \mathbb{A}_i$, $i \in I$. Such families will be denoted by Gothic letters: $\mathfrak{M} = \{M_i\}$, $i \in I$ or, if necessary, the index set will be specified: $(\mathfrak{M}, I)$. A *semialgebraic (semirational) map* of two families $(\mathfrak{M}, I)$, $(\mathfrak{N}, J)$ is a map $\chi : I \to J$, together with a family of semialgebraic (semirational) maps $\psi_i : \mathbb{A}_i \to \mathbb{A}_{\chi(i)}$, such that $\psi_i(M_i) \subset N_{\chi(i)}$.

DEFINITION. An *inductive family* (IF) of semialgebraic sets is a family $(\mathfrak{M}, I): M_i \subset \mathbb{A}_i$ and semialgebraic (semirational) maps $\varphi_{ij} : \mathbb{A}_i \to \mathbb{A}_j$ such that $\varphi_{jk}\circ\varphi_{ij} = \varphi_{ik}$ and $\varphi_{ij}(M_i) \subset M_j$, $i, j \in I$. Maps of inductive families are defined in a natural way.

The inductive limit (IL) of a family, defined in the usual way, will be denoted by $\underrightarrow{\lim}(\mathfrak{M}, I) = \mathbb{M}$. In the majority of cases $\mathbb{M}$ is not just an abstract limit: it can be constructed as a subset of the infinite-dimensional space $\mathbb{A}^\infty$ (see §2).

3. The MEMBERSHIP problem and polynomial-time computable families.
Let $\mathfrak{M} = \{M_i\}$; $M_i \subset \mathbb{A}_i$, $i \in I$, be a family of semialgebraic sets. We will study the following problem (MEMBERSHIP $\mathfrak{M}$):

Let $x = (x_1, \ldots, x_{n_i}) \in \mathbb{A}_i$ be a point with given coordinates. Is it true that $x \in M_i$?

We will be interested in the complexity of the solution to this problem.

It is clear from the definition of a semialgebraic set that the above problem can be solved in finitely many arithmetical operations on the coordinates of x. We will define the complexity of an algorithm that works with real numbers as in [1], where the notion of a real Turing machine capable of performing the operations of addition, subtraction, multiplication, and division of real numbers and comparison with zero was introduced. The input of the MEMBERSHIP problem is the set of coordinates $(x_1, \ldots, x_{n_i})$ of a point $x \in \mathbb{A}_i$ (and possibly also the index $i \in I$); the output is "yes" if $x \in M_i$, and "no" if $x \notin M_i$. By the complexity of an algorithm we will mean, roughly speaking, the number of operations performed [1].

We give two different definitions of polynomial-time computability. The first, from [1], is stronger.

DEFINITION. A *uniformly polynomial-time computable family* (UPCF) is a family of semialgebraic sets $(\mathfrak{M}, I) : I = \mathbb{Z}_+^k$ for which there is a polynomial $t(i_1, \ldots, i_k)$ such that

(1) $\dim \mathbb{A}_i \leq t(i)$, $i \in I$.
(2) There exists a Turing machine which, upon receiving input (i, x), $i \in I$, $x \in \mathbb{A}_i$, solves the MEMBERSHIP problem for M_i in time at most $t(i)$.

If we replace condition (2) by

(2′) There exists a Turing machine which, upon receiving input (i, x), $i \in I$, $x \in \mathbb{A}_i$, solves the MEMBERSHIP problem for M_i, and for any $i \in I$ there exists a Turing machine which, upon receiving input $x \in \mathbb{A}_i$, solves the MEMBERSHIP problem for M_i in time at most $t(i)$,

then we obtain the definition of a *polynomial-time computable family* (PCF).

Similarly, a *polynomial-time computable map* of two families $(\mathfrak{M}, I)$ and $(\mathfrak{N}, J)$ is a semirational map $s : \mathfrak{M} \to \mathfrak{N}$ for which there exist polynomials $t_1(i)$, $t_2(i)$, $i \in I$, $j \in J$ such that

(1) $\dim \mathbb{A}_i \leq t_1(i)$, $\dim \mathbb{A}_j \leq t_2(j)$.

(2) For any $i \in I$ there exists a Turing machine that computes the coordinates of the image $s_i(x) \in \mathbb{A}_{j(i)}$ in time at most $t_1(i)t_2(j)$.

The definition of a uniformly polynomial-time computable map (UPCM) is obvious. All subsequent notions in the category of polynomial-time computable families can obviously be defined for uniformly polynomial-time computable families as well. The authors do not know to what extent these categories are distinct. Up to §4 it makes no difference which of them is being considered. In §4 we consider only PCF's. We must emphasize that this notion is used only in the theory of real complexity.

4. The category of equinomial families and universality. The image of a polynomial-time computable family of semialgebraic sets under a polynomial-time computable map (PCM) is called an equinomial family (EF).

DEFINITION. Let $(\mathfrak{M}, I)$ be a PCF, $(\mathfrak{N}, J)$ a family of semialgebraic sets and $p : \mathfrak{M} \to \mathfrak{N}$ a polynomial-time computable map. If for any $j \in J$ the set N_j is the image of some set $M_i \in \mathfrak{M}$, then $\mathfrak{N}$ is called an *equinomial family*.

This clearly defines a category $\mathscr{K}$ whose objects are equinomial families and whose morphisms are polynomial-time computable maps.

DEFINITION. An equinomial family $(\mathfrak{N}, J)$ is *universal* if for any EF $(\mathfrak{M}, I)$, there exists a PCM $p = \{p_i\}$, $p_i : \mathbb{A}_i \to \mathbb{A}_{j(i)}$ such that $p_i(M_i) = N_{j(i)}$ and $p_i^{-1}(N_{j(i)}) = M_i$.

According to this definition, if there exists a polynomial-time algorithm that solves the MEMBERSHIP problem for a universal family, a similar algorithm exists for any equinomial family. The notion of a PCF is analogous to the class P in complexity theory [9], [10]. Equinomial families correspond to the class NP, and universal families to the class of NP-complete problems. We may consider the counterparts of these notions in terms of UPCF's. This yields a narrower class of EF's: uniformly equinomial families (UEF's) and consequently a different notion of universality. Given an EF $\mathfrak{N}$ and a point $x \in N_j$, we also have a "certificate" $y \in M_i$, where $\mathfrak{M} = \{M_i\}$ is a PCF, $p : \mathfrak{M} \to \mathfrak{N}$ a polynomial-time computable map and $p(y) = x$.

The problem arises as to whether a universal (uniformly) equinomial family is polynomial-time computable (uniformly polynomial-time computable)? This problem is a counterpart of the $P \stackrel{?}{=} NP$ problem.

All definitions given above obviously extend to the category of inductive families of semialgebraic sets. The new element in the system of definitions is the appearance of limit objects: inductive limits $\mathbf{N} = \varinjlim(\mathfrak{N}, I)$. We intend to show elsewhere how topological and analytical structures of $\mathbf{N}$ are related to the complexity of the MEMBERSHIP problem for the inductive family $\mathfrak{N}$.

DEFINITION. Let $\mathfrak{N}$ be an EF, $\mathfrak{M}$ a PCF, and $p : \mathfrak{M} \to \mathfrak{N}$ a PCM. A map $s : \mathfrak{N} \to \mathfrak{M}$ is called a *cut* of the map p if $p \circ s = \mathrm{id}_{\mathfrak{N}}$.

An algorithm in the MEMBERSHIP problem for an EF $\mathfrak{N}$ often amounts to the construction of a cut s which, given points $y \in N_j$, assigns them "certificates" $x \in M_i$, $p(x) = y$. More precisely, if there exists a polynomial-time computable cut on an EF, then the EF is polynomial-time computable.

§2. Examples

In this section we present some basic examples of equinomial inductive families, which will be important. Other examples were considered in [3]–[5] or will be presented in future publications. As a rule, the "universes" for semialgebraic sets are not defined, since they are obvious. We will use the following notation: the inductive families themselves will be denoted by script Latin letters; p will always denote the projection of a PCF onto an EF, whose existence is required by the definition in §1.4. The elements of the PCF will then be prefixed C (for "certificate").

1. Linear programming ($\mathscr{LP}$). Let I be the index set $\mathbb{Z}_+ \oplus \mathbb{Z}_+$. Put $i = (m, n)$, $CLP_{m,n} = \{(A, x) : A \text{ is an } m \times n \text{ matrix}, x \text{ is an } n\text{-dimensional vector}, x \geq \mathbf{0}, Ax = \mathbf{0} \text{ and } \sum_{i=1}^{n} x_i = 1\}$.

Let p be the projection, $p(A, x) = A$,

$$LP_{m,n} = \left\{ A \in \text{Mat}(m, n) : \exists x \in \mathbb{R}_+^n, Ax = \mathbf{0}, \sum_{i=1}^{n} x_i = 1 \right\}.$$

$\{LP_{m,n}\}$ and $\{CLP_{m,n}\}$ are inductive families, since the embeddings $LP_i \to LP_j$, $CLP_i \to CLP_j$, $i \leq j$, are defined in an obvious way via augmentation of A and x by zeroes to reach the required dimensions. These embeddings commute with p.

The MEMBERSHIP problem for the EF $\mathscr{LP}$ is one of the many equivalent forms of the existence problem for solutions of a system of linear inequalities and of the following linear programming problem, which is polynomial-time reducible to it: for a given $m \times n$-matrix A, determine whether there exists a vector $x \geq \mathbf{0}$, $x \neq \mathbf{0}$ such that $Ax = \mathbf{0}$.

2. π-assignment problem ($\mathscr{AP}$). Let $I = \mathbb{N}$, let $\{S_n\}$, $n \in \mathbb{N}$, be a sequence of symmetric groups and $\{\pi_n\}$ a sequence of real representations in spaces V_{π_n}. The algebraic group S_n is identified with the subset $\text{GL}(n; \mathbb{R}) \subset \mathbb{R}^{n^2}$ consisting of the permutation matrices.

We will assume that $\dim \pi_n \leq t(n)$, where t is a polynomial. Suppose that $L_n = V_{\pi_n}^* \otimes V_{\pi_n}$ is the operator space of the representation π_n, equipped with a scalar product $\langle\,,\,\rangle$.

Define

$CAP_n = \{(x, g) : x \in L_n,\ g \in S_n \text{ and } \langle x, \pi_n(g) \rangle \geq 1\}$, $p(x, g) = x$,
$AP_n = \{x \in L_n : \exists g \in S_n \langle x, \pi_n(g) \rangle \geq 1\}$.

The MEMBERSHIP problem for the EF $\{AP_n\}$ is what is known as the problem of justification of a π-assignment [3]–[5], which is equivalent to the determination of whether a point x lies in the interior of the polar to the polyhedron $P_{\pi_n} = \text{conv}\{\pi_n(g)\} \subset L_n$. It is also equivalent to the following problem:

Find
$$\min_{g \in S_n} \langle x, \pi_n(g) \rangle. \tag{2.1}$$

The condition $\dim \pi_n \leq t(n)$ imposes the following constraints on the representations π_n.

From some n on, the irreducible representations of S_n that appear as direct summands in π_n have the form $(\lambda_1, \ldots, \lambda_s, 1^{n-k})$ or $(n-k, \lambda_2, \ldots, \lambda_s)$, where $k \leq \deg t$. (We are using the standard notation for irreducible representations of S_n, indexing them by partitions of n, see, e.g., [11].) If we assume that S_{n-1} is a subgroup of S_n, the stabilizer of the point $\{n\}$, and $\pi_n \downarrow S_{n-1} \supset \pi_{n-1}$, then natural embeddings $V_{\pi_{n-1}} \to V_{\pi_n}$, $L_{n-1} \to L_n$ are defined, and $\{CAP_n\}$, $\{AP_n\}$ are inductive families of semialgebraic sets.

We now present some particular cases of problem (2.1):

(a) Let π_n be the natural representation of S_n in $V = \mathbb{R}\{e_1, \ldots, e_n\}$, $\pi_n(g)e_j = e_{g(i)}$. Then $x = \{x_j^i\}$ is an $n \times n$-matrix, $\pi_n(g)$ is a permutation matrix, $\langle x, \pi_n(g) \rangle = \sum_{i=1}^n x_{g(i)}^i$, and problem (2.1) is the ordinary assignment problem [9].

(b) If π_n is the representation of S_n in $V = \mathbb{R}\{e_{ij}\}$, $1 \leq i, j \leq n$, defined by
$$\pi_n(g)e_{ij} = e_{g(i)g(j)}, \qquad x = \{x_{j_1 j_2}^{i_1 i_2}\} \in V^* \otimes V,$$

$\langle x, \pi_n(g) \rangle = \sum_{i_1, i_2} x_{g(i_1)g(i_2)}^{i_1 i_2}$, then (2.1) is the quadratic assignment problem [9, 10].

(c) If x is specially chosen in the form
$$x = u_n^* \otimes v_n, \qquad \text{where } u_n^* \in V_{\pi_n}^*, \ v_n \in V_{\pi_n},$$

then (2.1) is the problem of maximizing the functional u_n^* on the orbit of the point v_n in the space V_{π_n}.

Thus we obtain the traveling salesman problem, the weighted matching problem, etc., see [3]–[5].

The general π-assignment problem can also be formulated for other series $\{G_n\}$ of algebraic groups (GL_n, SL_n) and semialgebraic groups (O_n, U_n) and polynomial (rational) representations π_n of these groups. If G_{n-1} is a subgroup of G_n and $\pi_n \downarrow G_{n-1} \supset \pi_{n-1}$, then, as before, we obtain an inductive family of semialgebraic sets $\{AP_n\}$.

3. Problem of the orbit of a set ($\mathscr{OL}$). Let π_n be a sequence of representations of (semi)algebraic groups G_n in $\mathbb{A}_n$, $\dim \pi_n \leq t(n)$, where t is a polynomial. $(\mathfrak{M}, \mathbb{N})$ is a polynomial-time computable family of semialgebraic sets such that $M_n \subset \mathbb{A}_n$.

Put

$$COS_n = \{(x, g) : x \in M_n, \ g \in G_n\}, \quad p(x, g) = \pi_n(g)x,$$
$$OS_n = \{x \in \mathbb{A}_n : \exists g \in G_n, \ \pi_n(g)x \in M_n\}.$$

The MEMBERSHIP problem for the family $\{OS_n\}$ is to determine whether the orbit of a given element intersects M_n. It is easy to show that the isomorphism problem for graphs ($G_n = S_n$), the π-assignment problem (M_n is a half-space $\langle \pi_n(e), x \rangle \geq 1$, where e is the unit of the group G_n), the problem of determining G_n-equivalence of multilinear forms, the problem of the rank of a tensor (see §4.2) and practically all combinatorial optimization problems can be reduced to this form.

Now we show that the problem of whether the solution set of a system of linear inequalities is nonempty can also be formulated as a MEMBERSHIP problem for the EF $\mathcal{OL}$ with $G_n = S_n$ and a suitable choice of $\mathfrak{M}$.

Consider a system of linear inequalities

$$Ax = e, \quad x \geq \mathbf{0}, \tag{2.2}$$

where A is a $(m+1) \times n$-matrix; $e = (1, 0, \ldots, 0)$; the first row of A is $(1, \ldots, 1)$; x is an n-dimensional vector. We assume that rank $A = m+1$.

If the set of solutions of (2.2) is nonempty, there exists a square $(m+1) \times (m+1)$-matrix A_δ such that $A_\delta^{-1} e \geq \mathbf{0}$.

Let $I = \mathbb{Z}_+ \oplus \mathbb{Z}_+$. The points of the space $\mathbb{A}_i$, $i = (m, n)$, will be interpreted as $(m+1) \times n$-matrices. The group S_n acts in $\mathbb{A}_i$ by permutations of the columns of the matrices in $\mathbb{A}_i$ (this representation $\pi_{m,n}$ is the sum of m natural representations of S_n in $\mathbb{R}^n$).

Set $\mathfrak{M} = \{M_n\}$:

$$M_{m,n} = \{A \in \mathbb{A}_{m,n} : A_{\delta_0}^{-1} e \geq \mathbf{0}\},$$

where A_{δ_0} is the submatrix formed by the first m columns of A. (If A_{δ_0} is a singular submatrix, we will assume that $A \notin M_{m,n}$.)

If (2.2) has a solution, then there exists a permutation g of the columns of A such that $g(A) \in M_{m,n}$.

§3. Approximations

1. Definition of polynomial-time computable approximations. Let $\{\mathfrak{N}, I\}$ be a family of semialgebraic sets. A collection of families of semialgebraic sets $\{(\mathfrak{X}(n), I)\}$, $n \in \mathbb{N}$, is called a *polynomial-time computable approximation* of $\{\mathfrak{N}, I\}$ if:

(1) $N_i \in \mathfrak{N}$ and $N_i \subset \mathbb{A}_i$ imply that $\mathfrak{X}_i(n) \subset \mathbb{A}_i$ for all $n \in \mathbb{Z}_+$ and N_i

is the limit of $\mathfrak{X}_i(n)$, $n \to \infty$, that is,

$$N_i = \bigcup_{n=1}^{\infty} \bigcap_{m \geq n} \mathfrak{X}_i(m) = \bigcap_{n=1}^{\infty} \bigcup_{m \geq n} \mathfrak{X}_i(m), \quad \forall i \in I.$$

(2) Each family $\mathfrak{X}(k) = \{X_i(k)\}_I$ is polynomial-time computable (the polynomials that measure the complexity may depend on k).

The following special cases are important:

(a) For each k, $\mathfrak{X}(k) \subset \mathfrak{X}(k+1) \subset \mathfrak{N}$ and $\bigcup_k \mathfrak{X}(k) = \mathfrak{N}$ (monotonic approximation from inside);

(b) For each k, $\mathfrak{N} \subset \mathfrak{X}(k+1) \subset \mathfrak{X}(k)$ and $\bigcap_k \mathfrak{X}(k) = \mathfrak{N}$ (monotonic approximation from outside).

Inclusion relations between families of sets should be understood as inclusion relations between the corresponding semialgebraic sets for all $i \in I$.

Polynomial-time computable approximations always exist. However, our aim is to construct PCA's that satisfy certain metric estimates.

2. Methods for solving linear programming problems and PCA. As a first example we will consider a PCA for the family $\mathscr{LP}$, and in §4 we will consider a polynomial-time computable approximation for $\mathscr{AP}$ and some other families.

The classical Simplex Method and the Ellipsoid Method, which has recently become well known [2] (see also [12]), are in fact means for constructing approximations in the above sense. Let us discuss the second method.

Unlike §2.1, we will consider the family of interiors of the sets:

$$L^{\circ}P_{m,n} = \left\{ A \in \mathrm{Mat}(m,n) : \exists x \in \mathbb{R}^n, \ Ax = \mathbf{0}, \ x > \mathbf{0}, \ \sum_{i=1}^n x_i = 1 \right\}.$$

Call the family thus obtained $\mathscr{L}^{\circ}\mathscr{P}$. We will construct a PCA using the Ellipsoid Method in both standard and shallow cut versions. To solve a system of linear inequalities

$$\begin{aligned} Ax &= \mathbf{0}, \\ \sum_{i=1}^n x_i &= 1, \quad x_i > 0, \end{aligned} \tag{3.1}$$

the basic Ellipsoid Method constructs a sequence of ellipsoids $E_A(k)$, $k = 1, 2, \ldots$, each of which includes the solution set of (3.1). Let $c(k) \in \mathbb{R}^n$ be the center of the k th ellipsoid $E_A(k)$. Denote $\Delta_n = \{(x_1, \ldots, x_n) : \forall i, \ x_i > 0 \text{ and } \sum_{i=1}^n x_i = 1\}$. Fix some polynomial $p(m,n)$ with positive integer coefficients and let

$$X_{m,n}(k) = \{A \in \mathrm{Mat}(m,n) : c(k \cdot p(m,n)) \in \Delta_n\}.$$

It is easy to check that $\{\mathfrak{X}(k)\} = \{X_i(k)\}$, $i = (m,n) \in I$, $k \in \mathbb{N}$, is indeed a polynomial-time computable approximation of $\mathscr{L}^{\circ}\mathscr{P}$, provided we allow the Turing machine to compute square roots of nonnegative real numbers as well; this is necessary to calculate the centers of the successive ellipsoids. Before discussing the case in which the extraction of square roots is forbidden, let us study the metric properties of the approximation thus obtained.

Let P_A be the polyhedron of solutions of (3.1) (if it is nonempty) and $\text{vol}(P_A)$ its $(n-m-1)$-dimensional volume. The Ellipsoid Method constructs an interior point of P_A in time $O(n^3 m |\log \text{vol}(P_A)|)$.

The following assertion is true.

Let $\varepsilon > 0$ be an arbitrary real number, $q(m, n)$ a polynomial. Consider the set

$$Y_{m,n}(\varepsilon) \subset L^\circ P_{m,n}; \quad Y_{m,n}(\varepsilon) = \{A : \text{vol}(P_A) \geq \exp\{q(m,n)\log\varepsilon\}\}.$$

Then for some $k \in \mathbb{N}$ and some polynomial $p(m, n)$, for all m, n,

$$Y_{m,n}(\varepsilon) \subset X_{m,n}(k).$$

Define the (ε, q)-interior $U_{\varepsilon, q}$ of the set $LP_{m,n}$ by the condition

$$A \in U_{\varepsilon, q} \Leftrightarrow \rho(A, \partial LP_{m,n}) \geq \exp\{\log\varepsilon \cdot q(m,n)\},$$

where $\partial LP_{m,n}$ is the boundary of $LP_{m,n}$ (obviously made up of polyhedrons of zero volume), ρ is the ordinary Euclidean metric in the space $\text{Mat}(m, n)$ and q is an arbitrary polynomial. Then for suitable k and $p(m, n)$,

$$U_{\varepsilon, q} \subset \mathfrak{X}(k).$$

Thus, the Ellipsoid Method constructs a mapping

$$s : \text{Mat}(m, n) \to \text{Mat}(m, n) \times \mathbb{R}^n,$$
$$A \to (A, x),$$

which turns out to be a polynomial-time computable cut of the projection $p : CLP_{m,n} \to LP_{m,n}$ over the interior $U_{\varepsilon, q}$ of $LP_{m,n}$.

We now consider an alternative approximation, corresponding to the case when square roots cannot be extracted. In that case the ellipsoid method can be modified by computing the center $c(k)$ of the k th ellipsoid $E(k)$ to within $\exp\{-r_1(m, n)\}$ and performing the next truncation if for some $1 \leq i \leq n$,

$$c_i(k) < \exp\{-r_2(m, n)\},$$

where r_1, r_2 are polynomials with positive coefficients.

Suppose we know that a polyhedron P_A contains an interior point x such that $x_i \geq \exp\{-q(m, n)|\log\varepsilon|\}$ for some $\varepsilon > 0$ and polynomial q. Then there exist polynomials r_1, r_2, p, definable in terms of q and ε, such that $c(k)$ will lie in the interior of P_A no later than on the $p(m, n)$-th step.

Let $X_{m,n}(k) = \{A \in \text{Mat}(m, n) : c(k \cdot p(m, n)) > 0\}$ and let $\mathfrak{X}(k) = \{X_{m,n}(k)\}$ be a polynomial-time computable approximation of $\mathscr{L}^\circ \mathscr{P}$.

Note that in this case there also exist polynomials p and $k \in \mathbb{N}$ for the ε, q-interior of $\mathscr{L}°\mathscr{P}$ such that $U_{\varepsilon,q} \subset \mathfrak{X}(k)$. The question of whether one can construct a polynomial-time computable cut over the larger interior of $\mathscr{L}\mathscr{P}$ (in particular, over $\mathscr{L}°\mathscr{P}$) remains unsolved.

3. Topology of the space of linear programming problems. We now return to the basic Ellipsoid Method which permits extraction of square roots from real numbers.

Let $A \in L°P_{m,n}$. It is easy to show that there exist a neighborhood U of A and a modification of the Ellipsoid Method (based on the "shallow cut" version) which constructs a smooth cut over U: $s: U \to CLP_{m,n}$ of the projection $p: CLP_{m,n} \to LP_{m,n}$. Now let $A \in \partial LP_{m,n}$ and let U be an arbitrary small punctured neighborhood of A in $LP_{m,n}$. It is easy to give examples of matrices A for which there is no continuous cut over U, owing to obstacles of a topological nature.

§4. Approximation in group-theoretic problems

1. Approximation in π-assignment problems. Recall that irreducible representations of the symmetric group S_n are parametrized by the Young diagrams Λ_n, that is, the partitions of the number n. We will use the same notation for an irreducible representation corresponding to a Young diagram $\Lambda_n = (\lambda_1, \ldots, \lambda_s)$ as in [11]. Let $\mu_i = \sum_{j=1}^{i} \lambda_j$, $\mu_1 = \lambda_1$. Let

$$S_{\Lambda_n} = S_{\{1,\ldots,\lambda_1\}} \times S_{\{\lambda_1+1,\ldots,\mu_2\}} \times \cdots \times S_{\{\mu_{s-1}+1,\ldots,\mu_s\}}$$

be the Young subgroup corresponding to the partition Λ_n. The representation of S_n induced by the identity representation S_{Λ_n} will be denoted by $\pi(\Lambda_n)$. A partial order is defined on Young diagrams by $\Lambda \leq K \Leftrightarrow \forall j \sum_{i=1}^{j} \lambda_i \leq \sum_{i=1}^{j} \kappa_i$. As in §2.2, let S_n be the sequence of symmetric groups, K_n the sequence of their real irreducible representations $(n-k, \kappa_2, \ldots, \kappa_s)$, $\sum_{i=2}^{s} \kappa_i = k$ in spaces V_n. Instead of the EF $\mathscr{A}\mathscr{P}$ it is more convenient to consider the family of polars $\mathscr{P}\mathscr{O}\mathscr{L} = \{POL_n\}$, where

$$POL_n = \{c \in L_n = V_n^* \otimes V_n : \forall g \in S_n \langle c, K_n(g) \rangle \leq 1\}.$$

The first author has established the following result.

THEOREM 1. *To every Young diagram Λ_n ($\Lambda_n \leq K_n$) one can associate a polyhedron $P(\Lambda_n) \subset L_n$ such that*

(1) *the MEMBERSHIP problem for the family $\{P(\Lambda_n)\}$ is solvable in polynomial time with respect to $\dim \pi(\Lambda_n)$, that is, there exists a polynomial $t(x)$ such that, for any n, there is an algorithm solving the MEMBERSHIP problem for $P(\Lambda_n)$ with complexity at most $t(\dim \pi(\Lambda_n))$.*

(2) *Let $\alpha(n) = (\dim K_n)/(\pi(\Lambda_n) : K_n)$, where the denominator of the fraction is the multiplicity of the irreducible representation K_n in the representation $\pi(\Lambda)$.*

Then $POL_n \supset P(\Lambda_n) \supset \alpha(n)^{-1} POL_n$.

For the proof, see [3], [5].

It is easy to extract a PCA for the family $\mathscr{POL}$ from the family of polyhedra $P(L_n)$. In fact, put $\mathfrak{X}(n) = \{X_i(n)\}$,

$$X_i(n) = P(\Lambda_i), \quad \text{where } \Lambda_i = \begin{cases} 1^i & \text{if } n \geq i \\ (i-n, 1^n) & \text{if } n \leq i. \end{cases}$$

By Theorem 1, for any $i \in \mathbb{N}$, the MEMBERSHIP problem for $X_i(n)$ is solvable in polynomial time with respect to i, and for any i we have $X_i(n) = POL_i$ for $n \geq 1$ (since if $\Lambda_n = (1^n)$, then $\dim K_n/(\pi(\Lambda_n) : K_n) = 1$, see, e.g., [11]).

A metric characterization of the approximation also follows from Theorem 1. Indeed, for any $\varepsilon > 0$ there exists $m_0 \in \mathbb{N}$ such that, for all $m > m_0$ and any i,

$$X_i(m) \subset POL_i \subset \varepsilon i^k X_i(m).$$

(Recall that $\dim K_n$ is a polynomial in n of degree k, $(\pi(i-n, 1^n) : K_i) \geq C_n^k$ [11].)

To end this subsection, we briefly describe a method for constructing the polyhedra $P(\Lambda_n)$.

Let $\mathbb{R}S_n$ be the real group algebra of S_n, and let I_{K_n} be the two-sided ideal of $\mathbb{R}S_n$ in which the representation K_n acts via left multiplication by elements $g \in S_n$. The linear functionals on L_n are in one-to-one correspondence with the elements of I_{K_n}:

$$c \leftrightarrow \sum_{g \in S_n} \langle c, K_n(g) \rangle g,$$

and the space of affine functionals $\langle c, x \rangle + b$ is isomorphic to $I_{K_n \oplus (n)}$, where (n) is the identity representation. Then POL_n is identified with the polyhedron

$$I_{K_n \oplus (n)} \cap \operatorname{conv}\{g : g \in S_n\},$$

$$\operatorname{conv}\{g : g \in S_n\} = \left\{ \sum_{g \in S_n} z(g)g : \sum_{g \in S_n} z(g) = 1; \; z(g) \geq 0 \right\}.$$

The polyhedron $P(\Lambda_n)$ is defined as follows:

$$P(\Lambda_n) = \operatorname*{conv}_{h_1, h_2 \in S_n} \left\{ |S_{\Lambda_n}|^{-1} \sum_{g \in S_{\Lambda_n}} h_1 g h_2 \right\} \cap I_{K_n \oplus (n)}.$$

The proof of Theorem 1 is based on the fact that the number of the vertices of the polyhedron

$$\tilde{P}(\Lambda_n) = \operatorname*{conv}_{h_1, h_2 \in S_n} \left\{ |S_{\Lambda_n}|^{-1} \sum_{g \in S_{\Lambda_n}} h_1 g h_2 \right\} \subset \mathbb{R}S_n$$

does not exceed $(n!/|S_{\Lambda_n}|)^2 = \dim^2 \pi(\Lambda_n)$.

Another topic discussed in [3] is the combinatorics of the approximating polyhedra $P(\Lambda_n)$. Note that if $K_n = (n-1, 1)$, then $P(\Lambda_n) = POL_n$, where $\Lambda_n = (n-1, 1)$. This assertion is equivalent to the Birkhoff-von Neumann Theorem (see, e.g., [9]).

2. The method of statistical sums. Let X be a finite set, μ a signed measure on X and $f : X \to \mathbb{R}$ a real-valued function. We will consider sums

$$S_\mu(f; t) = \int_X \exp\{tf\}\, d\mu = \sum_{x \in X} \exp\{tf(x)\}\mu(x),$$

where $t \in \mathbb{R}$. The following result is obvious.

PROPOSITION. *Let f be a function such that*

$$f(x) = f(y) \Rightarrow \mu(x) = \mu(y).$$

Then

$$\lim_{t \to +\infty} t^{-1} \log |S_\mu(f; t)| = \max_{x \in X \,:\, \mu(x) \neq 0} f.$$

This proposition can obviously be modified to hold for infinite sets X. If f and μ satisfy the assumptions of the proposition, we say that f is in general position relative to μ. We show how the result yields an approximation in the following very general situation.

Let $\{X_n\}$ be a sequence of sets, F_n a sequence of linear spaces of functions $f : X_n \to \mathbb{R}$, M_n the family of sets

$$M_n = \{f \in F_n : \forall x \in X_n\, f(x) \leq a_n\}, \qquad a_n \in \mathbb{R}.$$

Suppose we have been able to find signed measures $\mu_n : X_n \to \mathbb{R}$ such that

(a) $\mu_n(x) \neq 0$, $\forall x \in X_n$;

(b) the functions $f \in F_n$ in general position relative to μ_n form a dense open set.

Let $M_n(k) = \{f \in F_n : k^{-1} \log |S_{\mu_n}(f; k)| \leq a_n\}$. In view of the above proposition, the collection of families of sets $M_n(k)$ satisfies condition (1) in the definition of a polynomial-time computable approximation for all functions f in a dense open subset of F_n.

What is the complexity of computation of the sums $S_\mu(f; t)$? We will have to extend the set of operations of the real Turing machine by adding the operation of exponentiation. If we now define the complexity of an algorithm as the number of operations, including exponentiation, then we obtain a broader concept of polynomial complexity. We do not know to what extent it differs from the original concept.

The sums $S_\mu(f; t)$ have in fact been considered (though in another form) in the optimization literature. Unlike the "annealing" and "spin glass" methods [13], computation of the sums $S_\mu(f; t)$ will be based on algebraic rather than probabilistic considerations, in particular on the existence of nontrivial symmetries of the functional for a suitable choice of the signed measure μ. The examples given below will be discussed in more detail in future publications of Barvinok, where the necessary proofs will also be presented. All our examples are intimately connected with the classical problems of combinatorial optimization and are often just rephrased versions of such problems; this will be reflected in the headings.

In examples (a), (b), and (c), $X_n = S_n$ is a symmetric group, ε denotes the signed measure $\varepsilon(g) = \operatorname{sgn} g$, where $\operatorname{sgn} g$ is the sign of a permutation g; $\operatorname{sgn} g = 1$ if g is even, $\operatorname{sgn} g = -1$ if it is odd.

(a) *The assignment problem* [9, 14]. Let $\|a(i, j)\|$, $1 \leq i, j \leq n$, be a real matrix (from now on, for convenience, we will write the indices of matrices and tensors in parentheses), and

$$f(g) = \sum_{i=1}^{n} a(i, g(i)) \quad \text{for } g \in S_n.$$

Let $A_t(i, j) = \exp\{ta(i, j)\}$ and $A_t = \|A_t(i, j)\|$. Then

$$S_\varepsilon(f; t) = \det A_t.$$

We will consider S_n as an algebraic group, represented by permutation matrices $\|x_g(i, j)\| \in \mathbb{R}^{n^2}$. Let $p(x(i, j))$ be a polynomial of fixed degree k in n^2 variables. Let $A_n \subset S_n \subset \mathbb{R}^{n^2}$ be the Zariski-closed subset of S_n defined by the condition $p(x) = 0$. Then, setting $\mu(g) = \operatorname{sgn} g \cdot p(x_g(i, j))$, we obtain

$$\lim_{t \to +\infty} t^{-1} \log |S_{\mu_n}(f; t)| = \max_{g \in S_n - A_n} f(g).$$

The sums $S_{\mu_n}(f; t)$ are polynomial-time computable. Such signed measures turn out to be useful in finding several first values (in decreasing order) of f.

(b) *Weighted partition into l-tuples* [9], [10]. Let $l \in \mathbb{N}$ be fixed, $n = ml$, $m \in \mathbb{N}$, and consider a given l-valent tensor $b = \|b(i_1, \ldots, i_l)\| \in (\mathbb{R}^n)^{\otimes l}$.

Define a function $f : S_n \to \mathbb{R}$ by

$$f(g) = \sum_{i=0}^{m-1} b(g(li + 1), g(li + 2), \ldots, g(li + l)).$$

$f(g)$ is the total weight of the partition of $\{1, 2, \ldots, n\}$ into m ordered l-tuples

$$\{g(1), \ldots, g(l)\}, \{g(l + 1), \ldots, g(2l)\}, \ldots, \{g(n - l + 1), \ldots, g(n)\},$$

where the weight of a tuple $\{i_1, \ldots, i_l\}$ is defined as $b(i_1, \ldots, i_l)$. If g ranges over S_n, then $f(g)$ ranges over the weights of all such partitions. Define

$$B_t(i_1, \ldots, i_l) = \exp\{tb(i_1, \ldots, i_l)\}, \qquad B_t \in (\mathbb{R}^n)^{\otimes l}.$$

$S_\varepsilon(f;t)$ is a polynomial $P(B)$ in the coordinates of the tensor B_t. It turns out that $P(B)$ is a relative invariant of the action of the group $\mathrm{GL}(n;\mathbb{R})$ (the group $\mathrm{GL}(n;\mathbb{R})$ acts in the tensor space as the lth tensor power of the natural action in the space $\mathbb{R}^n$):

$$P(G(B)) = \det G \cdot P(B).$$

For odd l we have $P(B) = 0$, so we assume that l is even. In that case the set of tensors b for which f is in general position relative to the signed measure ε is open and dense in $(\mathbb{R}^n)^{\otimes l}$. If $l = 2$ (the weighted matching problem), then $P(B) = \mathrm{Pf}(B)$ is the Pfaffian of B, which is polynomial-time computable [14].

DEFINITION. The *rank* of an l-valent tensor B is the least number $r = \mathrm{rank}\,B$ such that $B \in (\mathbb{R}^n)^{\otimes l}$ can be represented as

$$B = \sum_{i=1}^r u_i^1 \otimes u_i^2 \otimes \cdots \otimes u_i^l, \quad \text{where } u_i^j \in \mathbb{R}^n. \qquad (4.1)$$

We define the *2-rank* of a tensor $B \in (\mathbb{R}^n)^{\otimes l}$, $l = 2q$, to be the least number $r = \mathrm{rank}_2 B$ such that B is representable as

$$B = \sum_{i=1}^r a_i^1 \otimes \cdots \otimes a_i^q, \quad \text{where } a_i^j \in \mathbb{R}^n \otimes \mathbb{R}^n. \qquad (4.2)$$

THEOREM 2. *Let k be a fixed natural number (k does not depend on n) and assume that at least one of the following conditions is satisfied*:

(1) $\mathrm{rank}\,B \leq n/l + k$, *and B is given by* (4.1);
(2) $\mathrm{rank}_2 B \leq k$ *and B is given by* (4.2).

Then there exists a polynomial-time algorithm that computes $P(B)$.

Note that if we weaken condition (1) by stipulating only $\mathrm{rank}\,B \leq 2n/l$, the problem of computing $P(B)$ will be polynomially equivalent to computing the invariant of P for an arbitrary l-valent tensor. Theorem 2 yields several examples of special tensors $b \in (\mathbb{R}^n)^{\otimes l}$ for which the sums $S_\varepsilon(f;t)$ are polynomial-time computable for all $t \in \mathbb{R}$.

(c) *Weighted covering by l-tuples* [10]. Fix numbers $k, l \in \mathbb{N}$ and set $X_n = S_{kn}$. We will assume that $n = lm$, $m \in \mathbb{N}$. Let $b \in (\mathbb{R}^n)^{\otimes l}$ be a given l-valent tensor; $\overline{a}$ will denote the residue of $a \pmod n$, where $1 \leq \overline{a} \leq n$. Define f by

$$f(g) = \sum_{i=0}^{km-1} b(\overline{g(li+1)}, \overline{g(li+2)}, \ldots, \overline{g(li+l)}).$$

$f(g)$ is the total weight of a covering of $\{1, 2, \ldots, n\}$ by ordered l-tuples

$$\{\overline{g(1)}, \ldots, \overline{g(l)}\},\ \{\overline{g(l+1)}, \ldots, \overline{g(2l)}\},\ \ldots,\ \{\overline{g(n-l+1)}, \ldots, \overline{g(n)}\},$$

such that any point lies exactly in k tuples, and the weight of a tuple is defined, as before, as $b(i_1, \ldots, i_l)$. Let μ be a signed measure on S_{kn} such

that $\sum \mu(g)g$ is an element of the ideal $I_{(k^n)}$ in the group algebra $\mathbb{R}S_{kn}$.
Define $B_t(i_1, \ldots, i_l) = \exp\{tb(i_1, \ldots, i_l)\}$.

Then $S_\mu(f; t)$ is a polynomial $P_\mu(B)$ in the coordinates of the tensor B_t, and with μ chosen as above we have

$$P_\mu(G(B)) = \det{}^k G \cdot P_\mu(B), \quad G \in \mathrm{GL}(n, \mathbb{R}).$$

If $l = 2$, we obtain what is known as the k-matching problem [14]. In that case for any choice of a sequence of signed measures $\mu_n \in I_{(k^n)}$ there exists a polynomial-time algorithm that computes $S_{\mu_n}(f; t)$. An analogue of Theorem 2 can also be proved.

Note also that if $k > 1$, then $P_\mu(B)$ need not always vanish even for odd l.

(d) *Other π-assignment problems* (the traveling salesman problem). In other particular cases of the π-assignment problem, different from those considered above, the sums $S_\mu(f; t)$ are not relative invariants of the tensors. Nevertheless, these sums are interrelated in ways that enable us, in some special cases, to devise effective algorithms for the computation of $S_\mu(f; t)$. We will show one way to obtain such relationships in the traveling salesman problem [9], [10], [14].

Let $b = \|b(i, j)\|$ be a square $n \times n$ real matrix; for all $g \in S_n$, put $f(g) = \sum_{i=1}^n b(g(i), g(i+1))$ (we stipulate $n+1 = 1$). Let $B_t(i, j) = \exp\{tb(i, j)\}$. We interpret the matrices B_t as points of the space $\mathbb{R}^n \otimes \mathbb{R}^n$, $B(i, j) = \langle B, e_i \otimes e_j \rangle$. The expression $S_\varepsilon(f; t)$ is a polynomial $Q(B)$ in the elements of B_t, which can be written in the following form:

$$Q(B) = \langle B^{\otimes n}, y \rangle,$$

where

$$y = \sum_{g \in S_n} \mathrm{sgn}\, g\, e_{g(1)} \otimes e_{g(2)} \otimes e_{g(2)} \otimes \cdots \otimes e_{g(n)} \otimes e_{g(n)} \otimes e_{g(1)},$$

$\langle , \rangle$ being the scalar product, defined in the standard way in $(\mathbb{R}^n)^{\otimes 2n}$.

Consider the action of the universal enveloping Lie algebra $U = \mathrm{gl}(n; \mathbb{R})$ in $(\mathbb{R}^n)^{\otimes 2n}$. Suppose there exists $D \in U$, $Dy = 0$. Then

$$\langle B^{\otimes n}, Dy \rangle = \langle D^* B^{\otimes n}, y \rangle = 0.$$

Passing from the enveloping algebra U to the group algebra $\mathbb{R}\mathrm{GL}(n; \mathbb{R})$ and replacing $D^* \in U$ by a suitable linear combination $\sum \alpha_i G_i$, $G_i \in \mathrm{GL}(n; \mathbb{R})$, we obtain the condition

$$\sum \alpha_i Q(G_i(B)) = 0,$$

or, in matrix notation,

$$\sum \alpha_i Q(G_i B G_i^\mathrm{T}) = 0.$$

Let D_{ij} be differentiation in $\mathbb{R}^n$: $D_{ij}e_k = \delta_{jk}e_i$, I the identity operator on $(\mathbb{R}^n)^{\otimes 2n}$. It is easy to verify that

(1) $D_{ij}^2 y = 0$, if $i \neq j$, $(D_{ii} - 2I)y = 0$;
(2) $(D_{ij}D_{ji} - 2I)y = 0$, if $i \neq j$.

Let E_{ij}^α denote the matrix $I + \alpha e_{ij}$, where $e_{ij}(k, s) = \delta_{jk}\delta_{is}$ is the identity matrix. The following identities ($i \neq j$) are consequences of conditions (1) and (2):

$$\beta Q(E_{ij}^\alpha B E_{ji}^\alpha) - \alpha(E_{ij}^\beta B E_{ji}^\beta) = (\beta - \alpha)Q(B),$$

$$Q(E_{ij}^\alpha E_{ji}^\beta B E_{ij}^\beta E_{ji}^\alpha) - Q(E_{ij}^\alpha B E_{ji}^\alpha) - Q(E_{ji}^\beta B E_{ij}^\beta)$$
$$= (2\alpha\beta - 2)Q(B) + Q(E_{ij}^{\alpha^2 \beta} B E_{ji}^{\alpha^2 \beta}).$$

In some special cases, these conditions enable us to construct a polynomial-time algorithm for the computation of $S_\varepsilon(f; t)$.

(e) *Search for the lattice point nearest a given point in $\mathbb{R}^n$* [12]. Let $X_n = \mathbb{Z}_n \subset \mathbb{R}^n$ be the lattice of integer points, A a nondegenerate operator in $\mathrm{GL}(n; \mathbb{R})$, $y \in \mathbb{R}^n$, and $\|\ \|$ the Euclidean norm in $\mathbb{R}^n$. Put $\mu_n(x) = 1$ and define a function f by

$$f_{A,y}(x) = \|Ax - y\|^2.$$

The problem of minimizing $f_{A,y}$ is the problem of finding a point in the integer lattice generated by the vectors $Ae_1, \ldots, Ae_n$ that is the closest to a given point y in $\mathbb{R}^n$. This NP-hard problem is perhaps the most typical problem of integer linear programming. The sum $S_\mu(f; -t)$ is easily expressible in terms of the theta function. In the standard notation [15],

$$S_\mu(f; -t) = \sum_{x \in \mathbb{Z}^n} \exp\{-t\|Ax - y\|^2\} = \exp\{-t\|y\|^2\}\theta(z, \Omega),$$

where $\Omega = (it/\pi)AA^*$; $z = -(it/\pi)A^*y$.

The computation of theta functions is a separate problem, which will be discussed elsewhere. Here we will only outline possible solutions. Since the theta function is a modular form, it will suffice to compute it for at least one point of each orbit of the group $\Gamma_{1,2} \subset \mathrm{Sp}(2n; \mathbb{Z})$ (in the notation of [15]), just as in examples (b) and (c) it was necessary to evaluate $S_\mu(f; t)$ at only one point of each orbit of the group $\mathrm{GL}(n; \mathbb{R})$. In addition, the computation of theta functions can be reduced to the computation of a certain meromorphic function on an algebraic variety. The rich algebraic properties of the theta function undoubtedly imply a diversity of means for its computation.

References

1. L. Blum, M. Shub, and S. Smale, *On theory of computations over the real numbers: NP-completeness, recursive functions and universal machines*, Bull. Amer. Math. Soc. **21** (1989), 1–46.

2. L. G. Khachiyan, *Polynomial algorithms in linear programming*, Zh. Vychisl. Mat. i Mat. Fiz. **20** (1988), no. 1, 51–68; English transl. in U.S.S.R. Comput. Math. and Math. Phys. **20** (1988).
3. A. I. Barvinok, *Combinatorial theory of polyhedra with symmetry and its applications to the optimization problems*, Thesis, Leningrad, 1988. (Russian)
4. A. I. Barvinok and A. M. Vershik, *Convex hulls of orbits of representations of finite groups and combinatorial optimization*, Funktsional. Anal. i Prilozhen. **22** (1988), no. 3, 66–67; English transl. in Functional Anal. Appl. **22** (1988).
5. _____, *Methods of representation theory in combinatorial optimization problems*, Izv. Akad. Nauk SSSR Tekhn. Kibernet. (1988), no. 6, 64–71; English transl. in Soviet J. Comput. Systems Sci. **27** (1989), no. 5, 1–7.
6. A. M. Vershik and A. G. Chernyakov, *Fields of convex polyhedra and the Pareto-Smale optimum*, Optimizatsiya, vol. 27(45), Akad. Nauk SSSR, Sibirsk. Otd., Inst. Mat., Novosibirsk, 1982, pp. 112–146. (Russian)
7. A. M. Vershik and P. V. Sporyshev, *Estimation of the mean number of steps of the Simplex Method and problems of asymptotic integral geometry*, Dokl. Akad. Nauk SSSR **271** (1983), no. 5, 1044–1048; English transl. in Soviet Math. Doklady **28** (1983).
8. E. Becker, *On the real spectrum of a ring and its application to semialgebraic geometry*, Bull. Amer. Math. Soc. **15** (1986), 19–60.
9. C. H. Papadimitriou and K. Steiglitz, *Combinatorial optimization*, Prentice-Hall, Englewood Cliffs, N.J., 1982.
10. M. R. Garey and D. S. Johnson, *Computers and intractability*, Freeman, San Francisco, 1979.
11. G. D. James, *The representation theory of the symmetric groups*, Lecture Notes in Math., vol. 682, Springer-Verlag, Berlin, 1978.
12. L. Lovasz, *An algorithmic theory of numbers, graphs and complexity*, SIAM, Philadelphia, 1986.
13. S. Kirkpatrick and G. Toulouse, *Configuration space analysis of traveling salesman problems*, J. Physique **46** (1985), no. 8, 1277–1292.
14. L. Lovasz and M. D. Plummer, *Matching theory*, North Holland, Amsterdam, New York; Akademiai Kiado, Budapest, 1986.
15. D. Mumford, *Tata lectures on theta*, Progress in Mathematics **28**, Birkhäuser, Boston, 1983.

Translated by A. BOCHMAN

Construction of Invariant Tori of Systems of Differential Equations with a Small Parameter

YU. N. BIBIKOV

Let n be an integer. Consider a d-dimensional system of differential equations depending on a small nonnegative parameter

$$\dot{x} = X(x, \varepsilon) + X^*(x, \varepsilon). \tag{1}$$

Here $X^* : \mathfrak{M} \to \mathbb{R}^d$, $\mathfrak{M} = \{(x, \varepsilon) : \|x\| < H, \ 0 \le \varepsilon < \varepsilon_0\}$, is a continuous function satisfying the condition

$$X^*(\sqrt{\varepsilon}x, \varepsilon) = \sqrt{\varepsilon^{3n+2}}\tilde{X}(x, \varepsilon), \tag{2}$$

where $\tilde{X}(x, \varepsilon)$ is a function of class $C^{10}_{x\varepsilon}(\mathfrak{M})$; $X(x, \varepsilon)$ is a vector polynomial; $X(0, 0) = 0$ ($\|x\|$ is the Euclidean norm of x).

Set $P = D_x X(0, 0)$, and let $\det P \ne 0$. Invoking (2), we have

$$D_x X^*(\sqrt{\varepsilon}x, \varepsilon) = \sqrt{\varepsilon} D_y X^*(y, \varepsilon)|_{y=\sqrt{\varepsilon}x} = \sqrt{\varepsilon^{3n+2}} D_x \tilde{X}(x, \varepsilon);$$

hence $D_x X^*(0, 0) = 0$. By the Implicit Function Theorem, for every sufficiently small $\varepsilon \ge 0$ the system (1) has an equilibrium state $x^*(\varepsilon)$, $x^*(0) = 0$. For such ε, we would like to find invariant manifolds of system (1) that are homeomorphic to tori of positive dimension (such manifolds will be called briefly invariant tori). If such manifolds exist, we say that bifurcation to invariant tori from the equilibrium state $x = 0$ has occurred at $\varepsilon = 0$. If P is a noncritical matrix, that is, P has no eigenvalues on the imaginary axis, it is well known that there are no invariant tori of positive dimension. We shall assume, therefore, that all the elements of the spectrum of P are numbers $\pm i\lambda^1, \ldots, \pm i\lambda^m, \lambda^{2m+1}, \ldots, \lambda^d$, where $\operatorname{Re}\lambda^k \ne 0$ for $k = 2m+1, \ldots, d$, and $m > 0$.

Observe that a system

$$\dot{x} = f(x, \varepsilon), \qquad f(0, 0) = 0$$

1991 *Mathematics Subject Classification.* Primary 34A47.

©1993 American Mathematical Society
0065-9290/93 $1.00 + $.25 per page

can be reduced to the form (1) if $f \in C^{3n+3}(\mathfrak{M})$, and in that case

$$X(x, \varepsilon) = \sum_{k=0}^{N} X_k(x)\varepsilon^k,$$

where N is the integral part of $(3n+1)/2$ and the X_k are polynomials of degree at most $3n+1-2k$. This can be derived from the Taylor expansion of $f(x, \varepsilon)$ with the integral form of the remainder.

We propose an n-step procedure for the construction of invariant tori. The first step was discussed in [1]–[5] (in [5], bifurcation to invariant tori was studied in the case of quasiperiodic motions). The main results of this paper were presented in [6]–[8].

In §1 we describe the first step of the procedure and establish two theorems on bifurcation to an invariant torus at this step. If $m = 1$, this becomes the known phenomenon of bifurcation to a limit cycle [9]. In §2 we describe an arbitrary step of the procedure for the construction of invariant tori of maximal dimension (we are not referring here to the maximal possible dimension of an invariant torus in general, but to the maximal dimension obtained in this particular procedure). In §3 we prove the main existence theorem for invariant tori. The proof is a generalization of the proof of Lemma 2.1 in [10], which, in the author's words, was a "natural generalization" of proofs given in [11]. Some generalizations of the main result will be given in §§4–6. In §4 we consider the formation of an invariant torus not from an equilibrium state, but from an invariant torus of an arbitrary dimension. It turns out that under certain additional assumptions the procedure described in §§1–3 can be implemented in a neighborhood of the torus and will produce a torus of higher dimension. These additional assumptions are not restrictive if bifurcation from a closed orbit is considered. The equilibrium state can be replaced by an invariant torus at any step of the procedure, not only at the first one. In §5, we will see how to obtain invariant tori of fewer dimensions. In §§1–5 the small parameter ε is assumed to be a positive scalar. In general, the small parameter must vary in a neighborhood of a point in a multidimensional Euclidean space. In this case we set $\varepsilon = \|\mu\|$, $\mu = \varepsilon\delta$, where μ is a small multidimensional parameter, and apply the above procedure on fixed rays (see §6).

§1. First step of the procedure

The first step of the procedure for constructing invariant tori proceeds as follows.

Given m-vectors y, $\bar{y}$ (bar denotes complex conjugation) and a $(d-2m)$-vector z, we let $\mathrm{col}(y, \bar{y}, z)$ denote the column vector whose components are those of y, $\bar{y}$, z. Then the transformation

$$x = A\,\mathrm{col}(y, \bar{y}, z) + \varepsilon a, \qquad (3)$$

where A is a nonsingular $d \times d$ matrix and a is a d-vector, bring system (1) to the form

$$\dot{y} = i\lambda \cdot y + Y(y, \bar{y}, z, \varepsilon) + Y^*(y, \bar{y}, z, \varepsilon),$$
$$\dot{z} = Qz + Z(y, \bar{y}, z, \varepsilon) + Z^*(y, \bar{y}, z, \varepsilon) \qquad (4)$$

We denote $\lambda = \text{col}(\lambda^1, \ldots, \lambda^m)$, $\lambda \cdot y = \text{col}(\lambda^1 y^1, \ldots, \lambda^m y^m)$, where $y^1, \ldots, y^m$ are the coordinates of y. The asterisk $*$ means that the function in question possesses the property expressed by (2); for example,

$$Y^*(\sqrt{\varepsilon} y, \sqrt{\varepsilon} \bar{y}, \sqrt{\varepsilon} z, \varepsilon) = \sqrt{\varepsilon^{3n+2}} \tilde{Y},$$

where $\tilde{Y}(y, \bar{y}, z, \varepsilon)$ is a continuous function of class $C^1_{y,\bar{y},z}$. In (4), Y and Z are nonlinearities, i.e., functions whose expansions do not contain linear terms; $\lambda^{2m+1}, \ldots, \lambda^d$ are the eigenvalues of the matrix Q; the equation conjugated to the first one in system (4) is omitted. These conventions and notation will be used throughout the paper.

We require $\lambda^1, \ldots, \lambda^m$ to be incommensurable in the following sense:

$$\sum_{k=1}^{m} q_k \lambda^k \neq 0 \quad \text{for } 0 < \sum_{k=1}^{m} |q_k| \leq 3n + 2,$$

where all the q_k's are integers. Since Q is noncritical, under this condition, the polynomial change of variables

$$y = u + h(u, \bar{u}, \varepsilon), \qquad z = v + g(u, \bar{u}, \varepsilon) \qquad (5)$$

transforms the system

$$\dot{y} = i\lambda \cdot y + Y(y, \bar{y}, z, \varepsilon),$$
$$\dot{z} = Qz + Z(y, \bar{y}, z, \varepsilon)$$

to the form

$$\dot{u} = i\lambda \cdot u + u \cdot U(u \cdot \bar{u}, \varepsilon) + U^0(u, \bar{u}, v, \varepsilon) + U^*,$$
$$\dot{v} = Qv + V^0(u, \bar{u}, v, \varepsilon) + V^*, \qquad (6)$$

where U is a vector polynomial of degree at most N, $U(0, 0) = 0$; U^0, V^0 are nonlinearities, $U^0(u, \bar{u}, 0, \varepsilon) = 0$, $V^0(u, \bar{u}, 0, \varepsilon) = 0$. This assertion can be proved by the standard method of indeterminate coefficients (see [12] and the lemma in §2). Applying the change of variables (5) to system (4), we obtain a system of the same form as (6) but with different functions U^*, V^*, which still satisfy conditions of type (2).

We now transform system (6) to polar coordinates $\rho = \text{col}(\rho^1, \ldots, \rho^m)$, $\varphi = \text{col}(\varphi^1, \ldots, \varphi^m)$. Then

$$u = \rho \cdot e^{i\varphi}, \qquad \bar{u} = \rho \cdot e^{-i\varphi},$$

where $e^{i\varphi} = \mathrm{col}(e^{i\varphi^1}, \ldots, e^{i\varphi^m})$. We obtain the system

$$\dot{\rho} = \rho \cdot F(\rho^2, \varepsilon) + F^0(\rho, \varphi, v, \varepsilon) + F^*(\rho, \varphi, v, \varepsilon),$$
$$\dot{\varphi} = \lambda + \Phi(\rho^2, \varepsilon) + \rho^{-1} \cdot [\Phi^0(\rho, \varphi, v, \varepsilon) + \Phi^*(\rho, \varphi, v, \varepsilon)], \quad (7)$$
$$\dot{v} = Qv + V^0(\rho, \varphi, v, \varepsilon) + V^*(\rho, \varphi, v, \varepsilon),$$

where $\rho^2 = \rho \cdot \rho$; the vector ρ^{-1} is defined by the condition $\rho \cdot \rho^{-1} = (1, \ldots, 1)$; $F = \mathrm{Re}\, U(\rho^2, \varepsilon)$; $\Phi = \mathrm{Im}\, U(\rho^2, \varepsilon)$; F^0, Φ^0, V^0 are nonlinearities with respect to ρ, v; $F^0(\rho, \varphi, 0, \varepsilon) = 0$, $\Phi^0(\rho, \varphi, 0, \varepsilon) = 0$, $V^0(\rho, \varphi, 0, \varepsilon) = 0$; $F^*(\sqrt{\varepsilon}\rho, \varphi, \sqrt{\varepsilon}v, \varepsilon) = \sqrt{\varepsilon^{3n+2}}\tilde{F}$, $\Phi^*(\sqrt{\varepsilon}\rho, \varphi, \sqrt{\varepsilon}v, \varepsilon) = \sqrt{\varepsilon^{3n+2}}\tilde{\Phi}$, $V^*(\sqrt{\varepsilon}\rho, \varphi, \sqrt{\varepsilon}v, \varepsilon) = \sqrt{\varepsilon^{3n+2}}\tilde{V}$, where $\tilde{F}$, $\tilde{\Phi}$, $\tilde{V}$ are continuously differentiable in all arguments except ε. All the functions in system (7) are 2π-periodic with respect to the components of φ.

Obviously, $F(0, 0) = 0$. Denote the linear terms of F by $B\rho^2 + \varepsilon b$. We call

$$B\rho^2 + \varepsilon b = 0 \quad (8)$$

the bifurcation equation. If

$$\det B \neq 0, \quad (9)$$

we can find the solution $\rho^2 = -\varepsilon B^{-1} b$ of equation (8) and assume that $\rho^2 > 0$ in the sense that all its coordinates are positive. Rewrite the solution as $\rho^2 = \varepsilon \alpha^2$ and introduce new variables x_1, φ_1, v_1 in system (7) by

$$\rho = \sqrt{\varepsilon}(\alpha + x_1), \qquad v = \sqrt{\varepsilon} v_1, \qquad \varphi = \varphi_1.$$

This yields a system

$$\dot{x}_1 = \varepsilon X_1(x_1, \varepsilon) + \sqrt{\varepsilon} X_1^0 + \sqrt{\varepsilon^{3n+1}} \tilde{X}_1,$$
$$\dot{\varphi}_1 = \lambda_1(\varepsilon) + \varepsilon \Phi_1(x_1, \varepsilon) + \sqrt{\varepsilon} \Phi_1^0 + \sqrt{\varepsilon^{3n+1}} \tilde{\Phi}_1, \quad (10)$$
$$\dot{v}_1 = Qv_1 + \sqrt{\varepsilon} V_1^0 + \sqrt{\varepsilon^{3n+1}} \tilde{V}_1,$$

where X_1, Φ_1 are vector polynomials of degree at most $3n+1$, $X_1(0, 0) = 0$, $\Phi_1(0, \varepsilon) = 0$; X_1^0, Φ_1^0, V_1^0 vanish for $v_1 = 0$; $\tilde{X}_1$, $\tilde{\Phi}_1$, $\tilde{V}_1$ are continuous functions of the class $C_{x_1, \varphi_1, v_1}^1$ for $\|x_1\| < H_1$, $\varphi_1 \in \mathbb{R}^m$, $\|v_1\| < H_1$, $0 \leq \varepsilon < \varepsilon_0$, and in addition

$$P_1 = D_{x_1} X_1(0, 0) = 2(\mathrm{diag}\,\alpha) B (\mathrm{diag}\,\alpha)$$

where the matrix $\mathrm{diag}\,\alpha$ is defined by $\mathrm{diag}\,\alpha = \mathrm{diag}(\alpha^1, \ldots, \alpha^m)$.

THEOREM 1. *If P_1 is a noncritical matrix, then $n = 1$ and for every sufficiently small $\varepsilon > 0$, system (7) has an invariant torus defined by the equations*

$$\rho = \sqrt{\varepsilon}\alpha + \sqrt{\varepsilon^3}\psi(\varphi, \varepsilon), \qquad v = \sqrt{\varepsilon^5}\chi(\varphi, \varepsilon),$$

where ψ, χ are functions continuous for all φ and $0 \leq \varepsilon < \varepsilon_0$ and 2π-periodic in components of φ. In other words, we have bifurcation of an m-dimensional torus.

Theorem 1 is a special case of Theorem 2 for $p = 1$ and of Theorem 3 (§3) for $n = 1$.

When $m = 1$, the invariant torus is just a closed orbit. The bifurcation thus obtained is a Hopf bifurcation [9]. Condition (9) then means that when $\varepsilon = 0$ the first focal magnitude is not zero. We know that Hopf bifurcation also takes place in a more general situation when a focal magnitude of arbitrary index does not vanish if $\varepsilon = 0$. An analogue of this situation occurs when $m > 1$, if the polynomial $G(\rho^2)$ of terms of the smallest degree in the vector polynomial $F(\rho^2, 0)$ in system (7) is a homogeneous vector polynomial of order $p > 1$ in the components of ρ^2.

In this situation, we consider system (7) with $3n + 2$ replaced by $4p + 1$. As bifurcation equation we take

$$F(\rho^2, \varepsilon) = G(\rho^2) + \varepsilon b + \cdots = 0.$$

Setting $\delta = \varepsilon^{1/p}$, $\rho^2 = \delta z$, we write this equation as

$$G(z) + b + \mu(z, \delta) = 0, \quad (11)$$

where $\mu(z, 0) = 0$. Assume that (11) has a solution $z = \beta^2(\delta) > 0$. Introduce in (7) the variables r, w defined by

$$\rho = \sqrt{\delta}\beta + \varepsilon r, \qquad v = \varepsilon^2 w. \quad (12)$$

We obtain the system

$$\begin{aligned}
\dot{r} &= \varepsilon S r + \varepsilon \sqrt{\delta} R(r, \varphi, w, \varepsilon), \\
\dot{\varphi} &= \lambda(\varepsilon) + \varepsilon \sqrt{\delta} \Psi(r, \varphi, w, \varepsilon), \\
\dot{w} &= Q w + \sqrt{\delta} W(r, \varphi, w, \varepsilon),
\end{aligned} \quad (13)$$

where R, Ψ, W are continuous functions of class $C^1_{r, \varphi, w}$, and the constant matrix S is defined by

$$S = 2 \operatorname{diag} \beta(0) DG(\beta^2(0)) \operatorname{diag} \beta(0).$$

THEOREM 2. *Assume that condition* (2) *is satisfied with* $3n + 2$ *replaced by* $4p + 1$, *and that the equation* $G(z) + b = 0$ *has a coordinatewise positive solution* $z = \Delta$ *such that* $\det DG(\Delta) \neq 0$. *Then, if S is a noncritical matrix, system* (7) *has an invariant m-dimensional torus for every sufficiently small* $\varepsilon > 0$.

PROOF. By the Implicit Function Theorem, equation (11) has a solution $z = \beta^2(\delta)$ with $\beta^2(0) = \Delta$. The change of variables (12) transforms system (7) into system (13). The latter satisfies the conditions of Lemma 2.1 in [10], which states that for a sufficiently small $\varepsilon > 0$, system (13) has an invariant manifold defined by the equations $r = \sqrt{\delta}\psi(\varphi, \varepsilon)$, $w = \sqrt{\delta}\chi(\varphi, \varepsilon)$, where ψ, χ are continuous and 2π-periodic in the components of φ. Inserting these equations into (12), we obtain an invariant torus of system (7). Incidentally, the assertion of Theorem 2 can also be derived from Theorem 3 (§3).

REMARK. If $G(z)$ is a homogeneous vector polynomial of odd order and $\det DG(z) \neq 0$ for $z \neq 0$, then the equation $G(z) + b = 0$ has a solution.

This follows from the known fact that the vector fields $G(z)+b$ and $G(z)$ are homotopic on the sphere $\|z\| = R$, for a sufficiently large R.

The statements of Theorems 1 and 2 may be interpreted as follows. Consider the system

$$\dot{\rho} = \rho[G(\rho^2) + \varepsilon b], \qquad (14)$$

where, under the conditions of Theorem 1, $G(\rho^2) = B\rho^2$. We have proved that for any coordinatewise positive equilibrium state of system (14) such that the corresponding variational system has a noncritical matrix of coefficients, system (7) has an m-dimensional invariant torus. This torus can also be constructed using an invariant torus of system (14) (provided it exists) instead of an equilibrium state of the system; in particular, we can use a closed orbit (see §4).

§2. A general step of the procedure

We now drop the requirement that the matrix P_1 in system (10) is noncritical. By condition (9), $\det P_1 \neq 0$. Therefore, the system

$$\dot{x}_1 = \varepsilon X_1(x_1, \varepsilon)$$

is of the same type as

$$\dot{x} = X(x, \varepsilon),$$

but with a "slow" time εt and in at most half as many dimensions. Applying the transformations described in §1, we get a procedure that stops after a finite number of steps, which we denote by n.

We now describe the general step of the procedure. Suppose that after s steps ($1 \leq s < n$) we obtain the system

$$\dot{x}_s = \varepsilon^s X_s(x_s, \varepsilon) + \sum_{i=1}^{s} \sqrt{\varepsilon^{2i-1}} X_s^i + \sqrt{\varepsilon^{3n+2-s}} \tilde{X}_s,$$

$$\dot{\varphi}_{ks} = \varepsilon^{k-1}\lambda_{ks}(\varepsilon) + \varepsilon^s \Phi_{ks}(x_s, \varepsilon) + \sum_{i=1}^{s} \sqrt{\varepsilon^{2i-1}} \Phi_{ks}^i + \sqrt{\varepsilon^{3n+2-s}} \tilde{\Phi}_{ks}, \qquad (15)$$

$$\dot{v}_{ks} = \varepsilon^{k-1}Q_k v_{ks} + \sum_{i=1}^{s} \sqrt{\varepsilon^{2i-1}} V_{ks}^i + \sqrt{\varepsilon^{3n+2-s}} \tilde{V}_{ks}, \qquad k = 1, \ldots, s.$$

Here x_s, φ_{ks}, v_{ks} are vector variables, the sum of their dimensions being d. We denote the dimension of x_s by m_{s-1} ($m_0 = m$); φ_{ks} are the vectors of angular variables. Hence the dependence on the components of φ_{ks} is 2π-periodic, and $\dim \varphi_{ks} = \dim x_k = m_{k-1}$, $\dim v_{ks} + 2m_k = m_{k-1}$. The last equality implies that $m_s \leq m_{s-1}/2$.

We adopt the following assumptions:

I_s: Q_k are noncritical matrices, $P_s = D_{x_s} X_s(0, 0)$ is a nonsingular matrix; X_s, Φ_{ks} are polynomials in x_s^j and ε, $j = 1, \ldots, m_{s-1}$; $X_s(0, 0) = 0$, $\Phi_{ks}(0, \varepsilon) = 0$;

II_s: X_s^i, Φ_{ks}^i, V_{ks}^i, $\tilde{X}_s$, $\tilde{\Phi}_{ks}$, $\tilde{V}_{ks}$ are C^1-continuous functions of x_s, φ_{js}, v_{js}, ε in all variables except ε, in the domain

$$\mathfrak{M}_s = \{\|x_s\| < H_s, \ \varphi_{js} \in \mathbb{R}^{m_{j-1}}, \ \|v_{js}\| < H_s, \ 0 \leq \varepsilon < \varepsilon_0\}, \quad j = 1, \ldots, s;$$

X_s^i, Φ_{ks}^i, V_{ks}^i vanish for $v_{is} = 0$.

Clearly, system (10) with $Q = Q_1$ satisfies conditions I_1–II_1.

Assume that the matrix P_s has $m_s > 0$ pairs of imaginary eigenvalues $\pm i\lambda_s^1, \ldots, \pm i\lambda_s^{m_s}$, constituting a vector $i\lambda_s$, and that other eigenvalues (if they exist) do not lie on the imaginary axis. Consider the system

$$\dot{x}_s = \varepsilon^s X_s(x_s, \varepsilon).$$

Applying the nonsingular transformation

$$x_s = A_s \operatorname{col}(y_s, \bar{y}_s, \bar{z}_s) + \varepsilon a_s, \tag{16}$$

we reduce this system to the form

$$\begin{aligned} \dot{y}_s &= \varepsilon^s [i\lambda_s \cdot y_s + Y_s(y_s, \bar{y}_s, z_s, \varepsilon)], \\ \dot{z}_s &= \varepsilon^s [Q_{s+1} z_s + Z_s(y_s, \bar{y}_s, z_s, \varepsilon)]. \end{aligned} \tag{17}$$

We augment (17) by adding

$$\dot{\varphi}_{ks} = \varepsilon^{k-1} \lambda_{ks}(\varepsilon) + \varepsilon^s \Phi_{ks}(y_s, \bar{y}_s, z_s, \varepsilon), \quad k = 1, \ldots, s, \tag{18}$$

where $\Phi_{ks}(y_s, \bar{y}_s, z_s, \varepsilon)$ is obtained from Φ_{ks} in system (15) by the linear change of variables (16).

LEMMA. *If* $\lambda_s^1, \ldots, \lambda_s^{m_s}$ *satisfy the condition*

$$\sum_{k=1}^{m_s} q_k \lambda_s^k \neq 0 \quad \text{for } 0 < \sum_{k=1}^{m_s} |q_k| \leq 3(n-s) + 2, \tag{19}$$

then there exists a polynomial transformation

$$\begin{aligned} y_s &= u_s + h_s(u_s, \bar{u}_s, \varepsilon), \\ z_s &= w_s + g_s(u_s, \bar{u}_s, \varepsilon), \\ \varphi_{ks} &= \varphi_{k,s+1} + f_{ks}(u_s, \bar{u}_s, \varepsilon), \quad k = 1, \ldots, s, \end{aligned} \tag{20}$$

which reduces system (17), (18) *to the form*

$$\dot{u}_s = \varepsilon^s[i\lambda_s \cdot u_s + u_s \cdot U_s(u_s \cdot \bar{u}_s, \varepsilon) + U_s^0 + U_s^*],$$
$$\dot{w}_s = \varepsilon^s[Q_{s+1}w_s + W_s^0 + W_s^*], \qquad (21)$$
$$\dot{\varphi}_{k,s+1} = \varepsilon^{k-1}\lambda_{ks}(\varepsilon) + \varepsilon^s[\Psi_{ks}(u_s \cdot \bar{u}_s, \varepsilon) + \Psi_{ks}^0 + \Psi_{ks}^*], \quad k = 1, \ldots, s,$$

where U_s, Ψ_{ks} *are polynomials in powers of* ε *and* $u_s^j \bar{u}_s^j$, $j = 1, \ldots, m_s$, *of degree at most* N_s, *where* N_s *is the integral part of* $[3(n-s) + 1]/2$, $U_s(0, 0) = 0$, $\Psi_{ks}(0, 0) = 0$, U_s^0, W_s^0, Ψ_{ks}^0 *vanish for* $w_s = 0$, U_s^0, W_s^0 *are nonlinearities, by* (2), *the order of the functions* U_s^*, W_s^*, Ψ_{ks}^* *is* $3(n-s) + 2$.

PROOF. Differentiating (20) with respect to t, setting $w_s = 0$, dividing by ε^s, and using (17), (18), (21), we obtain the following equations for h_s, g_s, f_{ks}:

$$i\frac{\partial h_s}{\partial u_s}(\lambda_s \cdot u_s) - i\frac{\partial h_s}{\partial \bar{u}_s}(\lambda_s \cdot \bar{u}_s) - i\lambda_s \cdot h_s$$
$$= Y_s(u_s + h_s, \bar{u}_s + \bar{h}_s g_s, \varepsilon) - u_s \cdot U_s$$
$$- U_s^* - \frac{\partial h_s}{\partial u_s}(u_s \cdot U_s + U_s^*) - \frac{\partial h_s}{\partial \bar{u}_s}(\bar{u}_s \cdot \bar{U}_s + \bar{U}_s^*),$$

$$i\frac{\partial g_s}{\partial u_s}(\lambda_s \cdot u_s) - i\frac{\partial g_s}{\partial \bar{u}_s}(\lambda_s \cdot \bar{u}_s) - Q_{s+1}g_s$$
$$= Z_s(u_s + h_s, \bar{u}_s + \bar{h}_s g_s, \varepsilon) - W_s^*$$
$$- \frac{\partial g_s}{\partial u_s}(u_s \cdot U_s + U_s^*) - \frac{\partial g_s}{\partial \bar{u}_s}(\bar{u}_s \cdot \bar{U}_s + \bar{U}_s^*),$$

$$i\frac{\partial f_{ks}}{\partial u_s}(\lambda_s \cdot u_s) - i\frac{\partial f_{ks}}{\partial \bar{u}_s}(\lambda_s \cdot \bar{u}_s)$$
$$= \Phi_{ks}(u_s + h_s, \bar{u}_s + \bar{h}_s g_s, \varepsilon) - \Psi_{ks}$$
$$- \Psi_{ks}^* - \frac{\partial f_{ks}}{\partial u_s}(u_s \cdot U_s + U_s^*) - \frac{\partial f_{ks}}{\partial \bar{u}_s}(\bar{u}_s \cdot \bar{U}_s + \bar{U}_s^*),$$
$$k = 1, \ldots, s.$$

This system can be solved using indeterminate coefficients, by equating the homogeneous vector polynomials in u_s, $\bar{u}_s$, $\sqrt{\varepsilon}$ of order at most $3(n-s)+1$ on both sides. This simultaneously determines the polynomials h_s, g_s, f_{ks}

and U_s, Ψ_{ks} (see §4). The asterisked functions are not involved in the expressions for these polynomials. The lemma is proved.

We now apply (20) to the system derived from (15) by the linear substitution (16) and then transform to polar coordinates ρ_s, φ_s: $u_s = \rho_s \cdot e^{i\varphi_s}$, $\bar{u}_s = \rho_s \cdot e^{-i\varphi_s}$. It follows from the lemma that the system thus obtained has the form

$$\dot{\rho}_s = \varepsilon^s[\rho_s \cdot F_s(\rho^2, \varepsilon) + F_s^0 + F_s^*] + \sum_{i=1}^s \sqrt{\varepsilon^{2i-1}} F_s^i + \sqrt{\varepsilon^{3n+2-s}} \tilde{F}_s,$$

$$\dot{\varphi}_s = \varepsilon^s[\lambda_s + G_s(\rho^2, \varepsilon) + \rho_s^{-1} \cdot (G_s^0 + G_s^*)]$$
$$+ \rho_s^{-1}\left[\sum_{i=1}^s \sqrt{\varepsilon^{2i-1}} G_s^i + \sqrt{\varepsilon^{3n+2-s}} \tilde{G}_s\right],$$

$$\dot{w}_s = \varepsilon^s(Q_{s+1}w_s + W_s^0 + W_s^*) + \sum_{i=1}^s \sqrt{\varepsilon^{2i-1}} W_s^i + \sqrt{\varepsilon^{3n+2-s}} \tilde{W}_s, \quad (22)$$

$$\dot{\varphi}_{k,s+1} = \varepsilon^{k-1}\lambda_{ks}(\varepsilon) + \varepsilon^s[\Psi_{k,s}(\rho^2, \varepsilon) + \Psi_{ks}^0 + \Psi_{ks}^*]$$
$$+ \sum_{i=1}^s \sqrt{\varepsilon^{2i-1}} \Psi_{ks}^i + \sqrt{\varepsilon^{3n+2-s}} \tilde{\Psi}_{ks},$$

$$\dot{v}_{ks} = \varepsilon^{k-1} Q_k v_{ks} + \sum_{i=1}^s \sqrt{\varepsilon^{2i-1}} V_{ks}^i + \sqrt{\varepsilon^{3n+2-s}} \tilde{V}_{ks},$$

$$k = 1, \ldots, s,$$

where $F_s = \operatorname{Re} U_s(\rho^2, \varepsilon)$; $G_s = \operatorname{Im} U_s(\rho^2, \varepsilon)$; the properties of the other functions in the right-hand sides of the equations in (22) are determined by the corresponding indices. For example, F_s^0 is a nonlinearity in ρ_s and w_s, which vanishes for $w_s = 0$; $F_s^*(\sqrt{\varepsilon}\rho_s, \varphi_s, \sqrt{\varepsilon}w_s, \varepsilon) = \sqrt{\varepsilon^{3(n-s)+2}} F_s'$ where F_s' is a C^1-continuous function of ρ_s, φ_s, w_s.

Let $F_s = B_s \rho_s^2 + \varepsilon b_s + \cdots$. Assuming that

$$\det B_s \neq 0, \qquad (23)$$

let us consider the bifurcation equation

$$B_s \rho_s^2 + \varepsilon b_s = 0 \qquad (24)$$

under the assumption that it has a solution such that $\rho_s^2 = -\varepsilon B_s^{-1} b_s > 0$ in the sense that each coordinate is positive. Denote the solution by $\rho_s^2 = \varepsilon \alpha_s^2$. We introduce new variables x_{s+1}, v_{s+1}:

$$\rho_s = \sqrt{\varepsilon}(\alpha_s + x_{s+1}), \qquad w_s = \sqrt{\varepsilon} v_{s+1}. \qquad (25)$$

As a result, it follows from system (22) for $|x_{s+1}^j| < \alpha_s^j$, $j = 1, \ldots, m_s$, that

$$\dot{x}_{s+1} = \varepsilon^{s+1} X_{s+1}(x_{s+1}, \varepsilon) + \sum_{i=1}^{s+1} \sqrt{\varepsilon^{2i-1}} X_{s+1}^i + \sqrt{\varepsilon^{3n+1-s}} \tilde{X}_{s+1},$$

$$\dot{\varphi}_{k,s+1} = \varepsilon^{k-1} \lambda_{k,s+1} + \varepsilon^{s+1} \Phi_{k,s+1}(x_{s+1}, \varepsilon) + \sum_{i=1}^{s+1} \sqrt{\varepsilon^{2i-1}} \Phi_{k,s+1}^i$$

$$+ \sqrt{\varepsilon^{3n+1-s}} \tilde{\Phi}_{k,s+1},$$

$$\dot{v}_{k,s+1} = \varepsilon^{k-1} Q_k v_{k,s+1} + \sum_{i=1}^{s+1} \sqrt{\varepsilon^{2i-1}} V_{k,s+1}^i + \sqrt{\varepsilon^{3n+1-s}} \tilde{V}_{k,s+1},$$

$$k = 1, \ldots, s+1,$$

where we denote $\varphi_{s+1} = \varphi_{s+1,s+1}$, $v_{s+1} = v_{s+1,s+1}$, $v_{ks} = v_{k,s+1}$, $k = 1, \ldots, s$; note also that $P_{s+1} = D_{x_{s+1}} X_{s+1}(0, 0) = 2(\operatorname{diag} \alpha_s) B_s (\operatorname{diag} \alpha_s)$. It is easy to see that this system satisfies conditions $\mathrm{I}_{s+1} - \mathrm{II}_{s+1}$, i.e., the same conditions as imposed on system (15), but with s replaced by $s + 1$.

§3. Main theorem

It follows from the results of §§1, 2 that after the last, nth, step we obtain a system

$$\dot{x}_n = \varepsilon^n X_n(x_n, \varepsilon) + \sum_{i=1}^{n} \sqrt{\varepsilon^{2i-1}} X_n^i + \varepsilon^{n+1} \tilde{X}_n,$$

$$\dot{\varphi}_{kn} = \varepsilon^{k-1} \lambda_{kn}(\varepsilon) + \varepsilon^n \Phi_{kn}(x_n, \varepsilon) + \sum_{i=1}^{n} \sqrt{\varepsilon^{2i-1}} \Phi_{kn}^i + \varepsilon^{n+1} \tilde{\Phi}_{kn}, \qquad (26)$$

$$\dot{v}_{kn} = \varepsilon^{k-1} Q_k v_{kn} + \sum_{i=1}^{n} \sqrt{\varepsilon^{2i-1}} V_{kn}^i + \varepsilon^{n+1} \tilde{V}_{kn},$$

$$k = 1, \ldots, n,$$

which satisfies conditions $\mathrm{I}_n - \mathrm{II}_n$. Define $P_n = D_{x_n} X_n(0, 0)$.

THEOREM 3. *If P_n is a noncritical matrix, then for every sufficiently small $\varepsilon > 0$ system (26) has an invariant manifold, defined by the equations*

$$x_n = \varepsilon \psi_n(\varphi_{1n}, \ldots, \varphi_{nn}, \varepsilon),$$
$$v_{kn} = \varepsilon^{n-k+2} \psi_{k-1}(\varphi_{1n}, \ldots, \varphi_{nn}, \varepsilon), \qquad (27)$$
$$k = 1, \ldots, n,$$

where $\psi_0, \ldots, \psi_n$ are continuous and 2π-periodic functions of the components of $\varphi_{1n}, \ldots, \varphi_{nn}$. Using (25), we obtain an invariant torus of system (7) of dimension $m = m_0 + \cdots + m_{n-1}$.

PROOF. Replace x_n by $\sqrt{\varepsilon}x_n$ and v_{kn} by $\sqrt{\varepsilon^{2n-2k+3}}v_{kn}$ in system (26). The system becomes

$$\begin{aligned}\dot{x}_n &= \varepsilon^n P_n x_n + \sqrt{\varepsilon^{2n+1}}\tilde{X},\\ \dot{v}_{kn} &= \varepsilon^{k-1}Q_k v_{kn} + \sqrt{\varepsilon^{2k-1}}\tilde{V}_k,\\ \dot{\varphi}_{kn} &= \varepsilon^{k-1}\lambda_{kn}(\varepsilon) + \sqrt{\varepsilon^{2n+1}}\tilde{\Phi}_k,\\ & k = 1, \ldots, n,\end{aligned} \qquad (28)$$

where $\tilde{X}$, $\tilde{V}_k$, $\tilde{\Phi}_k$ are C^1-continuous in $\mathfrak{M}_n$ with respect to all variables except ε.

Define $y_{k-1} = v_{kn}$, $k = 1, \ldots, n$, $y_n = x_n$, $\varphi = \text{col}(\varphi_{1n}, \ldots, \varphi_{nn})$, and set $y_k = \sqrt{\varepsilon}z_k$, $k = 0, 1, \ldots, n$. Then system (28) can be rewritten as

$$\begin{aligned}\dot{z}_k &= \varepsilon^k P_k z_k + \varepsilon^k Y_k(\sqrt{\varepsilon}z_0, \ldots, \sqrt{\varepsilon}z_n, \varphi, \varepsilon),\\ \dot{\varphi} &= \mu(\varepsilon) + \varepsilon^{n+1/2}\Phi(\sqrt{\varepsilon}z_0, \ldots, \sqrt{\varepsilon}z_n, \varphi, \varepsilon),\\ & k = 0, 1, \ldots, n,\end{aligned} \qquad (29)$$

where P_k are noncritical matrices; Φ, Y_k are continuous and 2π-periodic in the components of φ, of class $C^1_{z,\varphi}$ for $\|z_k\| < \sqrt{\varepsilon_0^{-1}H_n}$, $\varphi \in \mathbb{R}^m$, $0 \leq \varepsilon < \varepsilon_0$. Denote

$$M(\varepsilon) = \max\left\{\sup_\varphi\|\Phi(0, \ldots, 0, \varphi, \varepsilon)\|, \sup_\varphi\|Y_k(0, \ldots, 0, \varphi, \varepsilon)\|, \right.$$
$$\left. k = 0, \ldots, n\right\}$$

and let $L(\varepsilon)$ be a Lipschitz constant for $Y_0, \ldots, Y_n, \Phi$ as functions of $y_0, \ldots, y_n, \varphi$.

We now follow the proof of Lemma 2.1 in [10], which deals with a special case of system (29). Represent P_k as $P_k = \text{diag}(P_k^+, P_k^-)$, where the real parts of the eigenvalues of P_k^+ and P_k^- are positive and negative, respectively. We introduce the following matrix functions:

$$J_k(t) = \begin{cases} -\begin{pmatrix} e^{-P_k^+ t} & 0 \\ 0 & 0 \end{pmatrix} & \text{for } t > 0,\\ \begin{pmatrix} 0 & 0 \\ 0 & e^{-P_k^- t} \end{pmatrix} & \text{for } t < 0,\end{cases}$$
$$k = 0, 1, \ldots, n.$$

Properties of P_k^+, P_k^- show that the following estimates hold for $t \in \mathbb{R}$:

$$\|J_k(t)\| \leq \beta e^{-\alpha|t|}, \qquad \alpha > 0, \beta > 0. \qquad (30)$$

Let $C_r(D, \Delta)$ be the class of r-dimensional continuous vector functions $F(\varphi, \varepsilon)$ that are 2π-periodic in the components of φ and satisfy the inequality $\|F\| \leq D(\varepsilon) < \sqrt{\varepsilon^{-1}H_n}$ and a Lipschitz condition with respect to

φ with Lipschitz constant $\Delta(\varepsilon)$. In what follows we will treat ε as a fixed number in the interval $0 \leq \varepsilon < \varepsilon_0$, so that the dependence of functions on ε will be omitted. Consider the differential equation

$$\dot{\varphi} = \mu + \varepsilon^{n+1/2} \Phi(\sqrt{\varepsilon} F, \varphi), \tag{31}$$

where $F = (F_0, \ldots, F_n)$, $F_k \in C_{m_k}(D, \Delta)$, $m_k = \dim z_k$. Let $\varphi_t = T^F(t, \varphi_0)$ denote the solution of equation (3.1) that equals φ_0 at $t = 0$. Similarly, we set $\varphi_t^* = T^{F^*}(t, \varphi^*)$. Then

$\|\varphi_t^* - \varphi_t\|$

$$\leq \|\varphi^* - \varphi_0\| + \varepsilon^{n+1/2} \left| \left(\int_0^t \|\Phi(\sqrt{\varepsilon} F^*(\varphi_\tau^*), \varphi_\tau^*) - \Phi(\sqrt{\varepsilon} F(\varphi_\tau), \varphi_\tau)\| \, d\tau \right) \right|$$

$$\leq \|\varphi^* - \varphi_0\| + \varepsilon^{n+1/2} L \left| \int_0^t \left(\|\varphi_\tau^* - \varphi_\tau\| + \sqrt{\varepsilon} \sum_{k=0}^n \|F_k^*(\varphi_\tau^*) - F_k(\varphi_\tau)\| \right) d\tau \right|.$$

Adding and subtracting $F_k(\varphi_\tau^*)$ in the norms of the differences in the sums, we obtain

$\|\varphi_t^* - \varphi_t\|$

$$\leq \|\varphi^* - \varphi_0\| + \varepsilon^{n+1/2} L(1 + \sqrt{\varepsilon} p \Delta) \left| \int_0^t \|\varphi_\tau^* - \varphi_\tau\| \, d\tau \right| + \varepsilon^{n+1} L R(F^*, F)|t|, \tag{32}$$

where $p = n+1$ and $R(F^*, F) = \sup_\varphi (\sum_{k=0}^n \|F_k^*(\varphi) - F_k(\varphi)\|)$. By the Gronwall Lemma,

$$\|\varphi_t^* - \varphi_t\| \leq \|\varphi^* - \varphi_0\| + \varepsilon^{n+1} LR|t| + \varepsilon^{n+1/2} L(1 + \sqrt{\varepsilon} p \Delta)$$

$$\times \left| \int_0^t \exp\{\varepsilon^{n+1/2} L(1 + \sqrt{\varepsilon} p \Delta)|t - \tau|\} (\|\varphi^* - \varphi_0\| + \varepsilon^{n+1} LR|\tau|) \, d\tau \right|.$$

Integrating by parts, we find

$$\|\varphi_t^* - \varphi_t\| \leq \|\varphi^* - \varphi_0\| \exp\{\varepsilon^{n+1/2} L(1 + \sqrt{\varepsilon} p \Delta)|t|\}$$
$$+ \frac{\sqrt{\varepsilon} R}{1 + \sqrt{\varepsilon} p \Delta} (\exp\{\varepsilon^{n+1/2} L(1 + \sqrt{\varepsilon} p \Delta)|t|\} - 1). \tag{33}$$

Now set $S_\varphi(F) = (S_\varphi^0(F), \ldots, S_\varphi^n(F))$, where

$$S_\varphi^k(F) = \varepsilon^k \int_{-\infty}^{+\infty} J_k(\varepsilon^k t) Y_k(\sqrt{\varepsilon} F(T^F(t, \varphi)), T^F(t, \varphi)) \, dt.$$

Since $T^F(t, \varphi + 2\pi I) = T^F(t, \varphi) + 2\pi I$, where $I = (1, \ldots, 1)$, the function $S_\varphi^k(F)$ is 2π-periodic in the components of φ. Using (30) and making ε_0

smaller if necessary, we obtain the following estimate for $k = 0, 1, \ldots, n$:

$$\|S_\varphi^k(F)\| \leq 2\alpha^{-1}\beta(M + \sqrt{\varepsilon}pLD) < D = 4\alpha^{-1}\beta M. \tag{34}$$

Furthermore, by (30),

$$\|S_{\varphi^*}^k(F^*) - S_{\varphi_0}^k(F)\|$$
$$\leq \varepsilon^k \beta L \int_{-\infty}^{+\infty} e^{-\varepsilon^k \alpha |t|} \left(\|\varphi_t^* - \varphi_t\| + \sqrt{\varepsilon} \sum_{k=0}^{n} \|F_k^*(\varphi_t^*) - F_k(\varphi_t)\| \right) dt.$$

Arguing as in the proof of (32), we obtain

$$\|S_{\varphi^*}^k(F^*) - S_{\varphi_0}^k\|$$
$$\leq \varepsilon^k \beta L \left[\sqrt{\varepsilon} R \int_{-\infty}^{+\infty} e^{-\varepsilon^k \alpha |t|} dt + (1 + \sqrt{\varepsilon}p\Delta) \int_{-\infty}^{+\infty} e^{-\varepsilon^k \alpha |t|} \|\varphi_t^* - \varphi_t\| dt \right].$$

Together with (33), this yields the estimate

$$\|S_{\varphi^*}^k(F^*) - S_{\varphi_0}^k(F)\|$$
$$\leq \varepsilon^k \beta L \left[\sqrt{\varepsilon} R \int_{-\infty}^{+\infty} \exp\{\varepsilon^{n+1/2}L(1 + \sqrt{\varepsilon}p\Delta) - \varepsilon^k \alpha)|t|\} dt \right.$$
$$\left. + (1 + \sqrt{\varepsilon}p\Delta)\|\varphi^* - \varphi_0\| \int_{-\infty}^{+\infty} \exp\{(\varepsilon^{n+1/2}L(1 + \sqrt{\varepsilon}p\Delta) - \varepsilon^k \alpha)|t|\} dt \right],$$
$$k = 0, 1, \ldots, n. \tag{35}$$

Let us stipulate that for $\varepsilon < \varepsilon_0$

$$\varepsilon^{n+1/2}L(1 + \sqrt{\varepsilon}p\Delta) - \varepsilon^k \alpha < -\varepsilon^k \alpha/2. \tag{36}$$

Then, by (35),

$$\|S_{\varphi^*}^k(F^*) - S_{\varphi_0}^k(F)\| \leq 4\alpha^{-1}\beta L[\sqrt{\varepsilon}R + (1 + \sqrt{\varepsilon}p\Delta)\|\varphi^* - \varphi_0\|].$$

Making ε smaller, if necessary, we can insure that for $\varepsilon < \varepsilon_0$

$$4\sqrt{\varepsilon}\alpha^{-1}\beta L < 1/(2p), \qquad 4\alpha^{-1}\beta L(1 + \sqrt{\varepsilon}p\Delta) < \Delta = 8\alpha^{-1}\beta L. \tag{37}$$

Then

$$\|S_{\varphi^*}^k(F^*) - S_{\varphi_0}^k(F)\| \leq R(F^*, F)/(2p) + \Delta\|\varphi^* - \varphi_0\|, \tag{38}$$
$$k = 0, 1, \ldots, n.$$

In particular, if $F = F^*$,

$$\|S_{\varphi^*}^k(F^*) - S_{\varphi_0}^k(F)\| \leq \Delta\|\varphi^* - \varphi_0\|.$$

It follows that S maps the class of functions $\prod_{k=0}^{n} C_{m_k}(D, \Delta)$ into itself, provided that D, Δ, ε_0 are chosen subject to (34), (36), (37). Moreover, this mapping is a contraction. Indeed, summing inequalities (38) over $k = 0, 1, \ldots, n$ and setting $\varphi^* = \varphi_0 = \varphi$, we obtain

$$R(S_\varphi(F^*), S_\varphi(F)) \leq R(F^*, F)/2.$$

By the contraction mapping principle, S has a fixed point $\psi = (\psi_0, \ldots, \psi_n)$, where

$$\psi_k(\varphi) = \varepsilon^k \int_{-\infty}^{+\infty} J_k(\varepsilon^k \tau) Y_k(\sqrt{\varepsilon}\psi(T^\psi(\tau, \varphi)), T^\psi(\tau, \varphi)) \, d\tau.$$

Now put $\varphi = \varphi_t = T^\psi(t, \varphi_0)$, $\tau + t = u$. Since equation (31) is autonomous, it follows that

$$T^\psi(\tau, \varphi_t) = T^\psi(u - t, T^\psi(t, \varphi_0)) = T^\psi(u, \varphi_0) = \varphi_u.$$

Therefore,

$$z_k(t) = \psi_k(\varphi_t) = \varepsilon^k \int_{-\infty}^{t} J_k(\varepsilon^k u - \varepsilon^k t) Y_k(\sqrt{\varepsilon} z(u), \varphi_u) \, du$$

$$+ \varepsilon^k \int_{t}^{\infty} J_k(\varepsilon^k u - \varepsilon^k t) Y_k(\sqrt{\varepsilon} z(u), \varphi_u) \, du.$$

Differentiating with respect to t, we see that the functions $z_k(t)$ satisfy system (29) for all φ_0. This means that the equation $z = \psi(\varphi, \varepsilon)$ defines an invariant manifold of system (29). Expressing the variables x_n, v_{kn} of (26) in terms of z_k as in (29), we arrive at equation (27), and the theorem follows.

§4. Formation of an invariant torus from an invariant torus of smaller dimension

In §§1–3 we studied the situation when the one-parameter vector field defined by system (1) in the chosen coordinates has a singular point $\varepsilon = 0$. Now we assume that at $\varepsilon = 0$ the vector field has an l-dimensional invariant torus T with angular coordinates $(\theta^1, \ldots, \theta^l) = \theta$. Suppose that in suitable coordinates $(x^1, \ldots, x^d) = x$, θ, the torus T is defined in its neighborhood by the equation $x = 0$, while for $\varepsilon \geq 0$ the vector field is determined in a neighborhood of T by the following expression:

$$\begin{aligned} \dot{x} &= Px + X(x, \theta, \varepsilon) + X^*(x, \theta, \varepsilon), \\ \dot{\theta} &= \omega + \Theta(x, \theta, \varepsilon) + \Theta^*(x, \theta, \varepsilon), \end{aligned} \quad (39)$$

where P is a constant $d \times d$ matrix; ω is the constant l-vector of frequencies of T; $X^* : \mathfrak{N} \to \mathbb{R}^d$, $\Theta^* : \mathfrak{N} \to \mathbb{R}^d$, $\mathfrak{N} = \{(x, \theta, \varepsilon) : \|x\| < H, \theta \in$

$\mathbb{R}^l$, $0 \le \varepsilon < \varepsilon_0\}$; X^*, Θ^* are continuous functions satisfying the conditions

$$X^*(\sqrt{\varepsilon}x, \theta, \varepsilon) = \sqrt{\varepsilon^{3n+2}}\tilde{X}, \quad \Theta^*(\sqrt{\varepsilon}x, \theta, \varepsilon) = \sqrt{\varepsilon^{3n+2}}\tilde{\Theta}, \qquad (40)$$

where $\tilde{X}$, $\tilde{\Theta}$ are functions of class $C^{1,1,0}_{x,\theta,\varepsilon}$; X, Θ are polynomials in x, ε whose coefficients are sufficiently smooth functions of θ; X is nonlinearity in x, ε; $\Theta(0, \theta, 0) = 0$. All functions are periodic of period 2π in $\theta^1, \ldots, \theta^l$.

Let us consider the case $\det P \ne 0$. If P is noncritical, it follows from Theorem 3 (since (39) corresponds to system (26) with $n = 0$) that for sufficiently small $\varepsilon > 0$ system (39) has an invariant l-dimensional torus T_ε, and $T_\varepsilon \to T$ as $\varepsilon \to 0$. What can we say about the existence of an invariant torus of more than l dimensions for small $\varepsilon > 0$? Assume that P possesses $m > 0$ pairs of pure imaginary eigenvalues $\pm i\lambda^1, \ldots, \pm i\lambda^m$, while other eigenvalues of P stay away from the imaginary axis. Under this assumption, we can no longer assert that system (39) has an invariant l-dimensional torus T_ε, so the existence of an invariant torus of higher dimension does not necessarily imply the occurrence of bifurcation phenomena in system (39).

We will now generalize the construction of invariant tori described in §§1–3 to this new situation. For $m = 1$, the first step of the procedure was described in [14], and for arbitrary m in [15]; the first two steps for $m = 2$ were described in [16].

What distinguishes system (39) from (1) is the presence of an angular variable. However, since the procedure described in §§1, 2 introduces angular variables after the first step, the difference is not crucial. The only essential fact is that X and Θ depend on θ. This creates additional difficulties in implementing the first step, and in order to overcome them we must impose additional restrictions on X and Θ. After that, the second and later steps will be identical to those described in §2.

The linearization (3) brings system (39) to the form

$$\begin{aligned}
\dot{y} &= i\lambda \cdot y + Y(y, \bar{y}, z, \theta, \varepsilon) + Y^*, \\
\dot{z} &= Qz + Z(y, \bar{y}, z, \theta, \varepsilon) + Z^*, \\
\dot{\theta} &= \omega + \Theta(y, \bar{y}, z, \theta, \varepsilon) + \Theta^*,
\end{aligned} \qquad (41)$$

where Q is a noncritical matrix; $*$ means that a condition of type (40) is fulfilled; Y and Z are nonlinearities in y, $\bar{y}$, z, ε; the functions Θ and Θ^* are obtained from the functions Θ, Θ^* of system (39) by (3). We want to find a change of variables

$$\begin{aligned}
y &= u + h(u, \bar{u}, \psi, \varepsilon), \\
z &= \omega + g(u, \bar{u}, \psi, \varepsilon), \\
\varphi &= \psi + f(u, \bar{u}, \psi, \varepsilon),
\end{aligned} \qquad (42)$$

where h, g, and f are polynomials in u, $\bar{u}$, ε whose coefficients are

2π-periodic in the components of ψ, which brings system (41) to the form

$$\dot{u} = i\lambda \cdot u + u \cdot U(u \cdot \bar{u}, \varepsilon) + U^0 + U^*,$$
$$\dot{w} = Qw + W^0 + W^*, \qquad (43)$$
$$\dot{\varphi} = \omega + \Psi(u \cdot \bar{u}, \varepsilon) + \Psi^0 + \Psi^*,$$

where $U(\eta, \varepsilon)$, $\Psi(\eta, \varepsilon)$ are polynomials in η, ε of degree at most N, N being the integral part of $(3n+1)/2$; $U(0, 0) = 0$; $\Psi(0, 0) = 0$; U^0, W^0 are nonlinearities in u, $\bar{u}$, w, ε; U^0, W^0, Ψ^0 vanish at $w = 0$; U^*, W^*, Ψ^* possess property (40); all these functions are 2π-periodic in the components of ψ. Differentiating (42) with respect to t along orbits of systems (41), (43) and setting $w = 0$, we find that the terms up to order $3n + 1$ in the expansions of h, g, f in terms of u, $\bar{u}$, $\sqrt{\varepsilon}$ satisfy the equations

$$\frac{\partial h}{\partial \psi}\omega + i\frac{\partial h}{\partial u}(\lambda \cdot u) - i\frac{\partial h}{\partial \bar{u}}(\lambda \cdot \bar{u}) - i\lambda h$$
$$= Y(u + h, \bar{u} + \bar{h}, g, \psi + f, \varepsilon)$$
$$- \frac{\partial h}{\partial u}(u \cdot U) - \frac{\partial h}{\partial u} - \frac{\partial h}{\partial \bar{u}}(\bar{u} \cdot \bar{U}) - \frac{\partial h}{\partial \psi}\Psi - u \cdot U,$$
$$\frac{\partial g}{\partial \psi}\omega + i\frac{\partial g}{\partial u}(\lambda \cdot u) - i\frac{\partial g}{\partial \bar{u}}(\lambda \cdot \bar{u}) - Qg$$
$$= Z(u + h, \bar{u} + \bar{h}, g, \psi + f, \varepsilon)$$
$$- \frac{\partial g}{\partial u}(u \cdot U) - \frac{\partial g}{\partial u} - \frac{\partial g}{\partial \bar{u}}(\bar{u} \cdot \bar{U}) - \frac{\partial g}{\partial \psi}\Psi,$$
$$\frac{\partial f}{\partial \psi}\omega + i\frac{\partial f}{\partial u}(\lambda \cdot u) - i\frac{\partial f}{\partial \bar{u}}(\lambda \cdot \bar{u})$$
$$= \Theta(u + h, \bar{u} + \bar{h}, g, \psi + f, \varepsilon)$$
$$- \Psi - \frac{\partial f}{\partial u}(u \cdot U) - \frac{\partial f}{\partial u} - \frac{\partial f}{\partial \bar{u}}(\bar{u} \cdot \bar{U}) - \frac{\partial f}{\partial \psi}\Psi.$$

Suppose that the coefficients of the expansions of h, g, f of order less than r in u, $\bar{u}$, $\sqrt{\varepsilon}$ have already been determined. Equating terms of order r, we obtain the following equation for r:

$$\frac{\partial g^{(r)}}{\partial \psi}\omega + i\frac{\partial g^{(r)}}{\partial u}(\lambda \cdot u) - i\frac{\partial g^{(r)}}{\partial \bar{u}}(\lambda \cdot \bar{u}) - Qg^{(r)} = G^{(r)},$$

where $g^{(r)}$ is the homogeneous vector polynomial formed by the terms in the expansion of g of order r in the sense indicated above; $G^{(r)}$ is a known polynomial with the same properties. Since Q is noncritical, this equation is uniquely solvable.

The corresponding equations for $h^{(r)}$ and $f^{(r)}$ split up into separate equations for the coefficients of the monomials $u_1^{q_1}\bar{u}_1^{\bar{q}_1}\cdots u_m^{q_m}\bar{u}_m^{\bar{q}_m}\varepsilon^q$ (here $q_1 + \bar{q}_1 +$

$\cdots + q_m + \bar{q}_m + 2q = r$). These equations are

$$\frac{\partial h_k^{(q_1,\ldots,q)}}{\partial \psi}\omega + i[(q_1 - \bar{q}_1)\lambda^1 + \cdots + (q_m - \bar{q}_m)\lambda^m - \lambda^k]h_k^{(q_1,\ldots,q)} \quad (44)$$
$$= a_k^{(q_1,\ldots,q)}(\psi) - U_k^{(q_1,\ldots,q_k-1,\ldots,q)}, \qquad k = 1,\ldots,m,$$

$$\frac{\partial f_j^{(q_1,\ldots,q)}}{\partial \psi}\omega + i[(q_1 - \bar{q}_1)\lambda^1 + \cdots + (q_m - \bar{q}_m)\lambda^m]f_j^{(q_1,\ldots,q)} \quad (45)$$
$$= c_j^{(q_1,\ldots,q)}(\psi) - \Psi_j^{(q_1,\ldots,q)}, \qquad j = 1,\ldots,l,$$

where $a_k^{(q_1,\ldots,q)}$, $c_j^{(q_1,\ldots,q)}$ are known functions; $U_k^{(q_1,\ldots,q)}$, $\Psi_j^{(q_1,\ldots,q)}$ are the coefficients of the polynomials U_k, Ψ_j; k, j number the components of the vectors h and f.

We require that

$$\sum_{j=1}^{l} k_j \omega^j + \sum_{k=1}^{m} l_k \lambda^k \neq 0 \quad (46)$$

for

$$\sum_{j=1}^{l}|k_j| + \sum_{k=1}^{m}|l_k| > 0, \qquad \sum_{j=1}^{l}|l_k| \leq 3n + 2.$$

Using equations (44), (45), we can determine the Fourier coefficients of $h_k^{(q_1,\ldots,q)}$, $f_j^{(q_1,\ldots,q)}$ and the constants $U_k^{(q_1,\ldots,q)}$, $\Psi_j^{(q_1,\ldots,q)}$. If $a_k^{(q_1,\ldots,q)}(\psi)$, $c_j^{(q_1,\ldots,q)}(\psi)$ are finite Fourier series, then the coefficients $h_k^{(q_1,\ldots,q)}$, $f_j^{(q_1,\ldots,q)}$ will also be determined; at the same time, inequality (46) must hold for a finite number of k_j. In the general case, the vectors ω and λ must satisfy certain arithmetical conditions which, however, hold for most vectors ω and λ in the sense of Lebesgue measure. In this case the solvability of equations (44), (45) follows from [17, p. 99].

We now proceed as in §1. Introducing polar coordinates $u = \rho \cdot e^{i\varphi}$, $\bar{u} = \rho \cdot e^{-i\varphi}$, we obtain the following system, corresponding to (7):

$$\begin{aligned}
\dot{\rho} &= \rho \cdot F(\rho^2, \varepsilon) + F^0 + F^*,\\
\dot{\varphi} &= \lambda + \Phi(\rho^2, \varepsilon) + \rho^{-1} \cdot (\Phi^0 + \Phi^*),\\
\dot{\psi} &= \omega + \Psi(\rho^2, \varepsilon) + \Psi^0 + \Psi^*,\\
\dot{w} &= Qw + W^0 + W^*,
\end{aligned} \quad (47)$$

where the symbols ^{0} and * indicate the same properties as before. The linear terms $F = B\rho^2 + \varepsilon b + \cdots$ determine the bifurcation equation

$$B\rho^2 + \varepsilon b = 0$$

for which we assume that $\det B \neq 0$ and $B^{-1}b < 0$ coordinatewise. Denote the solution of this equation by $\rho^2 = \varepsilon \alpha^2$. In the variables x_1, φ_1, v_1, where $p = \sqrt{\varepsilon}(\alpha + x_1)$, $w = \sqrt{\varepsilon}v_1$, $\varphi_1 = \text{col}(\varphi, \psi)$, system (47) becomes

$$\dot{x}_1 = \varepsilon X_1(x_1, \varepsilon) + \sqrt{\varepsilon}X_1^0 + \sqrt{\varepsilon^{3n+1}}\tilde{X}_1,$$
$$\dot{\varphi}_1 = \lambda_1(\varepsilon) + \varepsilon\Phi_1(x_1, \varepsilon) + \sqrt{\varepsilon}\Phi_1^0 + \sqrt{\varepsilon^{3n+1}}\tilde{\Phi}_1, \qquad (48)$$
$$\dot{v}_1 = Qv_1 + \sqrt{\varepsilon}V_1^0 + \sqrt{\varepsilon^{3n+1}}\tilde{V}_1,$$

where $\lambda_1(0) = \text{col}(\lambda, \omega)$. This system satisfies the same conditions as system (10), so that if $P_1 = D_{x_1}X_1(0, 0)$ is a noncritical matrix, we can apply Theorem 1; but if P_1 has pure imaginary eigenvalues, Theorem 3 shows that we can implement the procedure of §2, which leads at the nth step to an invariant torus. This torus is of dimension $l + m$, where m is defined as in Theorem 3.

SPECIAL CASE. Suppose that $\Theta = \Theta^* \equiv 0$ in system (39). Then system (39) is a system of type (1), whose right-hand side is a quasiperiodic function of t with the vector of basic frequencies ω.

REMARK. As we have already mentioned, the arguments of this section are effective only under two additional assumptions: (1) the existence of appropriate local coordinates in a neighborhood of T; (2) the solvability of equations (44), (45) by functions that are 2π-periodic in the components of ψ.

If $l = 1$, i.e., when one considers the bifurcation of a closed orbit T, the first assumption does not affect the generality of the arguments. In fact, in a neighborhood of a closed orbit one can always introduce local coordinates x, θ, where θ is a scalar, so that the vector field is described by a system just like (39), except that $P = P(\theta)$ is a periodic rather than a constant matrix [13, pp. 83–93]. Using the reducibility property for systems of linear homogeneous equations with periodic coefficients, we transform the system to a system of the required form. As for the second assumption, when Ψ is a scalar it follows from (46) that equations (44), (45) always have periodic solutions.

It should be noted that when $l = 1$ bifurcation indeed occurs, because when ε is small but positive, the invariant tori appear together with closed orbits. Indeed, eliminating t from (39), we obtain

$$\frac{dx}{d\theta} = \frac{1}{\omega}Px + S(\theta, x, \varepsilon), \qquad S(\theta, 0, 0) = 0, \qquad D_xS(\theta, 0, 0) = 0$$

and consider the equation $\beta(x_0, \varepsilon) = x(2\pi, x_0, \varepsilon) - x_0 = 0$, which is a necessary and sufficient condition for the solution $x(\theta, x_0, \varepsilon)$ with initial data $\theta = 0$, $x = x_0$ to be 2π-periodic. Since $\beta(0, 0) = 0$ and, by (46), $\det\frac{\partial\beta}{\partial x_0}(0, 0) = \det(\exp\{\frac{2\pi}{\omega}P\} - E) \neq 0$, the Implicit Function Theorem implies that the equation has a unique solution $x_0(\varepsilon)$ with $x_0(0) = 0$.

We have considered the case when the perturbation is applied not to an equilibrium state, as in §§1–3, but to an invariant torus of the original system. Each step of the procedure can be generalized in the same way.

Consider the first step. Instead of an equilibrium state of system (14) with $G(\rho^2) = B\rho^2$, let us take a closed orbit or an invariant torus. Set $\rho = \sqrt{\varepsilon} r$ in system (14). This gives the system

$$\frac{dr}{d\tau} = r(Br^2 + b), \qquad \tau = \varepsilon t, \tag{49}$$

which no longer depends on ε. Assume that (49) has an invariant l_1-dimensional torus T_1 in the domain $r > 0$. Assume also that in a neighborhood of T_1 one can introduce coordinates x_1, θ_1 in terms of which T_1 is defined by the equation $x_1 = 0$ and system (49) has the form

$$\frac{dx_1}{d\tau} = P_1 x_1 + C(x_1, \theta_1), \qquad \frac{d\theta_1}{d\tau} = \omega_1 + \Omega(x_1, \theta_1),$$

where P_1 is a constant $(m - l_1 \times m - l_1)$ matrix, $\det P_1 \neq 0$, ω_1 is a constant l_1-vector; C, Ω are polynomials in x_1, with 2π-periodic coefficients; $\Omega(0, \theta_1) = 0$, $C(0, \theta_1) = 0$, $D_{x_1} X_1(0, \theta_1) = 0$. As we have already noted, for $l_1 = 1$ such coordinates always exist.

In terms of x_1, θ_1, v_1, φ_1 system (47) has the form (48), with the following differences:

(1) $X_1 = X_1(x_1, \theta_1, \varepsilon)$, where $X_1(x_1, \theta_1, 0) = P_1 x_1 + C(x_1, \theta_1)$, so that $D_{x_1} X_1(0, \theta_1, 0) = P_1$ is a constant matrix;

(2) an additional equation

$$\dot{\theta}_1 = \varepsilon \omega_1 + \varepsilon \Theta_1(x_1, \theta_1, \varepsilon) + \sqrt{\varepsilon}\Theta_1^0 + \sqrt{\varepsilon^{3n+1}}\tilde{\Theta}_1,$$

where $\Theta_1(x_1, \theta_1, 0) = \Omega(x_1, \theta_1)$;

(3) $\lambda_1 = \lambda_1(0) + \varepsilon \mu_1(\theta_1, \varepsilon)$, $\Phi_1 = \Phi_1(x_1, \theta_1, \varepsilon)$.

The term $\mu_1(\theta_1, 0)$ in the equation for φ_1 in system (48) can be averaged by the change of variables $\varphi_1 = \psi_1 + h(\theta_1)$, where $h(\theta_1)$ is defined by the equation

$$\frac{\partial h}{\partial \theta_1}\omega_1 = \mu_1(\theta_1, 0) - \nu,$$

ν being the mean value of $\mu_1(\theta_1, 0)$.

If this averaging procedure is possible (as is always the case when $l_1 = 1$) and P_1 is a noncritical matrix, we can apply Theorem 1, which says that if $n = 1$ and $\varepsilon > 0$ is sufficiently small, there exists an $(m + l + l_1)$-dimensional invariant torus. But if P_1 has pure imaginary eigenvalues, the second step of our scheme is necessary.

Thus, at each step, replacement of the equilibrium state by an invariant torus has the same effect as the admission of pure imaginary eigenvalues corresponding to the equilibrium state: appearance of additional small frequencies on the bifurcating torus.

§5. Bifurcation of an invariant torus of arbitrary dimension

In §§1–3 we have constructed an invariant torus defined by an equilibrium state of the system

$$\dot{\rho}_s = \rho_s \cdot (B_s \rho_s^2 + \varepsilon b_s), \qquad s = 0, 1, \ldots, n-1. \tag{50}$$

In §4 we have shown how to obtain an invariant torus of a higher dimension than in the procedure of §§1–3. The basic idea was to replace equilibrium states of systems (50) by invariant tori.

In this section, we will analyze the constructing of invariant tori of smaller dimensions. When an equilibrium state of system (50) is being defined, we can equate an arbitrary number of components to zero, and then other components will be determined by a linear system similar to (24). As before, we assume that a condition of type (9) or (23) is satisfied, i.e., the determinant of the coefficient matrix of the system does not vanish and each component of the solution is positive.

Consider the first step. As in §1, we bring system (1) to the form (6). Let us assume that for a solution of system (8) we have $\alpha^k > 0$, $k = 1, \ldots, l$, and $\alpha^k = 0$, $k = l+1, \ldots, m$. Accordingly, we represent the vector u as $u = \text{col}(\xi, \eta)$, where $\xi = \text{col}(u^1, \ldots, u^l)$, $\eta = \text{col}(u^{l+1}, \ldots, u^m)$. Similarly, $\lambda = \text{col}(\mu, \nu)$. Transform from variables ξ, $\bar{\xi}$ to polar coordinates $\xi = \rho \cdot e^{i\varphi}$, $\bar{\xi} = \rho \cdot e^{-i\varphi}$. We obtain the system

$$\begin{aligned}
\dot{\rho} &= \rho \cdot F(\rho^2, \eta \cdot \bar{\eta}, \varepsilon) + F^0(\rho, \varphi, \eta, \bar{\eta}, w, \varepsilon) + F^*, \\
\dot{\varphi} &= \mu + \Phi(\rho^2, \eta \cdot \bar{\eta}, \varepsilon) + \rho^{-1} \cdot [\Phi^0(\rho, \varphi, \eta, \bar{\eta}, w, \varepsilon) + \Phi^*], \\
\dot{\eta} &= \eta \cdot [i\nu + \Gamma(\rho^2, \eta \cdot \bar{\eta}, \varepsilon)] + \Gamma^0(\rho, \varphi, \eta, \bar{\eta}, w, \varepsilon) + \Gamma^*, \\
\dot{w} &= Qw + W^0(\rho, \varphi, \eta, \bar{\eta}, w, \varepsilon) + W^*,
\end{aligned} \tag{51}$$

where F, Φ, Γ are polynomials that vanish when the arguments take zero values; F^0, Φ^0, Γ^0, W^0 are nonlinearities in ρ, η, $\bar{\eta}$, ε that vanish at $w = 0$; the functions F^*, Φ^*, Γ^*, W^* satisfy a condition of type (2), say,

$$F^*(\sqrt{\varepsilon}\rho, \varphi, \sqrt{\varepsilon}\eta, \sqrt{\varepsilon}\bar{\eta}, \sqrt{\varepsilon}w, \varepsilon) = \sqrt{\varepsilon^{2n+2}} \tilde{F},$$

where $\tilde{F}$ is C^1-continuous in all arguments except ε, $0 \leq \rho < H$, $\|\eta\| < H$, $\|\bar{\eta}\| < H$, $\|w\| < h$, $\varphi \in \mathbb{R}^l$, $0 \leq \varepsilon < \varepsilon_0$; Q is a noncritical matrix. According to our convention, the equation conjugated to the third equation of system (51) is omitted.

Let $F(\rho^2, 0, \varepsilon) = C\rho^2 + \varepsilon c + \cdots$. By the assumption stated at the beginning of this section, the bifurcation equation

$$C\rho^2 + \varepsilon c = 0$$

has a unique solution $\rho^2 = \varepsilon \beta^2 > 0$, where β is an l-dimensional vector.

First let $n = 1$. Setting $\rho = \sqrt{\varepsilon}\beta + \varepsilon z$, $\eta = \varepsilon y$, $w = \varepsilon^2 v$ in system (51) we obtain

$$\dot{z} = \varepsilon S z + \sqrt{\varepsilon^3} Z,$$
$$\dot{\eta} = \varepsilon \eta \cdot [i(\nu/3 + \delta + M\beta^2) + (\gamma + N\beta^2)] + \sqrt{\varepsilon^3} H, \qquad (52)$$
$$\dot{v} = Qv + \sqrt{\varepsilon} V,$$
$$\dot{\varphi} = \mu_1(\varepsilon) + \sqrt{\varepsilon^3} \Phi,$$

where $S = 2(\operatorname{diag}\beta)C(\operatorname{diag}\beta)$; $\gamma = \operatorname{Re}\frac{\partial \Gamma}{\partial \varepsilon}(0,0,0)$; $\delta = \operatorname{Im}\frac{\partial \Gamma}{\partial \varepsilon}(0,0,0)$; $M = \operatorname{Im}\frac{\partial \Gamma}{\partial \rho^2}(0,0,0)$; $N = \operatorname{Re}\frac{\partial \Gamma}{\partial \rho^2}(0,0,0)$; Z, H, V, Φ are C^1-continuous in all arguments except ε,

$$|z^j| < \frac{\beta^j}{\sqrt{\varepsilon_0}} \quad (j = 1, \ldots, l), \qquad \|\eta\| < \frac{H}{\varepsilon_0}, \qquad \|v\| < \frac{H}{\varepsilon_0^2},$$

$$\varphi \in \mathbb{R}^l, \qquad 0 \le \varepsilon < \varepsilon_0.$$

THEOREM 4. *Assume that the matrix S is noncritical and that the vector $\gamma + N\beta^2$ has no zero components. Then, for sufficiently small $\varepsilon > 0$, system (51) has an l-dimensional invariant torus, which is defined by the equations*

$$\rho = \sqrt{\varepsilon}\beta + \sqrt{\varepsilon^3}\psi(\varphi, \varepsilon), \qquad \eta = \sqrt{\varepsilon^3}\zeta(\varphi, \varepsilon),$$
$$\bar{\eta} = \sqrt{\varepsilon^3}\bar{\zeta}(\varphi, \varepsilon), \qquad v = \sqrt{\varepsilon^5}\chi(\varphi, \varepsilon),$$

where ψ, ζ, $\bar{\zeta}$, χ are 2π-periodic in the components of φ.

PROOF. Define $x_1 = \operatorname{col}(z, \eta, \bar{\eta})$ and $P_* = \operatorname{diag}(S, \operatorname{diag}(\gamma + N\beta^2))$. System (52) can be viewed as system (28) for $n = 1$, with $P_* + iQ_*$ in place of P_1, where $Q_* = \operatorname{diag}(0, \operatorname{diag}(\frac{\nu}{\varepsilon} + \delta + M\beta^2))$. As in §3, we assume that $S = \operatorname{diag}(S^+, S^-)$, in an analogous sense. Associate the positive components of the vector $\gamma + N\beta^2$ with S^+ and the negative components with S^-. Then $P_* = \operatorname{diag}(P_*^+, P_*^-)$ and $Q_* = \operatorname{diag}(Q_*^+, Q_*^-)$. Since

$$e^{(P_*^+ + iQ_*^+)t} = e^{P_*^+ t} e^{iQ_*^+ t},$$

where the matrix $e^{iQ_*^+ t}$ is bounded (a similar assertion with $-$ replacing $+$ is also valid), it follows that

$$\|J(t)\| \le b e^{-a|t|} \quad (a > 0,\ b > 0)$$

with

$$J(t) = \begin{cases} -\begin{pmatrix} e^{-(P_*^+ + iQ_*^+)t} & 0 \\ 0 & 0 \end{pmatrix} & \text{for } t > 0, \\ \begin{pmatrix} 0 & 0 \\ 0 & e^{-(P_*^- + iQ_*^-)t} \end{pmatrix} & \text{for } t < 0. \end{cases}$$

Hence the proof of Theorem 3 for $n = 1$ is applicable to system (52), and the assertion of Theorem 4 follows.

Turning to the general case $n > 1$, we confine ourselves to the case $n = 2$. Setting $\rho = \sqrt{\varepsilon}(\beta + x_1)$, $\eta = \sqrt{\varepsilon}\eta_1$, $\bar{\eta} = \sqrt{\varepsilon}\bar{\eta}_1$, $w = \sqrt{\varepsilon}v$ in system (51) we obtain

$$\dot{x}_1 = \varepsilon X_1(x_1, \eta_1 \cdot \bar{\eta}_1, \varepsilon) + \sqrt{\varepsilon} X_1^0 + \sqrt{\varepsilon^7} \tilde{X}_1,$$

$$\dot{\eta}_1 = \varepsilon \eta_1 \cdot \left[i\left(\frac{\nu}{\varepsilon} + \delta + M\beta^2\right) + (\gamma + N\beta^2) + \Gamma_1(x_1, \eta_1 \cdot \bar{\eta}_1, \varepsilon)\right]$$
$$+ \sqrt{\varepsilon}\Gamma_1^0 + \sqrt{\varepsilon^7}\tilde{\Gamma}_1,$$

$$\dot{v} = Qv + \sqrt{\varepsilon}V^0 + \sqrt{\varepsilon^7}\tilde{V},$$

$$\dot{\varphi} = \mu_1(\varepsilon) + \varepsilon\Phi_1(x_1, \eta_1 \cdot \bar{\eta}_1, \varepsilon) + \sqrt{\varepsilon}\Phi_1^0 + \sqrt{\varepsilon^7}\tilde{\Phi}_1,$$

where X_1, Γ_1, Φ_1 are polynomials; $X_1(0, 0, 0) = 0$, $\Gamma_1(0, 0, 0) = 0$, $\Phi_1(0, 0, \varepsilon) = 0$; the functions X_1^0, Γ_1^0, Φ_1^0 vanish for $v = 0$; $S = D_{x_1}X_1(0, 0, 0)$; δ, γ, M, N are the same as in system (52).

When $n = 1$ we assumed that the real parts of the eigenvalues of S and the components of $\gamma + N\beta^2$ did not vanish. We now allow at least some of these numbers to be zero. Suppose that the matrix S is conjugate to a matrix $\mathrm{diag}(i\,\mathrm{diag}\,\lambda_1, Q_1)$, where Q_1 is noncritical, and that $\eta_1 = \mathrm{col}(\tau, \sigma)$, where τ and σ correspond to the zero and nonzero components of $\gamma + N\beta^2$, respectively. Accordingly, $\nu = \mathrm{col}(\omega, \zeta)$. A linear substitution similar to (16) enables us to rewrite system (53) as

$$\dot{y}_1 = \varepsilon y_1 \cdot i\lambda_1 + \varepsilon Y_1(y_1, \bar{y}_1, z_1, \tau \cdot \bar{\tau}, \sigma \cdot \bar{\sigma}, \varepsilon) + \sqrt{\varepsilon} Y_1^0 + \sqrt{\varepsilon^7}\tilde{Y}_1,$$

$$\dot{z}_1 = \varepsilon Q_1 z_1 + \varepsilon Z_1(y_1, \bar{y}_1, z_1, \tau \cdot \bar{\tau}, \sigma \cdot \bar{\sigma}, \varepsilon) + \sqrt{\varepsilon} Z_1^0 + \sqrt{\varepsilon^7}\tilde{Z}_1,$$

$$\dot{\tau} = \varepsilon\tau \cdot \left[i\left(\frac{\omega}{\varepsilon} + \delta_1 + M_1\beta^2\right) + T(y_1, \bar{y}_1, z_1, \tau \cdot \bar{\tau}, \sigma \cdot \bar{\sigma}, \varepsilon)\right]$$
$$+ \sqrt{\varepsilon}T^0 + \sqrt{\varepsilon^7}\tilde{T},$$

$$\dot{\sigma} = \varepsilon\sigma \cdot \left[i\left(\frac{\zeta}{\varepsilon} + \delta_2 + M_2\beta^2\right) + (\gamma_2 + N_2\beta^2) + \Sigma(y_1, \bar{y}_1, z_1, \tau \cdot \bar{\tau}, \sigma \cdot \bar{\sigma}, \varepsilon)\right]$$
$$+ \sqrt{\varepsilon}\Sigma^0 + \sqrt{\varepsilon^7}\tilde{\Sigma},$$

$$\dot{v} = Qv + \sqrt{\varepsilon}V^0 + \sqrt{\varepsilon^7}\tilde{V},$$

$$\dot{\varphi} = \mu_1(\varepsilon) + \varepsilon\Phi(y_1, \bar{y}_1, z_1, \tau \cdot \bar{\tau}, \sigma \cdot \bar{\sigma}, \varepsilon) + \sqrt{\varepsilon}\Phi^0 + \sqrt{\varepsilon^7}\tilde{\Phi},$$

(54)

where the properties of the functions on the right follow from those of system (53).

Consider the following subsystem of system (54):

$$\begin{aligned}
\dot{y}_1 &= \varepsilon y_1 \cdot i\lambda_1 + \varepsilon Y_1(y_1, \bar{y}_1, z_1, \tau \cdot \bar{\tau}, 0, \varepsilon), \\
\dot{z}_1 &= \varepsilon Q_1 z_1 + \varepsilon Z_1(y_1, \bar{y}_1, z_1, \tau \cdot \bar{\tau}, 0, \varepsilon), \\
\dot{\tau} &= \varepsilon \tau \cdot \left[i\left(\frac{\omega}{\varepsilon} + \delta_1 + M_1\right) + T(y_1, \bar{y}_1, z_1, \tau \cdot \bar{\tau}, 0, \varepsilon) \right], \\
\dot{\varphi} &= \mu_1(\varepsilon) + \varepsilon \Phi(y_1, \bar{y}_1, z_1, \tau \cdot \bar{\tau}, 0, \varepsilon).
\end{aligned} \quad (55)$$

To this system we apply a polynomial change of variables

$$\begin{aligned}
y_1 &= u + h(u, \bar{u}, a \cdot \bar{a}, \varepsilon), \\
z_1 &= w + g(u, \bar{u}, a \cdot \bar{a}, \varepsilon), \\
\tau &= a + a \cdot b(u, \bar{u}, a \cdot \bar{a}, \varepsilon), \\
\varphi &= \psi + f(u, \bar{u}, a \cdot \bar{a}, \varepsilon)
\end{aligned} \quad (56)$$

to bring it to the form

$$\begin{aligned}
\dot{u} &= \varepsilon u \cdot [i\lambda_1 + U(u \cdot \bar{u}, a \cdot \bar{a}, \varepsilon)] + \varepsilon U^1 + \varepsilon U^*, \\
\dot{w} &= \varepsilon Q w + \varepsilon W^1 + \varepsilon W^*, \\
\dot{a} &= \varepsilon a \cdot \left[i\left(\frac{\omega}{\varepsilon} + \delta_1 + M_1 \beta^2\right) + A(u \cdot \bar{u}, a \cdot \bar{a}, \varepsilon) \right] + \varepsilon A^1 + \varepsilon A^*, \\
\dot{\psi} &= \mu_1(\varepsilon) + \varepsilon \Psi(u \cdot \bar{u}, a \cdot \bar{a}, \varepsilon) + \varepsilon \Psi^1 + \varepsilon \Psi^*,
\end{aligned}$$

where the superscript "1" means that the function vanishes at $w = 0$, and the symbol $*$ means that property (2) holds with the exponent 5, say,

$$U^*(\sqrt{\varepsilon} u, \sqrt{\varepsilon} \bar{u}, \sqrt{\varepsilon} w, \varepsilon(a \cdot a), \varepsilon) = \sqrt{\varepsilon^5} \tilde{U};$$

U, A are second-degree polynomials of their arguments without constant terms.

Assuming that condition (19) holds, the existence of a transformation (56) possessing the required properties for $n = 2$, $s = 1$ can be proved in the standard way, by using indeterminate coefficients, as in the proof of Lemma 2 in §3. Conditions of type (19) do not impose restrictions on the components of the vector ω, because the polynomials involved depend on a, $\bar{a}$ only through the product $a \cdot \bar{a}$.

Applying (56) to the full system (54) and transforming to polar coordinates

r, θ where $u = r \cdot e^{i\theta}$, $\bar{u} = r \cdot e^{-i\theta}$, we get

$$\dot{r} = \varepsilon r \cdot F(r^2, a \cdot \bar{a}, \varepsilon) + \varepsilon F^1 + \varepsilon F^* + \sqrt{\varepsilon} F^0 + \sqrt{\varepsilon^7} \tilde{F},$$

$$\dot{\theta} = \lambda_1 + \varepsilon \Theta(r^2, a \cdot \bar{a}, \varepsilon) + r^{-1} \cdot (\varepsilon \Theta^1 + \varepsilon \Theta^* + \sqrt{\varepsilon} \Theta^0 + \sqrt{\varepsilon^7} \tilde{\Theta}),$$

$$\dot{w} = \varepsilon Q_1 w + \varepsilon W^1 + \varepsilon W^* + \sqrt{\varepsilon} W^0 + \sqrt{\varepsilon^7} \tilde{W},$$

$$\dot{a} = \varepsilon a \cdot \left[i \left(\frac{\omega}{\varepsilon} + \delta_1 + M_1^2 \right) + A(r^2, a \cdot \bar{a}, \varepsilon) \right] + \varepsilon A^1 + \varepsilon A^* + \sqrt{\varepsilon} A^0 + \sqrt{\varepsilon^7} \tilde{A},$$

$$\dot{\sigma} = \varepsilon \sigma \cdot \left[i \left(\frac{\zeta}{\varepsilon} + \delta_2 + M_2^2 \right) + (\gamma_2 + N_2^2) + \Sigma(r, \theta, w, a \cdot \bar{a}, \sigma \cdot \bar{\sigma}, \varepsilon) \right]$$
$$+ \sqrt{\varepsilon} \Sigma^0 + \sqrt{\varepsilon^7} \tilde{\Sigma},$$

$$\dot{v} = Qv + \sqrt{\varepsilon} V^0 + \sqrt{\varepsilon^7} \tilde{V},$$

$$\dot{\psi} = \mu_1(\varepsilon) + \varepsilon \Psi(r^2, a \cdot \bar{a}, \varepsilon) + \varepsilon \Psi^1 + \varepsilon \Psi^* + \sqrt{\varepsilon} \Psi^0 + \sqrt{\varepsilon^7} \tilde{\Psi},$$

where $F(0, 0, 0) = 0$, $\Theta(0, 0, 0) = 0$, $\Sigma(0, \theta, 0, 0, 0) = 0$; the superscript "1" means that the functions vanish at $w = 0$; the other symbols retain their previous meanings.

Let $F(r^2, 0, \varepsilon) = Cr^2 + \varepsilon c + \cdots$, where $\det C \neq 0$; and $-C^1 c = \kappa^2 > 0$ is componentwise positive. Set $r = \sqrt{\varepsilon}\kappa + \sqrt{\varepsilon} z_1$, $a = \varepsilon a_1$, $\sigma = \varepsilon^2 \sigma_1$, $w = \varepsilon^2 w_1$, $v = \varepsilon^3 v_1$.

It follows from the properties of the functions on the right of system (57) that in the new variables this system can be written as

$$\dot{z}_1 = \varepsilon^2 S_1 z_1 + \sqrt{\varepsilon^5} Z_1,$$

$$\dot{a}_1 = \varepsilon^2 a_1 \cdot \left[i \left(\frac{\omega}{\varepsilon^2} + \frac{\delta_1 + M_1 \beta^2}{\varepsilon} + L_1 \right) + L_2 \right] + \sqrt{\varepsilon^5} A_1,$$

$$\dot{\sigma}_1 = \varepsilon \sigma_1 \cdot \left[i \left(\frac{\zeta}{\varepsilon} + \delta_2 + M_2^2 \right) + (\gamma_2 + N_2^2) \right] + \sqrt{\varepsilon^3} \Sigma_1,$$

$$\dot{w}_1 = \varepsilon Q_1 w_1 + \sqrt{\varepsilon^3} W_1, \qquad \dot{v}_1 = Qv_1 + \sqrt{\varepsilon} V_1,$$

$$\dot{\theta}_1 = \lambda_1(\varepsilon) + \sqrt{\varepsilon^5} \Theta_1, \qquad \dot{\psi}_1 = \mu_2(\varepsilon) + \sqrt{\varepsilon^5} \Psi_1,$$

where $S_1 = 2(\operatorname{diag}\kappa) C(\operatorname{diag}\kappa)$; L_1, L_2 are constant vectors; the functions Z_1, A_1, Σ_1, W_1, V_1, Θ_1, Ψ_1 are C^1-continuous in all the variables except ε, for sufficiently small $\|z_1\|$, $\|w_1\|$, $\|a_1\|$, $\|\sigma_1\|$, $\|v_1\|$, and for $\theta \in \mathbb{R}^{l_1}$, $\psi \in \mathbb{R}^l$, $0 \leq \varepsilon < \varepsilon_0$.

THEOREM 5. *If S_1 is a noncritical matrix and the components of vector L_2 are not zero then, for sufficiently small $\varepsilon > 0$, system (57) has an invariant* *torus defined by the equations*

$$r = \sqrt{\varepsilon}\kappa + \sqrt{\varepsilon^3}\Delta_1(\theta, \psi, \varepsilon),$$
$$a = \sqrt{\varepsilon^3}\Delta_2(\theta, \psi, \varepsilon), \qquad \bar{a} = \sqrt{\varepsilon^3}\bar{\Delta}_2(\theta, \psi, \varepsilon),$$
$$\sigma = \sqrt{\varepsilon^5}\Delta_3(\theta, \psi, \varepsilon), \qquad \bar{\sigma} = \sqrt{\varepsilon^5}\bar{\Delta}_3(\theta, \psi, \varepsilon),$$
$$w = \sqrt{\varepsilon^5}\Delta_4(\theta, \psi, \varepsilon), \qquad v = \sqrt{\varepsilon^7}\Delta_5(\theta, \psi, \varepsilon),$$

where $\Delta_1 - \Delta_5$ are 2π-periodic continuous functions in the components of θ, ψ.

Theorem 5 is proved in the same way as Theorem 4, by reduction to the proof of Theorem 3 in §3 for $n = 2$.

§6. Multidimensional parameter

Suppose that system (1) depends not on a scalar parameter $\varepsilon \geq 0$ but on a multidimensional parameter μ, $\dim \mu = m$, where m is the number of pairs of pure imaginary eigenvalues of P. In order to apply our results to such a system, we define $\varepsilon = \|\mu\| = \sqrt{\mu_1^2 + \cdots + \mu_m^2}$, $\mu = \varepsilon \delta$ (in this section, the coordinates of vectors are subscripted). The conditions stating that the solutions of the bifurcation equation are positive and that the matrices P_s have a given number of pure imaginary eigenvalues define cones in the space $O\mu_1, \ldots, \mu_m$ with vertices at the origin O. Bifurcation to invariant tori of the corresponding dimensions will take place on rays of these cones.

We will consider a few simple examples.

EXAMPLE 1. $m = 1$. The system depends on a scalar parameter μ. Then $\varepsilon = |\mu|$, $\delta = \pm 1$. Equation (8) becomes a pair of scalar equations

$$B\rho^2 \pm \varepsilon b = 0.$$

Here B is a Lyapunov constant for $\mu = 0$, and $b = \frac{d(\operatorname{Re}\lambda)}{d\mu}\big|_{\mu=0}$, where $\lambda(\mu)$ is the eigenvalue of $D_x X(0, \mu)$ such that $\lambda(0) = i\lambda^1$. The condition that the solution is positive defines one of the two possible values of δ, and for this δ we have bifurcation to a closed orbit.

EXAMPLE 2. $m = 2$. The system depends on a two-dimensional parameter $\mu = (\mu_1, \mu_2)$. Define $\varepsilon = \sqrt{\mu_1^2 + \mu_2^2}$, $\delta = (\cos\tau, \sin\tau)$. The bifurcation equation (8) is a system of two equations, depending on the parameter τ:

$$B_{11}\rho_1^2 + B_{12}\rho_2^2 + \varepsilon(b_{11}\cos\tau + b_{12}\sin\tau) = 0,$$
$$B_{21}\rho_1^2 + B_{22}\rho_2^2 + \varepsilon(b_{21}\cos\tau + b_{22}\sin\tau) = 0.$$

Let $\rho_1^2 = \varepsilon\alpha_1^2$, $\rho_2^2 = \varepsilon\alpha_2^2$ be a solution of this system, where

$$\alpha_1^2 = \alpha_{11}\cos\tau + \alpha_{12}\sin\tau,$$
$$\alpha_2^2 = \alpha_{21}\cos\tau + \alpha_{22}\sin\tau.$$

The matrix P_1 of the system (10) is

$$P_1 = 2 \begin{pmatrix} \alpha_1^2 B_{11} & \alpha_1 \alpha_2 B_{12} \\ \alpha_1 \alpha_2 B_{21} & \alpha_2^2 B_{22} \end{pmatrix}.$$

The condition that α_1^2, α_2^2, be positive defines a sector of the circle about 0 in which μ is allowed to vary. The characteristic equation of the matrix $P_1/2$ is $\lambda^2 - a\lambda + A = 0$, where

$$a = \alpha_1^2 B_{11} + \alpha_2^2 B_{22}, \qquad A = \alpha_1^2 \alpha_2^2 \det B.$$

If $\det B < 0$ or $\det B > 0$ and $a \neq 0$, then P_1 is a noncritical matrix and by Theorem 1 there is bifurcation to a two-dimensional torus along a ray. Suppose now that $\det B > 0$, $a = 0$, and that the ray defined by the equation $a = 0$ lies in the sector in which α_1^2, α_2^2 are positive. Applying the second step of the procedure, we conclude from Theorem 3 that if the solution of the scalar equation (24) with $s = 1$ is positive, there is bifurcation to a three-dimensional invariant torus with one slow frequency of order ε along the ray in question. For $\dim \mu = 3$ a similar bifurcation will occur in a two-dimensional sector of the three-dimensional ball $\|\mu\| < \varepsilon_0$, without any additional conditions.

EXAMPLE 3. $m = 3$. Here $\delta_1 = \sin\tau_1 \cos\tau_2$, $\delta_2 = \sin\tau_1 \sin\tau_2$, $\delta_3 = \cos\tau_2$. A solution of the bifurcation equations is a linear combination of coordinates of the unit vector δ:

$$\rho_k^2 = \varepsilon \alpha_k^2; \qquad \alpha_k^2 = \alpha_{k1}\delta_1 + \alpha_{k2}\delta_2 + \alpha_{k3}\delta_3, \qquad k = 1, 2, 3,$$

$$P_1 = 2 \begin{pmatrix} \alpha_1^2 B_{11} & \alpha_1 \alpha_2 B_{12} & \alpha_1 \alpha_3 B_{13} \\ \alpha_1 \alpha_2 B_{21} & \alpha_2^2 B_{22} & \alpha_2 \alpha_3 B_{23} \\ \alpha_1 \alpha_3 B_{31} & \alpha_2 \alpha_3 B_{32} & \alpha_3^2 B_{33} \end{pmatrix}.$$

Let $\lambda^3 - a\lambda^2 + b\lambda - c = 0$ be the characteristic equation of $P_1/2$. If $c = ab$, $b > 0$, the characteristic equation has a pair of pure imaginary roots. This condition defines a two-dimensional cone K in the space $O\mu_1\mu_2\mu_3$. Along all the rays defined by the inequalities $\alpha_k^2 > 0$, $k = 1, 2, 3$, except those lying in K, there will be bifurcation to three-dimensional tori. On rays of K, assuming that the solution of the bifurcation equation is positive, we obtain a four-dimensional torus with one frequency of order ε. If $\dim \mu = 4$, a four-dimensional torus exists provided that μ lies in a certain three-dimensional cone.

When $m = 4$, three steps of the procedure can be implemented, yielding a seven-dimensional torus with two frequencies of order ε and one frequency of order ε^2.

Proceeding from examples to the general case, let us assume that the dimension of the parameter μ is equal to $m = m_0 + \cdots + m_{n-1}$, where n is the number of steps of the procedure, and m_{s-1}, $s = 1, \ldots, n$, is the number of pairs of pure imaginary eigenvalues at step s. As before, we set $\mu = \varepsilon\delta$, where $\varepsilon = \|\mu\|$. We confine attention to cases $n = 1$ and $n = 2$. Carrying out the first step as in §1, we obtain system (7), with

$$F(\rho^2, \mu) = B\rho^2 + b\mu + \cdots,$$

where B, b are matrices of order $m_0 \times m_0$ and $m_0 \times m$ respectively; the symbol $*$ means that, say,

$$F^*(\sqrt{\varepsilon}\rho, \varphi, \sqrt{\varepsilon}v, \varepsilon\delta) = \sqrt{\varepsilon^{3n+2}}\tilde{F}.$$

Considering the bifurcation equation

$$B\rho^2 + b\mu = 0,$$

we assume that $\det B \neq 0$, $\operatorname{rank} b = m_0$. The solution is $\rho^2 = A\mu = \gamma$, $\operatorname{rank} A = m_0$. We divide the components of μ into two groups $\mu = (\mu_1, \nu)$, where μ_1 represents components for which A has a nonzero minor of order m_0. The parameter will be the vector (γ, ν), each component of γ being positive. Setting $\gamma = \varepsilon\delta_1$, $\nu = \varepsilon\delta_2$, $\rho = \sqrt{\varepsilon}(\sqrt{\delta_1} + x_1)$, $v = \sqrt{\varepsilon}v_1$, $\varphi = \varphi_1$, we obtain the following corollary of system (10):

$$\begin{aligned}
\dot{x}_1 &= \varepsilon X_1(x_1, \varepsilon, \delta_1, \delta_2) + \sqrt{\varepsilon}X_1^0 + \sqrt{\varepsilon^{3n+1}}\tilde{X}_1, \\
\dot{\varphi}_1 &= \lambda_1(\varepsilon, \delta_1, \delta_2) + \varepsilon\Phi_1(x_1, \varepsilon, \delta_1, \delta_2) + \sqrt{\varepsilon}\Phi_1^0 + \sqrt{\varepsilon^{3n+1}}\tilde{\Phi}_1, \quad (58) \\
\dot{v}_1 &= Qv + \sqrt{\varepsilon}V_1^0 + \sqrt{\varepsilon^{3n+1}}\tilde{V}_1,
\end{aligned}$$

where X_1, Φ_1 are polynomials in x_1, ε with coefficients depending on δ_1, δ_2; $P_1 = D_{x_1}X_1(0, 0, \delta_1, \delta_2) = 2(\operatorname{diag}\sqrt{\delta_1})B(\operatorname{diag}\sqrt{\delta_1})$; Q is a noncritical matrix; X_1^0, Φ_1^0, V_1^0 vanish at $v_1 = 0$.

For $n = 1$ we have $\mu = \gamma$. From the ball $\|\mu\| < \varepsilon_0$ we delete the rays on which P_1 has pure imaginary eigenvalues. By Theorem 1, in the remaining domain there exist m_0-dimensional invariant tori with frequency vector λ_1.

Now, let $m = m_0 + m_1$. If there are rays to be deleted, we carry out the second step for $n = 2$, $s = 1$, as in §2. The subsystem of system (58)

$$\begin{aligned}
\dot{x} &= \varepsilon X_1(x_1, \varepsilon, \delta_1, \delta_2), \\
\dot{\varphi}_1 &= \lambda_1(\varepsilon, \delta_1, \delta_2) + \varepsilon\Phi_1(x_1, \varepsilon, \delta_1, \delta_2)
\end{aligned}$$

corresponding to systems (17), (18) contains a free parameter δ_2 so that, generally speaking, we can insure that the solution of the bifurcation equation (24) will be positive. We can thus determine a sector of the ball $\|\mu\| < \varepsilon_0$ on the rays of which there will be bifurcation of an $(m_0 + m_1)$-dimensional invariant torus.

References

1. Yu. I. Neimark, *On certain cases of dependence of periodic motions on a parameter*, Dokl. Akad. Nauk SSSR **129** (1959), no. 4, 376–379. (Russian)

2. R. Jost and E. Zehnder, *Generalization of the Hopf bifurcation theorem*, Helv. Phys. Acta **45** (1972), no. 2, 258-276.
3. Yu. A. Mitropol'skiĭ and A. M. Samoĭlenko, *Multifrequency oscillations of weakly nonlinear second-order systems*, Ukrain. Mat. Zh. **28** (1976), no. 6, 745-762; English transl. in Ukranian Math. J. **28** (1976).
4. N. V. Nikolenko, *Invariant asymptotically stable tori of the perturbed Korteweg-deVries equation*, Uspekhi Mat. Nauk **35** (1980), no. 5, 139-207; English transl. in Russian Math. Surveys **35** (1980).
5. Yu. N. Bibikov, *Hopf type bifurcation for quasiperiodic motions*, Differentsial'nye Uravneniya **16** (1980), no. 10, 1539-1544; English transl. in Differential Equations **16** (1980).
6. _____, *Bifurcation of invariant tori*, Uspekhi Mat. Nauk **40** (1985), no. 5, 197-198. (Russian)
7. _____, *Bifurcation of invariant tori for systems with degeneration*, Problems of Qualitative Theory of Differential Equations (V. M. Matrosov, ed.), "Nauka", Sibirsk. Otdel., Novosibirsk, 1988, pp. 4-8. (Russian)
8. _____, *Bifurcation to invariant tori with small frequencies*, Proc. 11th Internat. Conf. Nonlinear Oscillations, Janos Bolyai Math. Soc., Budapest, 1987, pp. 18-25.
9. J. Marsden and M. McCracken, *The Hopf bifurcation and its applications*, Springer-Verlag, New York, 1976.
10. J. K. Hale, *Integral manifolds of perturbed differential systems*, Ann. of Math.(2) **73** (1961), no. 3, 496-531.
11. N. N. Bogolyubov and Yu. A. Mitropol'skiĭ, *Asymptotic methods in the theory of nonlinear oscillations*, GITTL, Moscow, 1955; Gordon and Breach, New York, 1962.
12. Yu. N. Bibikov, *Local theory of nonlinear analytic ordinary differential equations*, Lecture Notes in Math., vol. 702, Springer-Verlag, Berlin and New York, 1979.
13. L. E. Reizin', *Local equivalence of differential equations*, "Zinante", Riga, 1971. (Russian)
14. G. Sell, *Bifurcation of higher dimensional tori*, Arch. Rat. Mech. Anal. **69** (1979), no. 2, 199-230.
15. Yu. N. Bibikov, *Bifurcation of stable invariant tori from invariant tori of lower dimensions*, Differentsial'nye Uravneniya **19** (1983), no. 2, 354-357, 368; English transl. in Differential Equations **19** (1983).
16. D. Flockerzi, *Persistence and bifurcation of invariant tori*, Oscillations, Bifurcation, and Chaos (F. V. Atkinson, W. F. Langford, and A. B. Mingarelli, eds.), Proc. Conf. Canadian Math. Soc., vol. 8, Amer. Math. Soc., Providence, R.I., 1988.
17. N. N. Bogolyubov, Yu. A. Mitropol'skiĭ, and A. M. Samoĭlenko, *The accelerated convergence method in nonlinear mechanics*, "Naukova Dumka", Kiev, 1969 (Russian); Springer-Verlag, Berlin and New York, 1976.

Translated by LEONID G. HANIN

Approximation of Nonlinear Operators by Volterra Polynomials

I. BAESLER AND I. K. DAUGAVET

The so-called method of Volterra series is widely used in nonlinear circuit theory. Within the context of the method, it is quite common to find operators being approximated by operators that are formally the same as partial sums of Volterra series. We will call such operators Volterra polynomials. A fundamental question here is whether an arbitrary continuous operator defined on a compact subset of some function space can be approximated by Volterra polynomials. In this paper we show that the answer is positive.

A similar problem, concerning the approximation of nonlinear operators by arbitrary polynomial operators, was solved by Prenter [1], [2] (for Hilbert spaces) and Istratescu [3]. The latter author showed that any continuous operator defined on a compact subset K of a Banach space X and taking values in another Banach space Y can be uniformly and arbitrarily closely approximated by polynomial operators. The stability of approximation by polynomial operators was considered in [4]. Stability here means that if the argument is perturbed then the value of the approximating operator will remain close to that of the approximated operator at the unperturbed argument, provided that the perturbation is small enough. We shall have this condition in mind in our analysis of the approximation of operators by Volterra polynomials.

If the approximated operator possesses certain special properties (belongs to a certain class), it is natural to require that the approximating Volterra polynomials possess the same properties. We shall consider three classes of operators: causal operators, operators with finite memory, and stationary operators with finite memory. The definitions are as follows.

Let F be an operator that carries any function $x(t)$ defined on $[a, b]$ and belonging to a certain set of functions $D(F)$ (the domain of F) onto

---

1991 *Mathematics Subject Classification.* Primary 41A10, 47A58.

a function $F(x; t)$ defined on $[a, b]$. We say that F is *causal* if, for any $t_0 \in [a, b]$, the restriction of $F(x)$ to $[a, t_0]$ depends only on the values of $x(t)$ at points in that interval. Now let $\Delta \in (0, b - a)$ be a given number and suppose that for all $x \in D(F)$ the functions $F(x; t)$ are defined in $[a + \Delta, b]$. Then F is said to have *finite memory* Δ if, for any $t_0 \in [a + \Delta, b]$, $F(x; t_0)$ depends only on the values of the function $x(t)$ in $[t_0 - \Delta, t_0]$. According to this definition, if F has finite memory Δ, then it also has any memory $\Delta' > \Delta$. An operator F with finite memory Δ is said to be *stationary* (with finite memory Δ)* if, for any $t', t'' \in [a + \Delta, b]$ and any functions $x', x'' \in D(F)$, the equality $x'(t' - \tau) = x''(t'' - \tau)$, $\forall \tau \in [0, \Delta]$, implies the equality $F(x'; t') = F(x''; t'')$.

DEFINITION. A functional of the form

$$p(x) = A + \int_a^b k_1(\tau_1)x(\tau_1)\,d\tau_1 + \cdots$$
$$+ \int_a^b \cdots \int_a^b k_n(\tau_1, \ldots, \tau_n)x(\tau_1)\cdots x(\tau_n)\,d\tau_1 \cdots d\tau_n \qquad (1)$$

is called a *Volterra functional*. An operator of the form

$$P(x; t) = k_0(t) + \int_a^b k_1(t, \tau_1)x(\tau_1)\,d\tau_1 + \cdots$$
$$+ \int_a^b \cdots \int_a^b k_n(t, \tau_1, \ldots, \tau_n)x(\tau_1)\cdots x(\tau_n)\,d\tau_1 \cdots d\tau_n \qquad (2)$$

is called a *Volterra polynomial*. An operator of the form

$$P(x; t) = k_0(t) + \int_a^t k_1(t, \tau_1)x(\tau_1)\,d\tau_1 + \cdots$$
$$+ \int_a^t \cdots \int_a^t k_n(t, \tau_1, \ldots, \tau_n)x(\tau_1)\cdots x(\tau_n)\,d\tau_1 \cdots d\tau_n \qquad (3)$$

is called a *causal Volterra polynomial*. An operator of the form

$$P(x; t) = k_0(t) + \int_{t-\Delta}^t k_1(t, \tau_1)x(\tau_1)\,d\tau_1 + \cdots$$
$$+ \int_{t-\Delta}^t \cdots \int_{t-\Delta}^t k_n(t, \tau_1, \ldots, \tau_n)x(\tau_1)\cdots x(\tau_n)\,d\tau_1 \cdots d\tau_n \qquad (4)$$

---

*For brevity the words "with finite memory Δ" will usually be omitted.

is called a *Volterra polynomial with finite memory* Δ. An operator of the form

$$P(x;t) = A + \int_0^\Delta k_1(\tau_1) x(t-\tau_1)\, d\tau_1 + \cdots$$
$$+ \int_0^\Delta \cdots \int_0^\Delta k_n(\tau_1, \ldots, \tau_n) x(t-\tau_1) \cdots x(t-\tau_n)\, d\tau_1 \cdots d\tau_n \quad (5)$$
$$= A + \int_{t-\Delta}^t k_1(t-\tau_1) x(\tau_1)\, d\tau_1 + \cdots$$
$$+ \int_{t-\Delta}^t \cdots \int_{t-\Delta}^t k_n(t-\tau_1, \ldots, t-\tau_n) x(\tau_1) \cdots x(\tau_n)\, d\tau_1 \cdots d\tau_n$$

is called a *stationary Volterra polynomial* (with memory Δ). In all these definitions the k_j are continuous real functions and A is a real number.

Volterra functionals are defined on the set of functions $L(a, b)$. Volterra polynomials and causal Volterra polynomials map functions in $L(a, b)$ to continuous functions on $[a, b]$. Volterra polynomials with finite memory and stationary Volterra polynomials map functions in $L(a, b)$ to continuous functions on $[a+\Delta, b]$.

We will use the notation $[A, k_1, \ldots, k_n]_{[a,b]}^F$, $[k_0, k_1, \ldots, k_n]$, $[k_0, k_1, \ldots, k_n]_c$, $[k_0, k_1, \ldots, k_n]_\Delta$, $[A, k_1, \ldots, k_n]_{st}$ for the Volterra functional and the Volterra polynomials (1)–(5), respectively. Obviously, a causal Volterra polynomial is a causal operator in the sense of the previous definition. A similar assertion is true for Volterra polynomials with finite memory and for stationary Volterra polynomials.

The following lemma is an obvious corollary of the Stone-Weierstrass Theorem (see, e.g., [5]).

LEMMA. *Let K be a compact set in $L_p(a, b)$, $1 \le p < \infty$, or in $C[a, b]$, and f a continuous functional on K. Then for every $\varepsilon > 0$ there exists a Volterra functional $p = [A, k_1, \ldots, k_n]_{[a,b]}^F$ such that $|f(x) - p(x)| < \varepsilon$ for all $x \in K$.*

We will also need the fact that functionals can be stably approximated by Volterra functionals. Let $\delta > 0$; the closed δ-neighborhood of a compact set $K \subset X$, where $X = L_p$ or $X = C$, will be denoted by K_δ.

THEOREM 1. *For any continuous functional $f : K \to R$ and any $\varepsilon > 0$, $\delta > 0$, there exist Volterra functionals p_1 and p_2 such that for all $x \in K$ and all $\Delta x \in X$ with $\|\Delta x\| < \delta$,*

$$|f(x) - p_1(x + \Delta x)| \le \frac{1}{2}\omega(f; 2\delta) + \varepsilon,$$

and for all $x \in K$, $u \in K_\delta$,

$$|f(x) - p_2(u)| \leq \omega(f; 2\|u - x\| + \varepsilon) + \varepsilon.$$

Here

$$\omega(f; \tau) = \sup_{x', x'' \in K : \|x' - x''\| \leq \tau} |f(x') - f(x'')|$$

is the modulus of continuity of f.

The proof of this theorem duplicates that of Theorem 1 of [4], except that the projection P is chosen differently, on the basis of the following easily proved proposition. For any compact set $K \subset X$ and any $\varepsilon > 0$ there exist a finite-dimensional subspace $\widetilde{X} \subset X$ and a projection $P : X \to \widetilde{X}$ such that

(a) for any $x \in K$, $\rho(x, \widetilde{X}) < \varepsilon$ (where ρ is the distance function);
(b) $\|P\| \leq 1 + \varepsilon$;
(c) P is an integral operator:

$$P(x; t) = \int_a^b k(t, \tau) x(\tau) \, d\tau,$$

where k is a continuous kernel.

REMARK 1. All the kernels of the Volterra functional

$$p_1 = [A, k_1, \ldots, k_n]_{[a,b]}^F \quad \text{or} \quad p_2 = [A, k_1, \ldots, k_n]_{[a,b]}^F$$

can be chosen so that if $\tau_j = a$ or $\tau_j = b$ for at least one $j = 1, \ldots, \ell$, then $k_\ell(\tau_1, \ldots, \tau_\ell) = 0$.

REMARK 2. It can be shown that the term ε in the second inequality of Theorem 1 is necessary both in the argument of the modulus of continuity and outside it. If either of these ε's is replaced by zero, the assertion is false.

Let us consider analogs of Theorem 1 for the approximation of continuous operators by Volterra polynomials. To fix ideas, we will dwell in detail on the approximation of operators with finite memory Δ by Volterra polynomials with the same memory, merely stating the results for the other cases.

Let $K \subset X$ be a compact set, where X is either $L_p(a, b)$, $1 \leq p < \infty$, or $C[a, b]$, and let $F : K \to Y$ be a continuous operator on K with range in $Y = C[a + \Delta, b]$, where $0 < \Delta < b - a$. Instead of the modulus of continuity of the function f which appears in Theorem 1, we shall use a certain characteristic $\omega_\Delta(F; \tau)$ of F, which we call the Δ-modulus of continuity of F.[*] To define ω_Δ, we consider the space $Z = L_p(0, \Delta)$ if $X = L_p(a, b)$ or $Z = C[0, \Delta]$ if $X = C[a, b]$. For each $t \in [a, \Delta]$, let $Q_t \in \mathscr{L}(X, Z)$ be the linear operator defined by $Q_t(x; \xi) = x(t - \Delta + \xi)$, $x \in X$, $t \in [a + \Delta, b]$, $\xi \in [0, \Delta]$.

---

[*]Strictly speaking, according to [6], [7], ω_Δ should have been called the R-modulus of continuity of F relative to the R-structure corresponding to operators with finite memory Δ.

DEFINITION. The Δ-modulus of continuity of F is the function defined for $\tau \in [0, +\infty)$ by

$$\omega_\Delta(F; \tau) = \sup_{t \in [a+\Delta, b]} \sup_{x', x''} |F(x', t) - F(x'', t)|,$$

where the second supremum is taken over all pairs of functions $x', x'' \in K$ such that $\|Q_t x'' - Q_t x'\| \leq \tau$.

It is easy to show that the Δ-modulus of continuity is a nonincreasing right-continuous function of τ [7]. Obviously, F is an operator with finite memory Δ if and only if $\omega_\Delta(F; 0) = 0$. In particular, if F is an operator with finite memory Δ, then $\omega_\Delta(F; \tau) \to 0$ as $\tau \to 0$.

THEOREM 2. *Let $F : K \to Y$ be a continuous operator with finite memory Δ. For any $\varepsilon > 0$ and $\delta > 0$ there exist Volterra polynomials P_1 and P_2 with finite memory Δ such that*

(1) *for all $x \in K$ and $\Delta x \in X$ such that $\|\Delta x\| \leq \delta$,*

$$\|F(x) - P_1(x + \Delta x)\| \leq \frac{1}{2}\omega_\Delta(F; 2\delta) + \varepsilon;$$

(2) *for all $x \in K$ and $u \in K_\delta$,*

$$\|F(x) - P_2(u)\| \leq \omega_\Delta(F; 2\|u - x\| + \varepsilon) + \varepsilon.$$

PROOF. We first construct P_1. Since $\omega_\Delta(F; \tau)$ is right continuous, we can choose $\varepsilon_1 > 0$ so small that

$$\frac{1}{2}\omega_\Delta(F; 2\delta + 2\varepsilon_1) + 2\varepsilon_1 \leq \frac{1}{2}\omega_\Delta(F; 2\delta) + \varepsilon.$$

Since $F(K)$ is a compact subset of $C[a + \Delta, b]$, there exists $\delta_1 > 0$ such that if $t', t'' \in [a + \Delta, b]$, $|t' - t''| < \delta_1$, then $|y(t') - y(t'')| < \varepsilon_1$ for any functions $y \in F(K)$. The operators Q_t introduced in the definition of ω_Δ depend continuously on t: if $t \to t_0$, then $Q_t u \to Q_{t_0} u$ for every $u \in X$. Since K is compact, there exists $\delta_2 > 0$ such that $\|Q_{t'} x - Q_{t''} x\| < \varepsilon_1$ for every $x \in K$ and any $t', t'' \in [a + \Delta, b]$ with $|t' - t''| < \delta_2$.

Choose a natural number n so large that $h < \delta_1$, $h < \delta_2$, where $h = (b - a - \Delta)/n$. Set $t_j = a + \Delta + jh$, $j = 0, 1, \ldots, n$, and for every j consider the compact set $K_j = \{Q_{t_j} x \mid x \in K\}$ in Z. Define a functional f_j on each K_j as follows: $f_j(z) = F(x; t_j)$, where $z = Q_{t_j} x \in K_j$ and $x \in K$. This definition is legitimate, since F is an operator with memory Δ: if $z = Q_{t_j} x_1 = Q_{t_j} x_2$, then $F(x_1; t_j) = F(x_2; t_j)$. The functional f_j is continuous; moreover in fact, its modulus of continuity obviously satisfies the inequality $\omega(f_j; \tau) \leq \omega_\Delta(F; \tau)$. Using Theorem 1, construct a Volterra functional $p_j = [A^{(j)}, k_1^{(j)}, \ldots, k_n^{(j)}]_{[0,\Delta]}^F$ such that for all $z \in K_j$ and all Δz with $\|\Delta z\| \leq \delta_3 = \delta + \varepsilon_1$,

$$|f_j(z) - p_j(z + \Delta z)| \leq \frac{1}{2}\omega(f_j; 2\delta_3) + \varepsilon_1.$$

We may assume (see Remark 1) that the kernels $k_\ell^{(j)}(\tau_1, \ldots, \tau_\ell)$ vanish at the boundary of the hypercube $[0, \Delta]^\ell$ and are defined as zero outside this hypercube. Extended in this way, the kernels are continuous throughout $\mathbb{R}^\ell$.

We now define a continuous function $k_0(t)$ for $t \in [a + \Delta, b]$:

$$k_0(t) = \frac{t - t_j}{h} A^{(j+1)} + \frac{t_{j+1} - t}{h} A^{(j)}, \qquad t \in [t_j, t_{j+1}],$$

and continuous kernels

$$k_\ell(t, \tau_1, \ldots, \tau_\ell) = \frac{t - t_j}{h} k_\ell^{(j+1)}(\tau_1 + \Delta - t, \ldots, \tau_\ell + \Delta - t)$$
$$+ \frac{t_{j+1} - t}{h} k_\ell^{(j)}(\tau_1 + \Delta - t, \ldots, \tau_\ell + \Delta - t),$$
$$t \in [t_j, t_{j+1}].$$

Finally, we define the Volterra polynomial $P_1 = [k_0, k_1, \ldots, k_n]_\Delta$ with finite memory Δ. It is easy to see that for any function $u \in X$ and $t \in [t_j, t_{j+1}]$

$$P_1(u; t) = \frac{t - t_j}{h} p_{j+1}(Q_t u) + \frac{t_{j+1} - t}{h} p_j(Q_t u).$$

Now we consider arbitrary functions $x \in K$ and $u = x + \Delta x$, where $\|\Delta x\| \le \delta$. Let us estimate $F(x; t) - P_1(x + \Delta x; t)$. For $t \in [t_j, t_{j+1}]$,

$$|F(x; t) - P_1(x + \Delta x; t)|$$
$$= \left| \frac{t - t_j}{h}(F(x; t) - F(x; t_{j+1})) + \frac{t_{j+1} - t}{h}(F(x; t) - F(x; t_j)) \right.$$
$$\left. + \frac{t - t_j}{h}(f_{j+1}(Q_{t_{j+1}} x) - p_{j+1}(Q_t u)) + \frac{t_{j+1} - t}{h}(f_j(Q_{t_j} x) - p_j(Q_t u)) \right|.$$

Since $|t - t_j| \le h < \delta_1$, it follows that $|F(x; t) - F(x; t_j)| < \varepsilon_1$; similarly, $|F(x; t) - F(x; t_{j+1})| < \varepsilon_1$. Next,

$$\|Q_{t_j} x - Q_t u\| \le \|Q_{t_j} x - Q_t x\| + \|Q_t x - Q_t u\| \le \varepsilon_1 + \|\Delta x\| \le \varepsilon_1 + \delta = \delta_3$$

(we have used the fact that $|t_j - t| \le \delta_2$ and $\|Q_t \Delta x\| \le \|\Delta x\|$). Hence, by the choice of p_j,

$$|f_j(Q_{t_j} x) - p_j(Q_t u)| < \frac{1}{2} \omega_\Delta(F; 2\delta_3) = \varepsilon_1.$$

The same estimate holds for the difference $f_{j+1}(Q_{t_{j+1}} x) - p_{j+1}(Q_t u)$. Taking all these estimates into account, we obtain

$$|F(x; t) - P_1(x + \Delta x; t)| \le \frac{1}{2} \omega_\Delta(F; 2\delta_3) + 2\varepsilon_1 \le \frac{1}{2} \omega_\Delta(F; 2\delta) + \varepsilon.$$

Since $t \in [a + \Delta, b]$ is arbitrary, it follows that

$$\|F(x) - P_1(x + \Delta x)\| \le \frac{1}{2} \omega_\Delta(F; 2\delta) + \varepsilon.$$

We have thus proved that P_1 is the desired Volterra polynomial.

The construction of P_2 is similar, except that the functionals p_j are chosen to satisfy the second part of Theorem 1.

COROLLARY 1. *Under the assumptions of Theorem 2,*

$$\inf_{P} \sup_{x \in K, \|\Delta x\| \leq \delta} \|F(x) - P(x + \Delta x)\| = \frac{1}{2}\omega_\Delta(F; 2\delta),$$

where the greatest lower bound is taken over all Volterra polynomials with finite memory Δ.

The proof follows at once from Theorem 2 and Theorem 1 of [7], according to which, for any continuous operator $G : K_\delta \to C[a + \Delta, b]$ with finite memory Δ,

$$\sup_{x \in K, \|\Delta x\| \leq \delta} \|F(x) - G(x + \Delta x)\| \geq \frac{1}{2}\omega_\Delta(F; 2\delta).$$

COROLLARY 2. *Under the assumptions of Theorem 2, for every* $\varepsilon > 0$ *there exists a Volterra polynomial* P *with finite memory* Δ *such that*

$$\|F(x) - P(x)\| < \varepsilon$$

for all $x \in K$.

Analogs of Theorem 2 and Corollary 1 and 2 are true for approximation of nonlinear operators by Volterra polynomials and by causal and stationary Volterra polynomials. First of all, we have the following theorem.

THEOREM 3. *Let* K *be a compact subset of* X, *where* $X = L_p(a, b)$ *or* $X = C[a, b]$, *and let* $F : K \to C[a, b]$ *be a continuous operator. Then for any* $\varepsilon > 0$, $\delta > 0$ *there exist Volterra polynomials* P_1 *and* P_2 *such that*

(1) *for all* $x \in K$ *and* $\Delta x \in X$ *such that* $\|\Delta x\| \leq \delta$ *we have*

$$\|F(x) - P_1(x + \Delta x)\| \leq \frac{1}{2}\omega_\Delta(F; 2\delta) + \varepsilon;$$

(2) *for all* $x \in K$ *and* $u \in K_\delta$ *we have*

$$\|F(x) - P_2(u)\| \leq \omega(F; 2\|u - x\| + \varepsilon) + \varepsilon.$$

Here

$$\omega(F; \tau) = \sup_{x', x'' \in K, \|x' - x''\| \leq \tau} \|F(x') - F(x'')\|$$

is the modulus of continuity of F.

COROLLARY 1. *Under the assumptions of Theorem 3,*

$$\inf_{P} \sup_{x \in K, \|\Delta x\| \leq \delta} \|F(x) - P(x + \Delta x)\| = \frac{1}{2}\omega(F; 2\delta),$$

where the greatest lower bound is taken over all Volterra polynomials P.

COROLLARY 2. *Under the assumptions of Theorem 3, for every* $\varepsilon > 0$ *there exists a Volterra polynomial* P *such that for all* $x \in K$

$$\|F(x) - P(x)\| < \varepsilon.$$

Now let us consider causal operators. We first refine somewhat the definition. When $X = L_p(a, b)$, we say that functions $x', x'' \in K$ coincide in $[a, t_0]$ if they coincide almost everywhere at this interval. It is therefore natural to assume that if $t_0 = a$, then all functions $x \in K$ coincide in $[a, t_0]$. Similarly, if the range Y is $L_q(a, b)$, it is natural to assume that if $t_0 = a$ all functions $y \in F(K)$ coincide in $[a, t_0]$. However, when $X = L_p(a, b)$ and $Y = C[a, b]$ the case $t_0 = a$ requires special consideration.

We assume that a causal operator F from $L_p(a, b)$ to $C[a, b]$ satisfies the following condition: the values $F(x, a)$ coincide for all x in the domain of F: $F(x, a) = \mathrm{const}$. When $X = Y = C[a, b]$ the causality condition means that $F(x; a) = \varphi(x(a))$, where φ is a continuous real function (since F is continuous). Any causal Volterra polynomial P will have the property: $P(x; a) = k_0(a) = \mathrm{const}$ for all functions x. Thus, when $X = Y = C[a, b]$, a necessary condition for a causal operator F to be approximable arbitrarily closely by causal Volterra polynomials is that the above function φ is a constant.

Let us introduce a special name. A continuous causal operator F on a compact set $K \subset C[a, b]$ with values in $C[a, b]$ is said to be *initial* if $F(x, a) = \mathrm{const}$, $x \in K$.

In approximations by causal Volterra polynomials, instead of the Δ-modulus of continuity used in Theorem 2, we shall use the causal modulus of continuity defined as follows. Let $X = L_p(a, b)$, $1 \le p < \infty$, or $X = C[a, b]$. To every point $t \in [a, b]$ we associate the linear operator $Q_t \in \mathscr{L}(X, X)$ defined by

$$Q_t(x; \xi) = \begin{cases} x(\xi) & \text{if } a \le \xi \le t, \\ 0 & \text{if } t < \xi \le b, \end{cases}$$

if $X = L_p(a, b)$; or

$$Q_t(x; \xi) = \begin{cases} x(\xi) & \text{if } a \le \xi \le t, \\ x(t) & \text{if } t < \xi \le b, \end{cases}$$

if $X = C[a, b]$. Now let $F : K \to C[a, b]$ be a continuous operator, where K is a compact subset of X. Define the causal modulus of continuity as follows:

$$\omega_c(F, \tau) = \sup_{t \in [a,b]} \sup_{x', x'' \in K \,:\, \|Q_t x' - Q_t x''\| \le \tau} |F(x'; t) - F(x''; t)|.$$

The causal modulus of continuity is a nondecreasing right-continuous function in $\tau \in [0, \infty)$. An operator F is causal if and only if $\omega_c(F; 0) = 0$.

THEOREM 4. *Let K be a compact subset of X, where $X = L_p(a, b)$ or $X = C[a, b]$, and let $F : K \to C[a, b]$ be a continuous causal operator. If $X = C[a, b]$, assume that F is also an initial operator. Then for any $\varepsilon > 0$*

and $\delta > 0$ there exist causal Volterra polynomials P_1 and P_2 such that

(1) for all $x \in K$, $\Delta x \in X$ such that $\|\Delta x\| \leq \delta$,

$$\|F(x) - P_1(x + \Delta x)\| \leq \frac{1}{2}\omega_c(F; 2\delta) + \varepsilon;$$

(2) for all $x \in K$ and $u \in K_\delta$,

$$\|F(x) - P_2(u)\| \leq \omega_c(F; 2\|u - x\| + \varepsilon) + \varepsilon.$$

COROLLARY 1. *Under the assumptions of Theorem 4,*

$$\inf_P \sup_{x \in K, \|\Delta x\| \leq \delta} \|F(x) - P(x + \Delta x)\| = \frac{1}{2}\omega_c(F; 2\delta),$$

where the greatest lower bound is taken over all causal Volterra polynomials P.

COROLLARY 2. *Under the assumptions of Theorem 4, for any* $\varepsilon > 0$ *there exists a causal Volterra polynomial* P *such that for all* $x \in K$,

$$\|F(x) - P(x)\| < \varepsilon.$$

Two further corollaries of Theorem 4 show what kind of approximation is nevertheless possible when $X = C[a, b]$ and F is not an initial operator.

COROLLARY 3. *Let* $K \subset C[a, b]$ *be a compact set and* $F : K \to C[a, b]$ *a continuous causal operator. Then for any* $\varepsilon > 0$ *there exist a polynomial* $q(\xi)$ *and a causal Volterra polynomial* P *such that for all* $x \in K$ *and* $t \in [a, b]$

$$|F(x; t) - q(x, (a)) - P(x; t)| < \varepsilon.$$

COROLLARY 4. *Under the assumptions of Corollary 3, for any* $\varepsilon > 0$ *and* $\Delta > 0$ *there exist a causal Volterra polynomial* P *such that for all* $x \in K$ *and* $t \in [a + \Delta, b]$

$$|F(x; t) - P(x; t)| < \varepsilon.$$

We now define the stationary modulus of continuity. Let $F : K \to C[a + \Delta, b]$ be a continuous operator, $\Delta > 0$, and let K be a compact set in $X = L_p(a, b)$ or $X = C[a, b]$. The *stationary modulus of continuity* of F for memory Δ is the function

$$\omega_{\mathrm{st}}(F; \tau) = \sup |F(x'; t') - F(x''; t'')|,$$

where the smallest upper bound is taken over all points $t', t'' \in [a + \Delta, b]$ and all pairs of functions $x', x'' \in K$ such that $\|Q_{t'}x' - Q_{t''}x''\| \leq \tau$; here the operators $Q_t \in \mathscr{L}(X, Z)$, $t \in [a + \Delta, b]$, where $Z = L_p(0, \Delta)$ if $X = L_p(a, b)$ and $Z = C[a, b]$ if $X = C[a, b]$, are the same operators as in the definition of the Δ-modulus of continuity: $Q_t(x; \xi) = x(t - \Delta + \xi)$.

THEOREM 5. *Let K be a compact subset of X and $F : K \to C[a+\Delta, b]$ a continuous stationary operator. For any $\varepsilon > 0$ and $\delta > 0$ there exist stationary Volterra polynomials P_1, P_2 such that*

(1) *for any $x \in K$ and $\Delta x \in K$ such that $\|\Delta x\| \leq \delta$,*

$$\|F(x) - P_1(x + \Delta x)\| \leq \frac{1}{2}\omega_{st}(F; 2\delta) + \varepsilon ;$$

(2) *for any $x \in K$ and $u \in K_\delta$,*

$$\|F(x) - P_2(u)\| \leq \omega_{st}(F; 2\|x - u\| + \varepsilon) + \varepsilon.$$

COROLLARY 1. *Under the assumptions of Theorem 5,*

$$\inf_{P} \sup_{x \in K, \|\Delta x\| \leq \delta} \|F(x) - P(x + \Delta x)\| = \frac{1}{2}\omega_{st}(F; 2\delta),$$

where the greatest lower bound is taken over all stationary Volterra polynomials.

COROLLARY 2. *Under the assumptions of Theorem 5, for any $\varepsilon > 0$ there exists a stationary Volterra polynomial P such that for all $x \in K$*

$$\|F(x) - P(x)\| < \varepsilon.$$

Note that in the last theorem and its corollaries the length of memory for both the operator being approximated and the approximating operators is some fixed number $\Delta > 0$.

Proofs of Theorems 3–5 are similar to the proof of Theorem 2 and involve no new ideas.

An essential point in the proofs of Theorems 2–5 is that the range of the operator is a C-space of continuous functions; moreover, it turns out that the definitions of operators with finite memory and stationary operators involve the value of the image-function at a point in an essential way. It is therefore rather difficult to consider these two classes of operators when the range is $L_q(a, b)$. Analogs of Corollary 2 to Theorems 3 and 4 are nevertheless true for continuous operators in general and causal operators with range $L_q(a, b)$. We now state these analogs, proving only the second.

THEOREM 6. *Let K be a compact set in X, where $X = L_p(a, b)$ or $X = C[a, b]$, and let $F : K \to L_q(a, b)$ be a continuous operator. Then for any $\varepsilon > 0$ there exists a Volterra polynomial P such that for all $x \in X$*

$$\|F(x) - P(x)\| < \varepsilon.$$

THEOREM 7. *Let K be a compact set in X, where $X = L_p(a, b)$ or $X = C[a, b]$, and let $F : K \to L_q(a, b)$ be a continuous causal operator. Then for any $\varepsilon > 0$ there exists a causal Volterra polynomial P such that for all $x \in X$*

$$\|F(x) - P(x)\| < \varepsilon.$$

PROOF. We assume that all functions in $L_q = L_q(a, b)$ vanish at the left of a. For sufficiently small $\alpha > 0$, we define J_α to be a linear operator that associates to every function in L_q the Steklov average

$$J_\alpha(y; t) = \frac{1}{\alpha} \int_{t-\alpha}^{t} y(\xi) d\xi.$$

J_α is a bounded linear causal operator from L_q to $C = C[a, b]$, i.e., $J_\alpha \in \mathscr{L}(L_q, C)$. For any $y \in L_q$ we have

$$\|J_\alpha y - y\|_{L_q} \to 0 \quad \text{as } \alpha \to 0,$$

where the convergence is uniform on the compact set $F(K)$. Using this fact, we fix a number $\alpha > 0$ so small that $\|J_\alpha F(x) - F(x)\|_{L_q} < \varepsilon_1$ for all $x \in K$, where ε_1 is such that $(1 + (b - a)^{1/q})\varepsilon_1 = \varepsilon$. Consider the operator $F_\alpha(x) = J_\alpha F(x)$ on K. It is a continuous operator with values in $C[a, b]$. In addition F_α is causal, as a composition of two causal operators. If $X = C[a, b]$, F_α is also an initial operator, since $F_\alpha(x; a) = 0$ for all $x \in K$. Using Corollary 2 of Theorem 4, we construct a causal Volterra polynomial P such that

$$\|F_\alpha(x) - P(x)\|_{C[a, b]} < \varepsilon_1$$

for all $x \in K$.

It now follows that for $x \in K$

$$\|F(x) - P(x)\|_{L_q} \leq \|F(x) - F_\alpha(x)\|_{L_q} + \|F_\alpha(x) - P(x)\|_{L_q}$$
$$\leq \varepsilon_1 + (b - a)^{1/q} \|F_\alpha(x) - P(x)\|_C \leq \varepsilon_1 (1 + (b - a)^{1/q}) = \varepsilon.$$

This completes the proof.

References

1. P. M. Prenter, *A Weierstrass theorem for real separable Hilbert spaces*, J. Approx. Theory **3** (1970), 341–351.
2. _____, *A Weierstrass theorem for real normed spaces*, Bull. Amer. Math. Soc. **75** (1969), 860–862.
3. V. I. Istratescu, *A Weierstrass theorem for real Banach spaces*, J. Approx. Theory **19** (1977), 118–122.
4. I. K. Daugavet, *Polynomial approximation of stationary operators*, Vestnik Leningrad. Univ. Mat. Mekh. Astronom. **22** (1989), 21–25; English transl. in Vestnik Leningrad. Univ. Math. **22** (1989), 32–37.
5. W. Rudin, *Principles of Mathematical Analysis*, McGraw-Hill, New York, 1964.
6. I. K. Daugavet, *Approximation of operators by causal operators and their generalizations* I. *Linear Case*, Numerical Methods of Analysis and Their Applications, Irkutsk, 1987, pp. 114–128. (Russian)
7. _____, *Approximation of operators by causal operators and their generalizations* II. *Nonlinear case*, Methods of Optimization and Their Applications, Irkutsk, 1988, pp. 166–178. (Russian)

Translated by V. OPERSTEIN

Subgroups of Chevalley Groups Containing a Maximal Torus

N. A. VAVILOV

This paper is a systematic survey of subgroups of those Chevalley groups over a field that contain a split maximal torus. It will be shown, for the first time, that subgroups of this class, in groups of all normal types over infinite fields, admit a standard description. Together with Seitz's result for the finite case, this gives an almost complete solution of the problem, thus completing 25 years of research by many authors. The main steps in the proof of this result will be outlined. The detailed calculations at each stage of the proof involve considerable technical difficulties: the full proofs for the exceptional groups occupy several hundred typewritten pages and will be published elsewhere. In addition, we will review results concerning subgroups that contain a not necessarily split maximal torus and state some unsolved problems.

This survey is a natural continuation of our previous survey [30], where this and similar problems for the classical Chevalley groups were discussed in detail. We therefore concentrate our attention on special groups and on the methods and aspects of the theory that were not considered in [30]. For the same reason, we have not tried to present a comprehensive bibliography on subgroups of classical groups over rings containing maximal tori. Additional references may be found in [42, 27, 30]. Complete proofs of all results formulated here for the classical groups (except, of course, SL_2, which was considered by O. King at the end of 1988), as well as an exhaustive bibliography, can be found in the author's thesis [24].

§1. Preliminaries

We assume the reader to be familiar with at least one of the following texts on the theory of Chevalley groups: [7, 43, 69, 73, 83, 107]. Here we

---

1991 *Mathematics Subject Classification.* Primary 22E40; Secondary 20D06.

recall only the basic notation and a number of general facts which will be used throughout.

1. Chevalley groups. Let Φ be a reduced irreducible root system, P a lattice that lies between the root lattice $Q(\Phi)$ and the weight lattice $P(\Phi)$, $G_P(\Phi)$ a Chevalley-Demazure group scheme of type Φ, P [86, 7], $T_P(\Phi)$ a split maximal torus in $G_P(\Phi)$. We usually omit the weight lattice from the notation, since it does not play any significant role in our context. Nevertheless, if we want to emphasize that we are considering a simply-connected group (which is usually convenient), we use the subscript sc; similarly, for adjoint groups, we use the subscript ad. Since the general case can be reduced by standard means to that of simple groups, we will assume Φ to be irreducible, but we will have to consider subsystems of Φ that are not irreducible. If R is a commutative ring with unit — we need only the case when R is a field K — we denote the corresponding Chevalley group and its split maximal torus simply by $G = G(\Phi, R)$ and $T = T(\Phi, R)$, respectively.

The construction of Chevalley groups is based on the fact that in every semisimple complex Lie algebra L of type Φ we can choose a Chevalley basis x_α, $\alpha \in \Phi$; h_i, $\alpha_i \in \Pi$, where Π denotes the set of simple roots in Φ in some ordering of Φ. A Chevalley basis is a normalization of a Weyl basis all of whose structure constants are integers. The integral lattice spanned by x_α, h_i is denoted by $L_{\mathbb{Z}}$ and is called a Chevalley order in L. If K is a field, then the Lie K-algebra $L_K = L_{\mathbb{Z}} \otimes K$ with Chevalley basis $x_\alpha = x_\alpha \otimes 1$, $e_i = e_i \otimes 1$ is also called a Chevalley algebra. The adjoint Chevalley group $G_{\text{ad}}(\Phi, K)$ may be described most simply as the subgroup of the automorphism group of the Chevalley group generated by certain special automorphisms. This is how Chevalley groups are treated in [73] (see also [43, 83, 157]).

Now let π be a representation of a Lie algebra L in a finite-dimensional vector space V. We omit the symbol π in the notation for the action of L on V, writing simply xu for $\pi(x)u$, where $x \in L$, $u \in V$. A lattice $V_{\mathbb{Z}}$ in V is said to be admissible with respect to $L_{\mathbb{Z}}$ if it is stable with respect to all divided powers $x_\alpha^{(m)} = x_\alpha^m/m!$, that is, if $x_\alpha^{(m)} V_{\mathbb{Z}} \subseteq V_{\mathbb{Z}}$ for all $\alpha \in \Phi$ and $m \in \mathbb{N}$ (in other words, $V_{\mathbb{Z}}$ is stable under the action of the $\mathbb{Z}$-form $U(L)_{\mathbb{Z}}$ of the universal enveloping algebra $U(L)$ associated with the Chevalley basis, that is, the Kostant form). Such a lattice always exists [7, 69, 86]. Thus, if the weight lattice of π is P, then the group $G_P(\Phi, K)$ is defined from the start together with a representation in the vector space $V_K = V_{\mathbb{Z}} \otimes K$, which we denote by the same letter π.

2. Root elements. For every root $\alpha \in \Phi$ there exist root unipotent elements $x_\alpha(\xi)$, $\xi \in K$, and root semisimple elements $h_\alpha(\varepsilon)$, $\varepsilon \in K^*$. Recall that $h_\alpha(\varepsilon) = w_\alpha(\varepsilon) w_\alpha(1)^{-1}$, where $w_\alpha(\varepsilon) = x_\alpha(\varepsilon) x_{-\alpha}(-\varepsilon^{-1}) x_\alpha(\varepsilon)$ [69, 83]. If we want to emphasize that the root elements in question correspond to a fixed choice of a maximal torus T, we will call them elementary. In general, however, the term root unipotent (or semisimple) element, will be used for any element of G conjugate to an elementary unipotent (or semisimple) element, i.e., to a certain $x_\alpha(\xi)$ or $h_\alpha(\varepsilon)$, respectively. These elements are called long if α is a long root and short if it is short.

All calculations in Chevalley groups are based on certain relations, called the Steinberg relations, among the elements $x_\alpha(\xi)$, $w_\alpha(\xi)$, $h_\alpha(\xi)$ [69, 83]. The most important are additivity of the x_α, multiplicativity of the h_α (the latter does not hold in the "Steinberg group"!) and the Chevalley commutation formula:

$$[x_\alpha(\xi), x_\beta(\zeta)] = \prod x_{i\alpha+j\beta}(N_{\alpha\beta ij}\xi^i\zeta^j),$$

where $[x, y] = xyx^{-1}y^{-1}$ denotes the commutator of two elements x, y of G; the product on the right is taken over all $i, j \in \mathbb{N}$ such that $i\alpha + j\beta \in \Phi$ in some fixed order, and $N_{\alpha\beta ij}$ are certain constants, independent of ξ and ζ and equal to ± 1, ± 2, ± 3, called the structure constants of G ($N_{\alpha\beta 11} = N_{\alpha\beta}$ are simply the structure constants of L in the Chevalley basis: $[x_\alpha, x_\beta] = N_{\alpha\beta}x_{\alpha+\beta}$).

For a root $\alpha \in \Phi$, let X_α denote the corresponding elementary root unipotent subgroup, consisting of all $x_\alpha(\xi)$, $\xi \in K$. The root subgroups are the subgroups conjugate in G to elementary root subgroups. The subgroup $E = E(\Phi, R)$ of G is called an elementary Chevalley group. Elementary groups are, of course, especially important in the theory of Chevalley groups over rings, but the groups G and E need not coincide even in the case of a field (counterexample: the orthogonal group). Nevertheless, in the most important case, that of simply-connected groups, $G_{sc}(\Phi, K) = E_{sc}(\Phi, K)$. Generally speaking, the difference between G and E is not very essential in the case of a field, being related only to the fact that the torus T may not coincide with the group

$$H = H(\Phi, K) = \langle H_\alpha(\varepsilon), \alpha \in \Phi, \varepsilon \in K^* \rangle,$$

generated by all the elementary root semisimple elements. As we just noted,

$$H_{sc}(\Phi, K) = T_{sc}(\Phi, K) = \text{Hom}(P(\Phi), K^*).$$

The center of G_{sc} consists of those elements of H_{sc} that are identically 1 on $Q(\Phi)$; hence it is isomorphic to $\text{Hom}(P(\Phi)/Q(\Phi), K^*)$.

3. Bruhat decomposition. Now let $N = N(\Phi, K)$ be the subgroup of G generated by T and all $w_\alpha(1)$, $\alpha \in \Phi$. As is well known, N is the normalizer of T in G, except for a very few minor exceptions such as $|K| = 2, 3$ [69]. The quotient group N/T is canonically isomorphic to the Weyl group $W = W(\Phi)$ of the root system Φ; hence, for every element $w \in W$, we can consider a preimage n_w of w in N. Usually, speaking of subgroups containing T, we shall write simply w instead of n_w. Thus, for example, we shall write BwB instead of Bn_wB to denote double cosets modulo Borel subgroups (see below).

Fix an ordering on Φ and let Φ^+, Φ^-, and $\prod = \{\alpha_1, \ldots, \alpha_l\}$ be the sets of positive, negative, and simple roots, respectively, with respect to this ordering. As usual, we set

$$U = U(\Phi, K) = \langle x_\alpha(\xi), \alpha \in \Phi^+, \xi \in K \rangle,$$
$$V = V(\Phi, K) = \langle x_\alpha(\xi), \alpha \in \Phi^-, \xi \in K \rangle,$$

where $\langle X \rangle$ denotes the subgroup of G generated by X. Then the subgroup $B = B(\Phi, K)$ — the product of T and U — is called the standard Borel subgroup of G corresponding to the given ordering in Φ, and U is called the unipotent radical of B. The subgroup $B^- = B^-(\Phi, K)$, which is equal to TV, is called the Borel subgroup opposite to B.

The Bruhat decomposition of G means that W is a system of representatives of the double cosets of G modulo B, i.e., any element $x \in G$ can be represented as $x = b_1 w b_2$ where $b_1, b_2 \in B$ and $w \in W$ is uniquely defined. The Bruhat decomposition can be made somewhat more precise. Indeed, given $w \in W$, we set $U_w^- = U \cap w^{-1} V w$. Then every element $x \in G$ can be uniquely represented as $x = uwvd$, where $u \in U$, $w \in W$, $v \in U_w^-$, $d \in T$. This decomposition, also called the canonical form, is a fundamental computational tool in the analysis of Chevalley groups over fields.

§2. Standard subgroups

In this section we consider certain subgroups of G that obviously contain T. The main result of the paper, formulated in the next section, asserts that, as a rule, there are no other subgroups containing T.

1. Standard subgroups. The groups U and V are special cases of a group $E(S)$ that we are now going to define for an arbitrary closed subset of roots S. Recall that a set of roots $S \subseteq \Phi$ is closed if, whenever $\alpha, \beta \in S$ and $\alpha + \beta \in \Phi$, then also $\alpha + \beta \in S$. Define $E(S)$ to be the subgroup generated by all X_α, $\alpha \in S$. Then U and V are the groups $E(\Phi^+)$ and $E(\Phi^-)$. The group $E(S)$ has a special meaning when S is a special set, that is, a closed subset such that if $\alpha \in S$, then $-\alpha \notin S$. In that case $E(S)$ is simply the product of all X_α, $\alpha \in S$, taken in any fixed order. Set $G(S) = T \cdot E(S)$.

If S is an arbitrary closed set of roots, let S^r denote its reductive or symmetric part, i.e., the set of all $\alpha \in S$ for which $-\alpha \in S$, and S^u its unipotent or special part, i.e., the set of all $\alpha \in S$ for which $-\alpha \notin S$. We avoid the notation S^+ and S^-, which is often used in this context, since it is more convenient to reserve it for the sets $S^+ = S \cap \Phi^+$, $S^- = S \cap \Phi^-$. Clearly, S is the disjoint union of S^r and S^u. It is easy to see that $E(S) = E(S^r) \cdot E(S^u)$ and $G(S) = G(S^r) \cdot E(S^u)$. The subgroup $E(S^u)$ is called the unipotent radical of $G(S)$, and $G(S^r)$ is called the Levi subgroup or Levi component.

An element $w \in W$ is said to normalize S if $wS = S$. The set $X(S)$ of all $w \in W$ that normalize S is called the normalizer of S in the Weyl group. It is clear that $X(S)$ contains the Weyl subgroup $W(S) = W(S^r)$; moreover (see, e.g., [84]), if $S = S^r$ is a root system, then $X(S)$ is the normalizer of $W(S)$ in W. Two sets S_1 and S_2 are said to be conjugate if there exists $w \in W$ such that $wS_1 = S_2$. Let $N(S)$ denote the subgroup of G generated by $G(S)$ and the elements n_w for all $w \in X(S)$. It follows from the Tits theorem (see [70] or §7 of this paper) that $N(S)$ is almost always equal to the normalizer of $G(S)$ in G. This equality may fail to hold only when $|K| = 2, 3$, and even then not for all S.

The next definitions play a major role in what follows. A subgroup F of a Chevalley group $G = G(\Phi, K)$ is called standard if there exists a closed set of roots $S \subseteq \Phi$ such that $G(S) \leq F \leq N(S)$. It is clear that the standard subgroups contain T. We will say that G admits a standard description of subgroups containing a split maximal torus T if the converse holds, i.e., every subgroup of G containing T is standard. The goal of this paper is to obtain such standard descriptions for all "sufficiently big" fields.

2. Parabolic subgroups. Recall that a closed set of roots containing Φ^+ is called a standard parabolic subset; any set conjugate to a standard parabolic subset is called a parabolic subset. It is clear that the only maximal standard parabolic subsets are the sets P_r, $1 \leq r \leq l$, defined as the smallest closed subsets of Φ that contain Φ^+ and all roots $-\alpha_i$ where $\alpha_i \in \Pi \setminus \{\alpha_r\}$. Since two different sets P_r and P_s are never conjugate, there exist exactly l classes of maximal parabolic subsets, up to conjugacy. In general, the standard parabolic subsets correspond in one-to-one fashion with all the subsets of the fundamental root system Π. Indeed, given an arbitrary subset $J \subseteq \Pi$, consider the parabolic set defined as the closure of $\Phi^+ \cup \{-\alpha |\, \alpha \in J\}$. This construction gives different parabolic subsets P_I and P_J for different subsets I and J, and every standard parabolic subset is a P_J for some $J \subseteq \Pi$. Thus, the total number of standard parabolic subsets of a root system of rank l is 2^l.

The subgroups of a Chevalley group $G = G(\Phi, K)$ that contain the standard Borel subgroup $B = B(\Phi, K)$ are called standard parabolic subgroups; any subgroup conjugate to a standard parabolic subgroup is called a parabolic subgroup. The classical Tits theorem (see, e.g., [69, 83]) states that the mapping $P_J \to G(P_J)$ is a bijection of the set of standard parabolic subsets onto the set of standard parabolic subgroups. If P is a parabolic subgroup, let U_P and L_P denote its unipotent radical and Levi subgroup, respectively. Naturally, the maximal parabolic subgroups are of the greatest importance for our purposes. If $P = G(P_r)$, we will also write $U_P = U_r$, $L_P = L_r$.

Parabolic subgroups play a special role because every nonreductive standard subgroup is contained in a parabolic subgroup. Namely, the Borel-Tits theorem [9] asserts that if $S^u \neq \varnothing$ then $N(S)$ is contained in some proper parabolic subgroup. Recall that parabolic subgroups coincide with their normalizers; in particular, $W(S) = X(S)$ for any parabolic set of roots S.

3. Subsystems of roots. For reduction to groups of lower rank, and for the classification of maximal subgroups of Chevalley groups containing a split maximal torus, we have to determine the maximal standard subgroups. Thanks to the Borel-Tits theorem, this can be done by finding all maximal standard subgroups $N(\Delta)$, where Δ is a subsystem of roots in Φ or, what is the same, by finding the pairs $(\Delta, X(\Delta))$ that are maximal with respect to inclusion in the set of pairs $(S, X(S))$, where S is a closed subset of Φ.

A description of all subsystems of reduced systems of roots was obtained independently by Borel-de Siebenthal and Dynkin, who used the following strikingly simple construction. (Dynkin [39] in fact solved the much more general problem of describing all the semisimple subalgebras of semisimple Lie algebras.) Let $\widetilde{\Pi}$ be an extended system of simple roots, obtained from Π by adding the root $\alpha_0 = -\delta$, where δ is the maximal root of Φ in some specific ordering. Let Δ_r, $1 \le r \le l$, be the smallest closed set of roots containing all roots $\pm\alpha$, $\alpha \in \widetilde{\Pi} \setminus \{\alpha_r\}$. Then Δ_r is a subsystem of roots of "maximal rank", which may be reducible. It may also coincide with Φ; this is always the case if $\Phi = A_l$. We can now repeat the process, starting from any irreducible component of Δ_r, and so on. In this way we eventually obtain all subsystems of maximal rank [39, Table 10]. Any root subsystem of Φ may now be obtained as follows. Let Δ be any subsystem of maximal rank in Φ and Σ its set of simple roots. Take an arbitrary subset $J \subseteq \Sigma$ and consider the closure of $(-J) \cup J$. As a result, we obtain all closed root subsystems of Φ. For the classical root systems, the classification of subsystems is completely elementary [39, Table 9]. In special systems, the following numbers of proper root subsystems (i.e., different from both Φ and $\varnothing$), up to conjugacy are as follows: 4 in G_2, 22 in F_4, 19 in E_6, 45 in E_7 and, finally, 75 in E_8 [39, Table 11].

As a rule, two subsystems of the same type are conjugate to each other in Φ. The only exceptions are as follows. If all roots of Δ are of the same length, it may be imbedded in Φ in two (generally different) ways, that is, as long roots or short roots. In order to distinguish between these two imbeddings, we write $\widetilde{\Delta} \subseteq \Phi$ for imbedding as short roots, reserving the notation $\Delta \subseteq \Phi$ for imbedding as long roots. Apart from this case, this occurs only in the systems B_l and D_l, owing to the difference between $2A_1$ and D_2, and also between A_3 and D_3, as well as in the systems D_l, E_7, and E_8 for certain subsystems $A_{k_1} + \cdots + A_{k_l}$ where all the k_i are odd.

4. Classification of maximal standard subgroups. The following theorem lists all the maximal standard subgroups of Chevalley groups. We use the following notation: $\Phi + \Delta$ denotes the orthogonal sum of root systems, $k\Delta$ the sum of k copies of Δ, $A_0 = B_0 = D_1$ the empty root system.

THEOREM 1. *Let Φ be a reduced irreducible root system, K a field. Then the Chevalley group $G = G(\Phi, K)$ contains the following conjugacy classes of maximal standard subgroups: l classes of maximal parabolic subgroups and the classes listed in the following table*:

A_l: $N\left(\dfrac{l+1}{k+1}A_k\right)$, $(k+1)|(l+1)$, $k \neq l$.

B_l: $N(B_k + D_{l-k})$, $0 \leq k \leq l-1$.

C_l: $N((l/k)C_k)$, $k|l$, $k \neq l$; $N(A_{l-1})$; $G(C_k + C_{l-k})$, $1 \leq k \leq l/2$.

D_l: $N((l/k)D_k)$, $k|l$, $k \neq l$; $N(A_{l-1})$; $N(D_k + D_{l-k})$, $1 \leq k \leq l/2$.

G_2: $N(A_2)$, $G(A_1 + \tilde{A}_1)$.

F_4: $G(B_4)$, $N(D_4)$, $G(A_1 + C_3)$, $N(A_2 + \tilde{A}_2)$.

E_6: $G(A_5 + A_1)$, $N(3A_2)$, $N(D_4)$, N.

E_7: $G(D_6 + A_1)$, $N(A_7)$, $N(A_5 + A_2)$, $N(D_4 + 3A_1)$, $N(7A_1)$, $N(E_6)$, N.

E_8: $G(E_7 + A_1)$, $G(D_8)$, $N(E_6 + A_2)$, $N(A_8)$, $N(2D_4)$, $N(2A_4)$, $N(4A_2)$, $N(8A_1)$, N.

Each of the types listed in the table gives one conjugacy class in G, except $N(A_{l-1})$, which gives two conjugacy classes in $G(D_l, K)$ if l is even and is not maximal for odd l. If K is an infinite field, the subgroups $G(S)$, and of the standard subgroups they alone, are connected in the Zariski topology. Thus, Theorem 1 gives, in particular, a classification of the connected maximal standard subgroups: just choose those Δ for which $G(\Delta)$ occurs in the table. An explicit list of maximal standard subgroups is also of interest, since subgroups that are not contained in any proper subgroup conjugate to a standard subgroup play the same role in the theory of Chevalley groups as the primitive irreducible subgroups in the theory of linear groups.

A theorem of this type for complex semisimple Lie groups was obtained by Golubitsky and Rothschild [103, 104]. Since the main calculations here involve only root systems and their Weyl groups, the result should be the same for all fields. However, the tables in the papers just cited contain three errors [21]. Theorem 1 was established by the author in [19, 21].

§3. Formulation of main results

We are now going to present the first really general formulations of two basic theorems about subgroups of $G = G(\Phi, K)$ containing $T = T(\Phi, K)$ — the classification theorem and the conjugacy theorem. Previously, such results have been stated either for special classes of fields or for groups of nonsymplectic types. The classification theorem states that the descriptions of subgroups in G that contain T are almost always standard. The conjugacy theorem states that under the same restrictions T is a pronormal subgroup of G. The proofs are outlined in §§4–9.

The assumption $\operatorname{char} K \neq 2$ for $\Phi = B_l, C_l, F_4$, $\operatorname{char} K = 2, 3$ for $\Phi = G_2$, is a necessary condition for a standard description. Otherwise, even the Borel subgroup $B = B(\Phi, K)$ will have nonstandard subgroups containing T. In fact, however, the situation is easy to remedy for all groups, except those of type C_l, by admitting not only closed but also quasiclosed sets of roots [8]. For the symplectic group $\operatorname{Sp}(2l, K) = G_{\operatorname{sc}}(C_l, K)$, in particular, $\operatorname{SL}(2, K) = \operatorname{Spin}(3, K) = G_{\operatorname{sc}}(C_1, K)$, $\operatorname{Spin}(4, K) = G_{\operatorname{sc}}(C_1, K)G_{\operatorname{sc}}(C_1, K)$, and $\operatorname{Spin}(5, K) = G_{\operatorname{sc}}(C_2, K)$, the situation is much more serious: If K is a nonperfect field of characteristic 2, there may be even infinitely many subgroups in G containing T. Taking this into consideration, we will assign the systems $A_1 = B_1 = C_1$ and $B_2 = C_2$ to the symplectic series. Thus, we assume $l \geq 2$ for A_l and $l \geq 3$ for B_l and D_l. Even so, it turns out that, apart from the forbidden characteristics, only extremely small fields may be exceptions as far as the standard description is concerned.

THEOREM 2. *Let Φ be a reduced irreducible root system and let K be a field. Assume that $|K| \geq 13$ and, moreover, $\operatorname{char} K \neq 2$ if $\Phi = B_l, C_l, F_4$, $\operatorname{char} K \neq 2, 3$ if $\Phi = G_2$. Then the Chevalley group $G = G(\Phi, K)$ admits a standard description of its subgroups that contain the split maximal torus $T = T(\Phi, K)$. In other words, for every intermediate subgroup F, $T \leq F \leq G$ there exists a unique closed set of roots $S \subseteq \Phi$ such that $G(S) \leq F \leq N(S)$.*

The main steps toward a proof of this theorem were as follows (see §6 for the very important case of extended groups). The result was proved for algebraically closed fields in the classical work of Borel and Tits [8]. Formally speaking, their proof concerned only subgroups that were connected and closed in the Zariski topology, which were proved to coincide with $G(S)$ for appropriate closed (or quasiclosed, in the exceptional characteristics) sets of roots S. However, the famous theorem of Chevalley immediately implies that in the case of an algebraically closed field, any subgroup that contains T is automatically closed in the Zariski topology; and another classical paper of Tits [70] states that the normalizers $G(S)$ in G coincide with $N(S)$, not only for algebraically closed fields but for all fields K with $|K| > 5$.

The next important step was a 1979 paper by Seitz [146], in which he established a standard description for finite fields K, $\operatorname{char} K \neq 2$, for all root systems, including the series $\Phi = A_l, D_l, E_l$ and $|K| \geq 13$. The proofs in [146] rely substantially on the theory of finite groups and they do not carry over to infinite fields.

In 1979, the author and Dybkova proved the following reduction theorem for symplectic groups: if $\operatorname{char} K \neq 2$ and $|K| \geq 7$, the description of the subgroups of $\operatorname{Sp}(2l, K)$ containing a split maximal torus is standard if and only if it is standard for $\operatorname{SL}(2, K)$ [33, 38, 24]. This result made it possible, in particular, to prove Theorem 2 for symplectic groups under the additional hypothesis $|K^*| > |K^*/K^{*2}|$ [33, Theorem 5].

In 1983, the author proposed a method that enabled him to prove Theorem

2 for all nonsymplectic groups over an infinite field (announced in [17]). The key idea was to reduce the treatment of the nonsymplectic groups, not to ordinary Chevalley groups, but to extended Chevalley groups of lower rank, for which the theorem had already been proved (see §6). This proof, for $SL(n, K)$, $n \geq 3$, over an infinite field, was published in [18, I]. Parts II–IV of [18] actually established a much stronger result, namely the truth of the theorem for $SL(l+1, K) = G_{sc}(A_l, K)$, $l \geq 2$, for all K, $|K| \geq 7$. Later we showed that the theorem also holds for the spinor group $Spin(2l, K) = G_{sc}(D_l, K)$, with the same bound $|K| \geq 7$ on the number of elements of the field (the proof will be published in a paper "On subgroups of the spinor group that contain a split maximal torus. I"). In [23] we lifted the restriction $\operatorname{char} K \neq 2$ for groups of type E_l over a finite field.

Thus, it remained only to consider the case $SL(2, K) = G_{sc}(A_l, K)$. Paradoxically, this case turned out to be difficult for infinite fields. In 1983 the author and Dybkova [33, II] verified that under the additional assumption $-1 \in K^{*2}$ the group $SN(2, K)$ of monomial matrices is maximal in $SL(2, K)$; in 1985 King proved this in full generality [118]. In 1986, analyzing King's proof, we noticed that an insignificant modification yields a proof for the group $SL(2, K)$ and, simultaneously, thanks to the results of [33], for all symplectic groups as well, provided that $-1 \in K^{*2}$ [32]. Finally, in the fall of 1988, King successfully completed the treatment of $SL(2, K)$ [119]; this was the last step in the proof of the classification theorem.

Theorem 2 is a substantial generalization of many previously known results, in particular, of the description of the parabolic subgroups of G. For example, Theorems 1 and 2 imply the following

COROLLARY. *Under the assumptions of Theorem 2, any maximal subgroup in G that contains T is either a maximal parabolic subgroup or conjugate to one of the groups listed in the table of Theorem 1.*

As particular cases, this corollary includes a great many previous results of Dye, Ki, King, Li Shangzhi and other authors, relating mostly to classical groups. We will not go into the details here, nor discuss geometrical versions of the corollary for classical groups; the reader is referred to [24, 27, 30] (in addition to the literature listed there, see also the recent work of Ton Dao-rong [167, 168]).

Our second main theorem, as we mentioned, is the conjugacy theorem. Recall that a subgroup T of a group G is said to be pronormal in G if any two subgroups conjugate to it are already conjugate in the subgroup that they generate. In other words, for every $x \in G$ there exists $y \in \langle T, xTx^{-1} \rangle$ such that $xTx^{-1} = yTy^{-1}$. The best known examples of pronormal subgroups are of course the Sylow subgroups in finite groups and the maximal tori in algebraic groups [6, 72]. Pronormality generalizes both normality and abnormality. Recall that a subgroup B of a group G is said to be abnormal if, for any $x \in G$, $x \in \langle B, xBx^{-1} \rangle$. In particular, the normalizer of a pronormal group is abnormal. It turns out [2] that pronormality is of paramount importance in describing the lattice of subgroups of an abstract group. The classical Tits theorem asserts that a Borel subgroup B of a Chevalley group G is abnormal. The following result is an analog — and essentially a substantial generalization — of this fact.

THEOREM 3. *Under the assumptions of Theorem 2, the split maximal torus $T = T(\Phi, K)$ is pronormal in the Chevalley group $G = G(\Phi, K)$.*

For the proof, see §7. The "prehistory" of the theorem, relating to extended groups, is described in §6. The following fairly useful assertion follows from Theorem 3.

COROLLARY. *Under the assumptions of Theorem 2, if two subgroups that contain T are conjugate to each other in G, they are conjugate by an element of N.*

In actual fact, pronormality of T is equivalent to a more precise assertion: if $F_2 = xF_1x^{-1}$ for two subgroups F_1, F_2 of G containing T, and for some $x \in G$, then $x = wy$ for a suitable $w \in N$ and $y \in F_1$.

§4. Long root semisimple elements

Underlying the proofs of Theorems 2 and 3 are calculations with long root semisimple elements. The starting point is the following well-known fact (see, e.g., [69]):

$$T_{\rm sc}(\Phi, K) = \langle h_\alpha(\varepsilon), \ \alpha \in \Phi_l, \ \varepsilon \in K^* \rangle,$$

where Φ_l is the set of long roots of Φ. Moreover, the long root semisimple elements (LRSE) not only generate a split maximal torus of a simply-connected group, but generally speaking, they are in fact the simplest elements of the torus. Certain microweight elements (see §5) may indeed have simpler structure, but it is well known that groups of types E_8, F_4, G_2 have no microweight representations. The use of LRSE's here is therefore dictated by hard necessity. But even for groups of types A_l, D_l, E_6 the use of LRSE's, as a rule, produces more precise results than that of microweight elements. In this section we assemble several results due to the author and Semenov (see, in particular, [28, 29, 31, 35, 66]) concerning the Bruhat decompositions of LRSE's; in §§7, 8 we shall show how this information is used to prove Theorems 2 and 3. In view of future applications, we are interested mainly not in the Bruhat decompositions of individual LRSE's, but in the Bruhat decompositions of "long root tori", i.e., of subgroups such as

$$Q_x = \{y(\varepsilon) = xh_\delta(\varepsilon)x^{-1}, \ \varepsilon \in K^*\},$$

where $x \in G$ and δ is a maximal root.

1. Reduction to D_4. The following fact is crucial: any problem concerning Bruhat decompositions of long root tori in Chevalley groups of all types can be reduced to the analogous problem in the group $\mathrm{Spin}(8, K)$. Let Φ, Δ be root systems. We will say that Δ is obtained from Φ if Δ is a twisting (possibly, trivial) of a subsystem of Φ (possibly of Φ itself).

THEOREM 4. *Let Q_x, $x \in G$, be a long root torus in a Chevalley group $G = G(\Phi, K)$. Then there exist a subsystem $\Delta \subseteq \Phi$ obtained from D_4 and an element $y \in U(\Phi, K)$ such that $yQ_x y^{-1} \leq G(\Delta, K)$.*

This is Theorem 1 of [35]. The idea of the reduction is contained (in a less rigorous form) in our papers [28, 29]. The main step in the proof consists in finding orbits of a Borel subgroup in certain special representations [28]; this was also done independently by A. G. Elashvili. Other applications of the theorem and another interpretation of the relevant orbit computations are discussed in [28, 29, 35]. As it happens, another problem which, at first glance, has nothing in common with our problem, also reduces to exactly the same computations: to find all possible configurations of triples of long root unipotent subgroups, two of which are opposite (see §9).

2. Number of degenerations. Using Theorem 4, in $\mathrm{Spin}(8, K)$, Semenov [35] obtained the following surprising result.

THEOREM 5. *A long root torus Q_x intersects at most four cosets in the Bruhat decomposition. Moreover, all elements $y(\varepsilon)$, except at most three, are in the same coset Bw_0B.*

In other words, there is at most one $\theta \neq 0$, 1 such that $y(\varepsilon) \in Bw_0B$ for $\varepsilon \neq 0$, 1, θ, θ^{-1}, but $y(\theta) \in BwB$ and $y(\theta^{-1}) \in Bw^{-1}B$ for some $w \neq w_0$. The elements $y(\varepsilon)$ in Bw_0B and the corresponding values of the parameter ε will be called typical; all others will be called degenerate. A comparison of the Bruhat decompositions of the elements $y(\varepsilon)$ and $y(\varepsilon^{-1}) = y(\varepsilon)^{-1}$ shows that w_0 is an involution. The author's own papers [28, 29] proved only that all but a finite set of the elements $y(\varepsilon)$ are in Bw_0B. Semenov previously proved Theorem 5 for the cases $\Phi = A_l$, C_l [66].

3. Factors from the Weyl group. The following corollary, relating to w_0, immediately follows from Theorem 4.

COROLLARY. *The element w_0 in the Bruhat decomposition of a long root torus is necessarily a product of reflections with respect to pairwise orthogonal roots. If r of these roots are long and s are short, then $r + 2s \leq m$, where $m = 4$ for $\Phi = B_l$, $l \geq 3$, D_l, $l \geq 4$, E_l, F_4; $m = 3$ for $\Phi = G_2$; $m = 2$ for $\Phi = A_l$, $l \geq 3$, C_l, $l \geq 2$, and, finally, $m = 1$ for $\Phi = A_1$, A_2.*

More detailed information about what collections of orthogonal roots occur here may be found in [28]. It is also of considerable interest to find all values that w may take in the Bruhat decomposition of a LRSE.

For $\mathrm{SL}(n, K)$, and hence also for $\mathrm{Sp}(2l, K)$, all possible $w \in S_n = W(A_{n-1})$ were found in [31]. It turns out that, besides transpositions and products of two independent transpositions, there may occur only 3-cycles and certain (but not all!) 4-cycles. A more detailed analysis of this problem was presented in [66], including a table of all possible w_0, w, corresponding to a long root torus in $SL(n, K)$. In particular, it was found that for a symplectic group $Sp(2l, K)$ degeneration may occur only at the points $\varepsilon = \pm 1$; moreover, if w_0 is a product of reflections with respect to two orthogonal long roots, degeneration will invariably occur at $\varepsilon = -1$, that is, $y(-1)$ lies outside the typical coset Bw_0B. We will make essential use of this fact in §8. Very recently, Semenov extended the results of [31, 66] to all Chevalley groups, but his results are too cumbersome to be presented here.

4. Factors from the Borel subgroup. It is also important to have information about the factors b_1 and b_2 from the Borel subgroup $B = B(\Phi, K)$ in the Bruhat decomposition of a LRSE: $y(\varepsilon) = b_1 w b_2$. The available results are too technical to state here in full detail. A rough impression may be gained from [20, 22, 24, 31], where explicit calculations of the factors from B are carried out for several microweight elements. Here we restrict ourselves to a simple proposition whose role will become clear in §7. Express the elements b_1 and b_2 as $b_1 = ud_1$, $b_2 = d_2v$, where $u, v \in U = U(\Phi, K)$ and $d_1, d_2 \in T$, and factorize the elements u and v into elementary factors:

$$u = \prod x_\alpha(u_\alpha), \qquad v = \prod x_\alpha(v_\alpha)$$

extending over all positive roots in an arbitrary but fixed order.

THEOREM 6. *The elements u_α and v_α are rational functions of ε, whose numerators and denominators are at most quadratic functions of ε.*

In actual fact, if the ordering of Φ^+ is suitably chosen, most of these coefficients do not depend at all upon the choice of a typical ε and satisfy the condition $u_\alpha = -v_\alpha$.

§5. Weight elements

The proofs of Theorems 2 and 3 for the classical groups and for types E_6 and E_7 may be somewhat simplified if one exploits the fact that in these cases T generally contains elements of simpler structure than LRSE's. The existence of these elements is closely bound up with the possibility of constructing nontrivial diagonal extensions of groups of these types — "the extended Chevalley groups", which stand in the same relation to the ordinary Chevalley groups as do the groups $GL(n, K)$ to $SL(n, K)$. The extended Chevalley groups are generally much easier to study than the ordinary Chevalley groups, and thay may serve as models. At the same time, they often appear naturally as a stage in the theory of Chevalley groups proper, if only because the Levi components of parabolic subgroups in G are almost always not semisimple but reductive: a maximal torus of G induces nontrivial diagonal automorphisms of their semisimple parts. We now recall some constructions and results due to Berman and Moody [81], and the author [20, 25, 28, 29].

1. Diagonal extensions of Chevalley groups. Assign a scalar $\chi(\alpha) \in K^*$ to each simple root $\alpha \in \Pi$. This correspondence extends linearly to a K-character, i.e., a homomorphism of the root lattice $Q(\Phi)$ into K^*:

$$\chi(m_1\alpha_1 + \cdots + m_l\alpha_l) = \chi(\alpha_1)^{m_1} \cdots \chi(\alpha_l)^{m_l}.$$

It is well known that there exists a unique automorphism φ_χ of the Chevalley group $G = G(\Phi, K)$ such that $\varphi_\chi(x_\beta(\xi)) = x_\beta(\chi(\beta)\xi)$. Steinberg [69] calls such automorphisms of G diagonal. The main part of diagonal automorphisms is of course the inner automorphisms. Indeed,

$$h_\alpha(\varepsilon)x_\beta(\xi)h_\alpha(\varepsilon)^{-1} = x_\beta(\varepsilon^{\langle\beta,\alpha\rangle}\xi),$$

where $\langle\beta,\alpha\rangle = 2(\beta,\alpha)/(\alpha,\alpha) = (\beta,\alpha^\vee)$ is the Cartan number (here, as usual, $\alpha^\vee = 2\alpha/(\alpha,\alpha)$ is the dual root to α). Thus, conjugation by $h_\alpha(\varepsilon)$ is equivalent to the diagonal automorphism φ_χ corresponding to the K-character $\chi = \chi_{\alpha,\varepsilon}$ defined by

$$\chi_{\alpha,\varepsilon}(\beta) = \varepsilon^{\langle\beta,\alpha\rangle} = \varepsilon^{(\beta,\alpha^\vee)}.$$

In fact, conjugation by any element $h \in H$ is equivalent to some diagonal automorphism. These are precisely the diagonal automorphisms corresponding to the K-characters of the root lattice $Q(\Phi)$ that extend to K-characters of the weight lattice $P(\Phi)$ [83]. This means that the quotient group of the group of all diagonal automorphisms by the subgroup of inner automorphisms is isomorphic (in the simply-connected case) to a product of groups K^*/K^{*m_j}, where the m_j are all the elementary divisors of the finite abelian quotient group $P(\Phi)/Q(\Phi)$. Thus, for any field K such that $K^{*m_j} = K^*$, all diagonal automorphisms are inner automorphisms.

One would naturally like to construct an extension of the Chevalley group (for any field K) in which all diagonal automorphisms are inner automorphisms. This is easy to do for adjoint groups: for every K-character of $Q(\Phi)$ there is an automorphism $h(\chi) = h_{\text{ad}}(\chi)$ of the Chevalley algebra L_K, defined by $h(\chi)h_i = h_i$, $h(\chi)x_\beta = \chi(\beta)x_\beta$. The automorphisms $h(\chi)$, where $\chi \in \text{Hom}(Q(\Phi), K^*)$ constitute a subgroup $\overline{T}_{\text{ad}} = \overline{T}_{\text{ad}}(\Phi, K)$ of the automorphism group of the Chevalley group $G_{\text{ad}}(\Phi, K)$ and, as one can easily see, conjugation by $h(\chi)$ defines a diagonal automorphism φ_χ of $G_{\text{ad}}(\Phi, K)$:

$$h(\chi)x_\beta(\xi)h(\chi)^{-1} = x_\beta(\chi(\beta)\xi).$$

In particular, in this notation, the root semisimple element $h_\alpha(\varepsilon)$ is just $h(\chi_{\alpha,\varepsilon})$. All this can be found, for example, in [83, 157]; the latter also presents an identification of the groups $\overline{G}_{\text{ad}} = \overline{T}_{\text{ad}}G_{\text{ad}}$ for the classical series.

It is much more difficult to define similar extensions for simply-connected groups (the dimension of the maximal torus must be increased by 1 or 2); this has been done only by Berman and Moody [81]. We shall not reproduce the details of their construction here (see also [20]); instead, in the next subsection we present a straightforward definition of weight elements — not as rigorous as might be based on [81], but quite sufficient for our purposes.

2. Weight elements. Let us return once more to the element $h_\alpha(\varepsilon)$. It corresponds to the character $\chi_{\alpha,\varepsilon}$ which is defined on $Q(\Phi)$ because $(\beta, \alpha^\vee) = \langle \beta, \alpha \rangle \in \mathbb{Z}$. But by definition, $(\beta, \omega) \in \mathbb{Z}$ for any $\beta \in \Phi$ and any $\omega \in P(\Phi^\vee)$, where $\Phi^\vee$ is the root system dual to Φ, i.e., $\Phi^\vee = \{\alpha^\vee \mid \alpha \in \Phi\}$. Thus, for any $\omega \in P(\Phi^\vee)$ and any $\varepsilon \in K^*$, there exists a K-character $\chi_{\omega,\varepsilon} \in \mathrm{Hom}(Q(\Phi), K^*)$; it is defined by $\chi_{\omega,\varepsilon}(\beta) = \varepsilon^{(\beta,\omega)}$. This character linearly depends on ω, i.e.,

$$\chi_{m_1\omega_1+m_2\omega_2,\varepsilon} = \chi_{\omega_1,\varepsilon}^{m_1} \cdot \chi_{\omega_2,\varepsilon}^{m_2}.$$

In addition, it is clear that $\chi_{m\omega,\varepsilon} = \chi_{\omega,\varepsilon^m}$. We can now consider an element $h_\omega(\varepsilon)$, which may be called a weight element, by analogy with the root semisimple element $h_\alpha(\varepsilon)$ (or as we should denote it now, $h_{\alpha^\vee}(\varepsilon)$), such that conjugation by $h_\omega(\varepsilon)$ defines a diagonal automorphism of $G = G(\Phi, K)$ corresponding to $\chi_{\omega,\varepsilon}$, i.e.,

$$h_\omega(\varepsilon) x_\beta(\xi) h_\omega(\varepsilon)^{-1} = x_\beta(\varepsilon^{(\beta,\omega)} \xi).$$

In the adjoint case, $h_\omega(\varepsilon)$ is uniquely defined by this condition, because then $h_\omega(\varepsilon) = h_{\mathrm{ad}}(\chi_{\omega,\varepsilon})$. However, in the simply-connected case $h_\omega(\varepsilon)$ is defined, generally speaking, only up to a factor in the center of the extended Chevalley group $\overline{G} = \overline{G}(\Phi, K)$; but this is not very important since the elements $h_\omega(\varepsilon)$ enter our calculations only in expressions like $[x, h_\omega(\varepsilon)]$. This essentially means that, in the context of extended groups, instead of subgroups containing $\overline{T}$ we could deal with subgroups invariant under diagonal automorphisms; nevertheless, the notation $h_\omega(\varepsilon)$ is rather convenient, especially since the construction of [8] can be used to refine the above definition so that the elements $h_\omega(\varepsilon)$ are uniquely defined in the simply-connected case, too, while all the necessary identifications for classical groups remain valid (see below).

In what follows we call any element $x h_\omega(\varepsilon) x^{-1}$, $x \in \overline{G}$, $\varepsilon \in K^*$, a weight element in the extended Chevalley group $\overline{G} = \overline{G}(\Phi, K)$ of type $\omega \in P(\Phi^\vee)$. Not all weight elements are equally important. The most common case in applications is when ω is $\overline{\omega}_i(\Phi)$, the fundamental weight of Φ. The following theorem, established in [25], defines the action of the elements $h_\omega(\varepsilon)$ in representations. As in §1, let π be a representation of $G = G(\Phi, K)$ in the K-vector space V_K defined as the tensor product of K and a certain admissible lattice in an irreducible representation V_L of the Lie algebra $L_\mathbb{C}$ (in other words, V_K is the Weyl module). Let $\Lambda(\pi)$ denote the set of weights of π.

THEOREM 7. *If $V_{\mathbb{C}}$ is irreducible, then $h_\omega(\varepsilon)$ acts diagonally on each weight subspace V^λ, $\lambda \in \Lambda(\pi)$ as multiplication by a certain scalar $c_\lambda \neq 0$, in such a way that for any two weights $\lambda, \mu \in \Lambda(\pi)$:*

$$c_\lambda c_\mu^{-1} = \varepsilon^{(\lambda,\omega)-(\mu,\omega)}.$$

We notice that the number in the exponent on the right is an integer, because the difference between two weights of an irreducible module belongs to $Q(\Phi)$. It is not always true that $c_\lambda = \varepsilon^{(\lambda,\omega)}$, since we have defined $h_\omega(\varepsilon)$ only up to a central factor and (λ, ω) need not be an integer. Considering an example, let us see what the weight elements look like in the usual representations (i.e., representations with highest weight $\overline{\omega}_1$) of the classical groups. For $\omega = \omega_k(A_l)$, we have

$$h_\omega(\varepsilon) = \mathrm{diag}(\varepsilon, \ldots, \varepsilon, 1, \ldots, 1),$$

where the number of ε's is k. For $\Phi = B_l$, C_l, D_l and $\omega = \overline{\omega}_k$, with the exception of $\overline{\omega}_{l-1}(D_l)$, $\overline{\omega}_l(D_l)$, $\overline{\omega}_l(C_l)$, we have

$$h_\omega(\varepsilon) = \mathrm{diag}(\varepsilon, \ldots, \varepsilon, 1, \varepsilon^{-1}, \ldots, \varepsilon^{-1}),$$

where the number of ε's and the number of ε^{-1}'s are both k. Finally, for the weight $\omega = \overline{\omega}_l$ in C_l, D_l, the element $h_\omega(\varepsilon)$ is the same as for the weight $\overline{\omega}_l(A_{2l-1})$. In general, the action of the weight elements may be quite surprising. Thus, a weight element of type $\overline{\omega}_1(E_6)$ in the representation of the highest weight $\overline{\omega}_1$ has one eigenvalue ε, ten eigenvalues ε^{-1}, and sixteen eigenvalues 1.

3. Bruhat decomposition of microweight elements. The weight elements of type ω have an especially simple structure when ω is a microweight. Recall that in that case the set $\Sigma_\omega = \{\alpha \in \Phi^+ | (\omega, \alpha) \neq 0\}$ is abelian, i.e., the sum of any two of its elements is not a root. All the fundamental weights for A_l; $\overline{\omega}_l$ for B_l; $\overline{\omega}_1$ for C_l; $\overline{\omega}_1, \overline{\omega}_{l-1}, \overline{\omega}_l$ for D_l; $\overline{\omega}_1, \overline{\omega}_6$ for E_6 and $\overline{\omega}_7$ for E_7 are microweights [10]. Microweight elements are particularly important because, as a rule, the torus $\overline{T}_{\mathrm{ad}}$ is generated by the microweight elements of the given type ω that it contains: it is sufficient that ω generate the weight lattice $P(\Phi^\vee)$ over the root lattice $Q(\Phi^\vee)$. Hence, in the general case, too, $\overline{T}$ is generated (under this assumption) by weight elements of type ω together with the center of the group $\overline{G}$. Thus the microweight elements play the same role in the theory of extended Chevalley groups that the LRSE's play in the theory of ordinary Chevalley groups. From the viewpoint of the Bruhat decomposition, the microweight elements are much simpler in structure than the LRSE's, as will be seen from the following theorem [22, 20, 25, 29]. Let $\overline{B} = \overline{B}(\Phi, K)$ denote a Borel subgroup of the extended Chevalley group $\overline{G}$, and let m denote the width of the partially ordered set Σ_ω. As usual, ω_α is reflection with respect to the root α.

THEOREM 8. *Let ω be a microweight; if $\Phi = B_l$, C_l, we also assume that* char$K \neq 2$. *Then every weight element of type ω belongs to one of the cosets*

$$\overline{B}w\overline{B} = \overline{B}w_{\gamma_1}\cdots w_{\gamma_{r+s}}\overline{B},$$

where $\gamma_1,\ldots,\gamma_{r+s}$ are pairwise strictly orthogonal roots, and if $\gamma_1,\ldots,\gamma_r$ are long roots and $\gamma_{r+1},\ldots,\gamma_{r+s}$ short roots, then $r+2s \leq m$. Moreover, for fixed $x \in G$, the element ω in the Bruhat decomposition of $xh_\omega(\varepsilon)x^{-1}$ does not depend upon the choice of $\varepsilon \neq 0, 1$.

In the above-mentioned papers one can also find quite detailed information about the factors in the Borel subgroup. We list here all the microweights together with the appropriate values of m:

A_l,	$\omega = \overline{\omega}_k$,	$m = \min(k, l+1-k)$;
B_l,	$\omega = \overline{\omega}_1$,	$m = 2$;
C_l,	$\omega = \overline{\omega}_l$,	$m = l$;
D_l,	$\omega = \overline{\omega}_1$,	$m = 2$;
	$\omega = \overline{\omega}_{l-1}, \overline{\omega}_l$,	$m = [l/2]$;
E_6,	$\omega = \overline{\omega}_1, \overline{\omega}_6$,	$m = 2$;
E_7,	$\omega = \overline{\omega}_7$,	$m = 3$.

It follows from this table that the "smallest" semisimple element in a group of type E_6 is approximately of the same complexity as in the classical groups, while the complexity of the "smallest" semisimple element in a group of type E_7 is higher. As we have already said, the LRSE's are the simplest semisimple elements in groups of the type E_8, F_4, G_2.

§6. Extended Chevalley groups

Here we formulate analogs of the classification and conjugacy theorems for extended Chevalley groups. These theorems are actually simpler than Theorems 2 and 3 and were proved earlier. Throughout this section we will assume that $\Phi \neq E_8, F_4, G_2$. Of course, going over from G and T to $\overline{G}$ and $\overline{T}$, one has to modify the notion "standard", that is, replace the groups $G(S)$ and $N(S)$ in the definition by the groups $\overline{G}(S)$ and $\overline{N}(S)$ obtained from them by multiplication by $\overline{T}$.

THEOREM 9. *Let $|K| \geq 7$; then also if $\Phi = B_l$, C_l, char$K \neq 2$. Then the extended Chevalley group $\overline{G} = \overline{G}(\Phi, K)$ admits a standard description of subgroups containing a split maximal torus $\overline{T} = \overline{T}(\Phi, K)$. In other words, for every intermediate subgroup F, $\overline{T} \leq F \leq \overline{G}$, there exists a unique closed set of roots $S \subseteq \Phi$ such that*

$$\overline{G}(S) \leq F \leq \overline{N}(S).$$

This theorem was known before Theorem 2 and is used in the proof of the latter. For $\mathrm{GL}(n, K)$, which is an extended Chevalley group of type A_{n-1}, the theorem was proved in 1976 by Borevich [1]. In 1978-1981 Borevich and the author proved a similar result for the general linear group over an arbitrary, not necessarily commutative semilocal ring Λ, (see, in particular, [3, 14] and a detailed survey of results of this kind in [42, 30]). This naturally requires redefining the notion "standard", to take the structure of the ideals of the underlying ring into account. The modification involved the notions of a net of ideals and net subgroups, first defined for the linear case in 1976 by Borevich and myself (see references in [1, 3, 30, 41, 42]), and, in the context of Chevalley groups, by Suzuki [164, 165] and the author and E. B. Plotkin [11, 34, 16]. The proofs relating to subgroups of $GL(n, \Lambda)$ were based on matrix technique, but in 1979 we found an invariant proof [13, 22].

Similar results were obtained in 1979, 1980 for other split classical groups: the general symplectic group $G\,Sp(2l, R)$ and the general orthogonal group $GO(2l, R)$, i.e., extended Chevalley groups of types C_l and D_l, respectively [33, 169]. In these papers R is any commutative semilocal ring such that $2 \in R^*$ and all the residue fields of R contain at least 7 elements. A similar result for the odd-dimensional general orthogonal group $GO(2l+1, R)$, which is an extended Chevalley group of type B_l, was also obtained at that time, but published only in [27]. When writing [169], we did not realize that the results are valid not only for extended but also for ordinary orthogonal groups $SO(n, R)$. This fact, noted only in [27], may be explained by the observation that the groups $SO(n, R)$ are already extended by means of certain, though not necessarily all, diagonal automorphisms. The correct analogs of the groups $SL(n, R)$ and $Sp(2l, R)$ are of course the spinor groups $Spin(n, R)$.

In 1981 we proved Theorem 9 for extended Chevalley groups of types E_6, E_7 too. The result was announced in [14], and parts of the proof were presented in [20, 25, 26]. The treatment of all these cases, except the case of symplectic groups, is based on the analysis of the microweight elements contained in F (in symplectic groups the structure of LRSE's is much simpler than that of microweight elements, so the latter are used in [33] only at the last stage).

An analog of the conjugacy theorem for extended groups is valid under somewhat weaker assumptions on the base field.

THEOREM 10. *Under the assumptions of Theorem 9, the split maximal torus $\overline{T} = \overline{T}(\Phi, K)$ is pronormal in the extended Chevalley group $\overline{G} = \overline{G}(\Phi, K)$. In particular, if two subgroups $\overline{T} \leq F_1$, $F_2 \leq \overline{G}$ are conjugate in $\overline{G}$, then they are already conjugate by an element of $\overline{N}$.*

This theorem was first proved for the general linear group over a local matrix ring [12, 170]. (Recall that a ring Λ is called a local matrix ring if its quotient ring over the Jacobson radical is the complete matrix ring over a division ring.) Koĭbaev then showed that the conjugacy theorem is valid for the complete linear group even when there are no standard descriptions, namely, for the fields F_4 and F_5. Moreover, he found a way to modify the formulation of the conjugacy theorem so that it also included the case of the field F_3. Of course, the group $\overline{T} = D(n, F_3)$ of diagonal matrices is not pronormal in $\overline{G} = \mathrm{GL}(n, F_3)$. In fact, if $n = 2$ the groups $\overline{T}$ and $x\overline{T}x^{-1}$, where

$$x = \begin{pmatrix} 1 & 1 \\ 1 & -1 \end{pmatrix},$$

are contained in the group $\overline{N} = N(2, F_3)$ of monomial matrices, but are certainly not conjugate in it. However, it was shown in [48] that this counterexample is, essentially, the only one. Theorem 10 was established for the symplectic and orthogonal cases in [33, 27]; in [26] one can find a general proof based on calculating the Bruhat decomposition for microweight elements (Theorem 8).

§7. Conjugacy and reduction theorems

We will now outline the proofs of Theorems 3 and 10, and create the basis for the induction on $|\Phi|$ used to prove Theorems 2 and 9.

1. Extraction of unipotent elements. The following theorem gives a description of unipotent subgroups normalized by a maximal split torus.

THEOREM 11. *Let* $|K| \geq 5$. *Then for any element* $u = \prod x_\alpha(u_\alpha) \in U$, *where the product is taken over all positive roots in arbitrary order,* $x_\alpha(u_\alpha) \in \langle H, uHu^{-1} \rangle$ *for any* $\alpha > 0$.

This result was actually proved in [145, 87] for finite ground fields, in a stronger form. It is clear that the proof for an infinite field can only be simpler (see, e.g., [79]). In fact, far more general results than Theorem 11, pertaining to arbitrary commutative rings, were proved in [164, 11].

COROLLARY. *Under the assumptions of Theorem* 11, *T is an abnormal subgroup of B*.

If the characteristic of K satisfies the same conditions as in Theorem 2, it follows immediately from Theorem 11 that $\langle T, uTu^{-1} \rangle = \langle T, u^{-1}Tu \rangle = \langle T, u \rangle$ is just $G(S)$, where S is the closure of the set of positive roots α such that $u_\alpha \neq 0$ in a certain fixed order. The following result is basic for the study of subgroups contained in a proper parabolic subgroup.

THEOREM 12. *Let* $|K| \geq 7$, *and let P be a standard parabolic subgroup of a Chevalley group G. Represent* $x \in P$ *as* $x = yu$, *where* $y \in L_P$ *and* $u \in U_P$. *Then* $u \in \langle T, x^{-1}Tx \rangle$.

2. Conjugacy theorems. The Tits theorem proved in [70] can be stated as follows in the present context.

TITS THEOREM. *Let $|K| \geq 4$. If there exist $S \subseteq \Phi$ and $x \in G$ such that $xTx^{-1} \leq G(S)$, then $x = yw$ for some $w \in N$ and $y \in G(S)$.*

Tits actually proved his theorem for any groups with "normal root data". The conjugacy theorem clearly follows immediately from the classification theorem and the following result, which we will refer to as the conjugacy theorem for standard subgroups. Being a generalization of the Tits theorem, it is one of the key steps in the proof of the classification theorem itself.

THEOREM 13. *Let $|K| \geq 7$. If there exist $S \subseteq \Phi$ and $x \in G$ such that $xTx^{-1} \leq N(S)$, then $x = yw$ for some $w \in N$ and $y \in G(S)$.*

Theorem 13, in turn, follows immediately from the Tits theorem and the following result, which we shall call the connectedness theorem. For an infinite ground field, this theorem follows immediately from the fact that $G(S)$ is a connected component of $N(S)$ in the Zariski topology.

THEOREM 14. *Let $|K| \geq 7$. For any $x \in G$, if $xTx^{-1} \leq N(S)$, then $xTx^{-1} \leq G(S)$.*

An analogous result for extended Chevalley groups was proved in [26]. However, the results of [26], applied to ordinary Chevalley groups, give a less sharp bound on $|K|$ than the bound in Theorem 14. Only Theorems 5 and 6 allow us to lower the bound to $|K| \geq 7$.

The connectedness theorem is proved by induction on $|\Phi|$. Theorem 12 at once reduces the case in which $N(S)$ is contained in a proper parabolic subgroup (as we know from the Borel-Tits theorem, this is always the case if $S^u \neq \varnothing$) to groups of lower rank. In what follows, therefore, we may assume that $S = \Delta$ is a subsystem of roots such that $N(\Delta)$ is not contained in a proper parabolic subgroup. We first examine the Bruhat decompositions of elements of $N(\Delta)$.

LEMMA. *Let $|K| \geq 4$ and $\Delta \subseteq \Phi$. Then every element of $N(\Delta)$ can be expressed uniquely as $x = uwvd$, where $u \in U(\Delta)$, $w \in X(\Delta)$, $v \in U_{w_0}^-(\Delta)$. Here $w_0 \in W(\Delta)$ is defined by w in the following way: $w_1 = ww_0^{-1}$ preserves the basis of Λ contained in Φ^+.*

It is now clear that a proof of the connectedness theorem can proceed as follows: we have to consider the Bruhat decomposition of the LRSE $y(\varepsilon) = xh_\delta(\varepsilon)x^{-1}$, calculated in §4, and check that, if it is of the type described in the lemma, then necessarily $w \in W(\Delta)$. A relatively simple inductive argument reduces everything to the case in which $N(\Delta)$ is maximal with respect to the standard subgroups and not connected, i.e., $N(\Delta) \neq G(\Delta)$. All such groups are listed in Theorem 1.

Indeed, let $y(\varepsilon) \in N(\Delta)$. By Theorem 5 we can assume that for all ε except $\varepsilon = 0, 1$, and at most two other values of ε, the element $y(\varepsilon)$ lies in a coset BwB, where $w = w_{\beta_1} \cdots w_{\beta_m}$ and $\beta = \beta_1, \ldots, \beta_m$ are pairwise strictly orthogonal roots, $m \leq 4$, and w does not depend on ε. Calculations of factors in B described in §4 imply that the coefficients $u(\varepsilon)_\beta$ and $v(\varepsilon)_\beta$ in the Bruhat decomposition $y(\varepsilon) = u(\varepsilon)wd(\varepsilon)v(\varepsilon)$ are nonzero rational functions of ε (their denominators may vanish at the excluded values of ε). By Theorem 6, the numerators and denominators of these functions are of degree at most two. Hence, excluding at most two more values (this gives at most 6 forbidden values, so ε exists if K contains at least 7 elements), we may assume, for example, that $u(\varepsilon) \neq 0$. But then by the lemma, $\beta \in \Delta$. As this argument may be applied to all β_i, it follows that $w \in W(\Delta)$, i.e., $y(\varepsilon) \in G(\Delta)$ for all typical ε. It remains to notice that, since $|K| \geq 7$, the root torus is generated by its typical elements, and therefore $y(\varepsilon) \in G(\Delta)$ for all ε, proving the connectedness theorem.

3. Reduction theorem. The following result reduces the description of all subgroups containing T to those subgroups that are not contained (up to conjugacy) in a proper standard subgroup.

THEOREM 15. *Assume that $|K| \geq 7$ and that the classification theorem for all proper root subsystems in Φ is valid. If F is a subgroup of the Chevalley group $G = G(\Phi, K)$ containing $T = T(\Phi, K)$ and contained in a subgroup conjugate to a proper standard subgroup, then F is itself standard.*

Therefore, a similar assertion holds for extended Chevalley groups. Since T in fact induces nontrivial diagonal automorphisms on most of the standard subgroups, it is usually sufficient to assume that, in the statement of Theorem 15, the standard description holds in the extended Chevalley group corresponding to a proper subsystem Δ of Φ. In particular, there is no nonsymplectic system for which SL_2 admits standard descriptions.

The proof of Theorem 15 may be sketched as follows. It follows easily from the results of the two preceding subsections that for $|K| \geq 7$ any subgroup that contains T and is conjugate to a standard subgroup, is itself standard. Thus, we may assume from the start that $F \leq N(S)$, where $S \subseteq \Phi$ (if desired, we may even assume that $N(S)$ is a maximal standard subgroup). Now, if $N(S) = P$ is a parabolic subgroup then, representing $x \in F$ as $x = yu$, where $y \in L_P$ and $u \in U_P$, we conclude by Theorem 12 that $u \in F$. But, by assumption, y together with the maximal torus of the group $[L_P, L_P]$ generate a standard subgroup of the latter. Therefore, y and T generate a standard subgroup of G. Hence, the subgroup generated by T and an arbitrary element $x \in F$ is standard (as the composite of standard subgroups $\langle T, u \rangle$ and $\langle T, y \rangle$). But then, of course, F itself is standard. If F is contained in a reductive standard subgroup, the proof is even simpler. Indeed, let $T \leq F \leq N(\Delta)$, where Δ is a root subsystem in Φ. As we have just pointed out, the assumptions of the theorem imply that the subgroup $F \cap G(\Delta)$ is a standard subgroup. But the conjugacy theorem for standard subgroups states, in particular, that the normalizer of a standard subgroup is standard.

It is convenient to use Theorem 15 in the following form.

COROLLARY. *Under the assumptions of the theorem, let $x \in G$ be an element of a proper standard subgroup. Then if $x \notin N$, $\langle T, x \rangle$ contains an elementary root unipotent subgroup X_α.*

In fact, we shall see in the following section that an analogous assertion is true without the assumption that x lies in a proper standard subgroup.

§8. Extraction of a root unipotent

The following results are the crucial — and the most difficult — steps in the proofs of Theorems 2 and 8. Throughout, we will continue to assume that the characteristic of the ground field satisfies the same conditions as in Theorem 2.

THEOREM 16. *Let K be an infinite field. Then any subgroup of the Chevalley group $G = G(\Phi, K)$ that contains the split maximal torus $T = T(\Phi, K)$ but is not contained in its normalizer $N = N(\Phi, K)$ contains an elementary root subgroup X_α, $\alpha \in \Phi$.*

THEOREM 17. *Let $\Phi = A_l, B_l, C_l, D_l, E_6, E_7$, and $|K| \geq 7$. Then any subgroup $\overline{G} = \overline{G}(\Phi, K)$ containing the split maximal torus $\overline{T} = \overline{T}(\Phi, K)$, but not contained in its normalizer $\overline{N} = \overline{N}(\Phi, K)$ contains an elementary root subgroup X_α, $\alpha \in \Phi$.*

In this section we shall only illustrate some very general ideas underlying the proof. The point is that all proofs known to me are technical in nature and need voluminous calculations, with each case examined separately. For the classical groups the proofs, formulated in Lie and matrix languages, are scattered over a large number of papers (see references in Subsections 3, 6 and the bibiliography in [30]). All the details are collected in my thesis [24]. The proofs for the exceptional groups have not yet been published in full because of their length.

1. General strategy. By the corollary to Theorem 15, it suffices to show that Φ contains an element $g \notin N$ that belongs to a proper standard subgroup. We will do this as follows. Since $F \not\leq N$, it follows from the connectedness theorem that we may choose a one-parameter weight subgroup of F,

$$Q = \{y(\varepsilon): xh_\omega(\varepsilon)x^{-1}, \ \varepsilon \in K^*\},$$

not contained in N. For ordinary Chevalley groups, we shall as a rule define $h_\omega(\varepsilon)$ as a LRSE $h_\alpha(\varepsilon)$; for extended Chevalley groups it will be a microweight element corresponding to the weight $\overline{\omega}_1$ (for the symplectic case, see subsection 4). Nevertheless, there are other alternatives; for example, in the proofs of Theorems 16 and 17 for groups of type B_l, it proves most convenient to use the elements $h_{\overline{\omega}_l}(\varepsilon)$ [24], which are not microweight elements.

We want to show that the required element g may already be found in the subgroup $\langle T, Q \rangle$ generated by T and Q (of course, we may assume that Q itself is not contained in a proper standard subgroup). As a rule, however, we will not be able to combine elements from T and Q at once so as to obtain an element with the required properties. A remarkable exception is the general linear group, for which g is always available in QTQ; this is yet another confirmation of the Verma principle (the famous Harish-Chandra principle claims that all reductive groups have the same structure, Verma stated that the structure of the general linear group is even more uniform than that of all others). In the general case one has to proceed by induction: starting with Q, one determines another one-parameter subgroup Q' of F, also not in a proper standard subgroup, which is in some sense simpler than Q itself.

Recall that almost all elements of Q lie in the same coset $Bw_0 B$ of the Bruhat decomposition, and that, moreover, w_0 is an involution. Our criteria for the complexity of the group Q are as follows, in order of decreasing priority: the number m of root reflections in the element $w_0 = w_{\beta_1}, \ldots, w_{\beta_m}$ (i.e., the number of eigenvalues of w_0 other than 1, denoted by $\bar{l}(w_0)$ in [84]); the length $l(w_0)$ of w_0; the height h of the highest of the roots $\beta_1, \ldots, \beta_m$; the number of nonzero coefficients in factors from the Borel subgroup; the validity of certain equations or inequalities for these coefficients. In each case one can show that $\langle T, Q \rangle$ contains a one-parameter subgroup $Q' \not\leq N$ of weight elements of simpler structure compared with Q, i.e., with smaller h, and so on. One then begins to analyze the group $\langle T, Q' \rangle$, eventually locating the required g in one of these subgroups. We shall now show approximately how this is done in a few simple cases (for the classical groups see also [1, 3, 15, 18, 22, 24, 27, 32, 33, 38]).

2. Group SL_2. The only groups of semisimple rank 1 among the groups of normal types are SL_2 and GL_2. We will reproduce a quite fantastic proof by King for the group SL_2 [119]; the much easier case of GL_2 was known long before and will be considered in the next subsection, in a far more general context.

Let $\Phi = A_1$ and $T \leq F \leq G = \mathrm{SL}(2, K)$, but $T \not\leq N$, B, B^-. In the case of a finite field, all the subgroups of G are known. We may therefore assume that K is infinite. Replacing F by a subgroup conjugate to it by a diagonal matrix in $\mathrm{GL}(2, K)$, we may assume that F contains an element

$$f = \begin{pmatrix} 1 & 1 \\ \alpha & 1+\alpha \end{pmatrix}.$$

Replacing F, if necessary, by $\langle T, fTf^{-1} \rangle$, we may also assume that α does not belong to a fixed finite subset in K^*. Denote the root element $h_\delta(\varepsilon)$ by

$$h(\varepsilon) = \begin{pmatrix} \varepsilon & 0 \\ 0 & \varepsilon^{-1} \end{pmatrix}.$$

If now char $K \neq 2, 3, 5$, King chooses $\alpha \neq 0, 1, \pm 2, \pm 3, -4, 5, -6$ and forms the following product:

$$g = f^{-1}h\left(\frac{2}{\alpha}\right)f^{-1}h\left(\frac{\alpha+4}{\alpha+2}\right)fh\left(\frac{3}{4}\right)f^{-1}h\left(\frac{3}{\alpha-3}\right)f^{-1}$$
$$\times h\left(\frac{\alpha+6}{\alpha+3}\right)fh\left(\frac{5}{6}\right)f^{-1}h\left(\frac{\alpha-2}{\alpha-5}\right)fh(3)f^{-1}h\left(\frac{\alpha+3}{\alpha+2}\right)f$$
$$\times h\left(\frac{2}{3}\right)f^{-1}h\left(\frac{\alpha-1}{\alpha-2}\right)fh(\alpha-3)fh\left(\frac{4}{3}\right)f^{-1}h\left(\frac{\alpha+2}{\alpha+4}\right)fh\left(\frac{\alpha}{2}\right)f.$$

Calculations show that $g_{21} = 0$, and g_{12} vanishes only for a finite number of values of α. Therefore, a certain group conjugate to F by a diagonal matrix contains an elementary transvection, and hence $F = G$. If char $K = 3, 5$, King constructs other, shorter formulas expressing a transvection in terms of f and $h(\varepsilon)$.

Of course, our search for an element g in groups of rank greater than 1 does not require such ingenuity as King's formula, but it is not reasonable to expect it to be substantially simpler.

3. Reflections with respect to a single root. Let us return to the notation of subsection 1. Let w be a microweight. We will outline a proof of Theorem 17 for the simplest case, when F contains a one-parameter subgroup Q of weight elements of type W, such that if $\varepsilon \neq 0, 1$ the element $\mu(\varepsilon)$ belongs to the class $\overline{B}w_\alpha \overline{B}$. In particular, this will prove Theorem 17 and also — by McLaughlin's theorem (see [141] and §9 of this paper) — Theorem 9 for the simplest case of the general linear group. In fact, our proof exactly models the original proof [1, 3, 15, 22] for this case.

Namely, we claim that elements $\varepsilon, \eta, \theta \in K \setminus \{0, 1\}$ can be chosen in such a way that

$$z = y(\varepsilon)h_\omega(\theta)y(\eta^{-1})$$

is contained in a proper parabolic subgroup, but not in N. We may of course assume that the support of the root α is Π (otherwise $\overline{B}w_\alpha \overline{B}$ is contained in a standard parabolic subgroup). As we know from §5 (see also [20, 22, 24, 25, 29] for the details), the Bruhat decomposition of $y(\varepsilon)$ is $y(\varepsilon) = u(\varepsilon)n(\varepsilon)v(\varepsilon)$, where

$$n(\varepsilon) = w_\alpha(\pm c(\varepsilon-1)^{-1})h_{w\omega}(\varepsilon) \in N,$$

and $u(\varepsilon), v(\varepsilon) \in U$. In fact, $v(\varepsilon) = x_\alpha(v(\varepsilon)_\alpha)v_1 v_2$ belongs to the Heisenberg group $Y_\alpha = U_{w_\alpha}^-$, where $v(\varepsilon)_\alpha$ is a rational function of ε, with denominator and numerator of degree at most 1; v_1 and v_2 do not depend on ε and are products of root elements $x_\gamma(v_\gamma)$ over all γ_1 and γ_2, respectively, where γ_1 and γ_2 are positive roots such that $\alpha = \gamma_1 + \gamma_2$, and $\gamma_1 \in \Sigma_\omega$ (see the definition of this set in §6).

Now choose ε and η, differing from 0, 1, such that $v(\varepsilon)_\alpha, v(\eta)_\alpha \neq 0$, $v(\varepsilon)_\alpha \neq v(\eta)_\alpha$, and then take $\theta = v(\varepsilon)_\alpha v(\eta)_\alpha^{-1}$. It is clear that $\theta \neq 0, 1$

and, by the very choice of θ,
$$x_\alpha(v(\varepsilon)_\alpha)h_\omega(\theta)x_\alpha(v(\eta)_\alpha)^{-1} = h_\omega(\theta).$$
Obviously, v_2 commutes with $h_\omega(\theta)$ and
$$[v_1, h_\omega(\theta)] = \prod x_\gamma(v_\gamma(1-\theta)),$$
where the product is taken over all $\gamma \in \Sigma_\omega$ such that $\alpha - \gamma \in \Phi^+$. Let $\omega = \overline{\omega}_i$; then the simple root α_i occurs in the factorization of each of these roots γ with the coefficient 1. Therefore, the support of each of the roots $w_\alpha \gamma = \gamma - \alpha$ is contained in $\Pi \setminus \{\alpha_i\}$. This means that
$$z_0 = n(\varepsilon)v(\varepsilon)h_\omega(\theta)v(\eta)^{-1}n(\eta)^{-1}$$
is an element of the subgroup $\overline{G}(\pm(\Pi \setminus \{\alpha_i\}), K)$, so that z lies in a proper standard parabolic subgroup $\overline{G}_i = \overline{G}(P_i)$.

Careful calculation of the factors from the Borel subgroup will now show that, by excluding at most six values of ε, we can choose η with the required properties, in such a way that z does not belong to $\overline{N}$; this concludes the proof. Note that the above argument goes through almost completely for LRSE's in the symplectic group, except that the numerator and denominator of $v(\varepsilon)_\alpha$ then depend on ε not as linear but as quadratic functions, so we have to impose two restrictions on ε instead of only one; hence the stronger requirements on the order of the field K.

4. Symplectic group. The case of Sp_{2l} was the last to be considered, because it finally reduces to the group SL_2 and not to GL_2, like the others. However, the actual reduction to a lower rank is much easier for Sp_{2l}, $l \geq 2$, than for all other groups, except GL_n. We will now translate the proof of [33] into our language (see also [38, 24]), with due attention to the simplifications suggested in [66].

We have to show that there exists a long root torus Q in F lying in the coset $Bw_\gamma B$; having done this, we will be able to refer to subsection 3. Let $Q = \{y(\varepsilon), \varepsilon \in K^*\}$ be a long root torus in F that is not contained in N. According to §4, all the elements with $\varepsilon \neq 0, \pm 1$ lie in the same coset $Bw_\alpha w_\beta B$, where α and β are two orthogonal long roots, but the element $y = y(-1)$ cannot be in this coset. A computation shows that $y(-1)$ belongs either to B or to $Bw_\gamma B$, where γ is a short root. Without loss of generality, we may assume that $\gamma = \varepsilon + \varepsilon_i$, where $2 \leq i \leq l$ (here we are using the ordinary notation for root systems, see [10]), for otherwise Q is contained in a proper standard parabolic subgroup G_1 or G_l.

Now it is easy to see that the root torus $\widetilde{Q} = \{yh_\delta(\varepsilon)y^{-1}, \varepsilon \in K^*\}$ is contained either in a proper parabolic subgroup G_1 or in $Bw_\gamma B$, depending on whether the coefficient $v(-1)_\gamma$ vanishes or not (this is because $w_\gamma X_\alpha w_\gamma^{-1} \leq G_1$ for all roots $\alpha \in \Sigma$, $\alpha \neq \gamma$). Thus, referring to §7 and subsections 2, 3, we can complete the proofs of Theorems 16 and 17 for the symplectic case, provided only that $\tilde{Q}$ can be chosen nondiagonal. If $\tilde{Q}$ is diagonal, it follows from what we know about the conjugacy classes of semisimple elements and their centralizers (see, e.g., [67, 85]) that y is of the form $y = wz$, where $w \in W$ and $z \in G(C_{l-1}, K)$ for an appropriate imbedding $C_{l-1} \leq C_l$. Since $\langle T, y^{-1}Ty \rangle = \langle T, z^{-1}Tz \rangle$, we may again refer to Theorem 15. Thus, the required $\tilde{Q}$ exists, provided the matrix y itself does not belong to N.

Now, if indeed $y \in N$, this actually imposes certain conditions on the coefficients of the factors in the Bruhat decomposition of the element x such that $y = xh_\delta(-1)x^{-1}$. We must therefore change x. Indeed, replace x by $y(\varepsilon)$ and, naturally, y by $z(\varepsilon) = y(\varepsilon)h_\delta(-1)y(\varepsilon^{-1})$. If $z(\varepsilon) \in N$ for all ε, then (since $|K| \geq 7$) in fact $z(\varepsilon) \in T$, and once again our information about the centralizers of semisimple elements implies that $y(\varepsilon)$ belongs to the subset $W \cdot G(C_{l-1}+A_1, K)$. Since, by assumption, Q itself is not contained in N, we may apply the inductive hypothesis, to conclude that $X_\alpha \leq F$ for some $\alpha \in \Phi$.

The proofs of Theorems 16 and 17 in other cases are based on repeated applications of the above and similar techniques, but even for SL_n, $Spin_{2l}$, and SO_{2l+1} — not to speak of the exceptional groups — they are far more complicated. Complete arguments at every stage of the proofs involve huge amounts of calculation.

§9. Subgroups generated by root subgroups

In this section we shall complete our outline of the proofs of Theorems 2 and 9 and discuss what must still be done to make this part of the proof completely natural.

1. Completing the proof of the classification theorems. We continue to assume that the characteristic of the ground field satisfies the usual conditions. The following result is the last step in the proof of the classification theorems.

THEOREM 18. *Let $|K| \geq 7$. Then there are no proper subgroups in G that contain TX_α for some $\alpha \in \Phi$ but are not contained in proper standard subgroups.*

It is clear that the classification theorems follow immediately from Theorem 18 modulo the previous results. Indeed, let F, $T \leq F \leq G$, be an intermediate subgroup. If F lies in a proper standard subgroup, then by Theorem 15 it is standard itself. Suppose, therefore, that F is not contained in any proper subgroup. Then, by Theorems 16 (or 17, for extended groups) and 18, F coincides with G.

Theorem 18 is proved as follows. Let S be the maximal set of roots such that $G(S) \leq F$. By assumption, $\alpha \in S$, so S is nonempty. If $S \neq \Phi$, we choose a unipotent root subgroup X in F, not lying in $G(S)$ (such a subgroup exists, since F is not contained in $N(S)$). Commuting X with elements of $G(S)$, one can construct an elementary root subgroup X_β in F not lying in $G(S)$, and this contradicts the definition of S. The proof of this fact imitates proofs of Theorems 16 and 17, but it is substantially simpler, since the computations are done with unipotent root subgroups instead of one-parameter subgroups of semisimple elements. The former have a much simpler structure, which is the same for all series; for example, a long root subgroup X always lies in one of the cosets $Bw_\alpha B$; to be precise, it is of the form $uX_\alpha u^{-1}$ for a suitable $u \in U(\Phi, K)$.

In reality, however, Theorem 18 imposes a needlessly stringent condition on F, demanding that it contains not just X_α but also T. It would be far more natural to list all the "irreducible" subgroups — in some sense of the word — that contain the root subgroups or the subgroups generated by the root subgroups, and then to show that the only group in the list that contains T is G itself. Suitable concepts of "irreducibility" might be the property of not being contained up to conjugacy in a proper standard subgroup, which is an analog of primitivity and irreducibility; or in a proper subgroup of the form $G(S)$ — an analog of ordinary irreducibility; or in a proper parabolic subgroup (in the linear case this is equivalent to the preceding property but it is generally weaker).

We shall now briefly discuss the current situation. More detailed information and references can be found in reviews by Zalesskiĭ [41, 42], Kantor [116], Kondrat'ev [54], and the author [29]. Of the more recent publications, not mentioned there, Brown and Humphries [82, 108–110], Li Shangzhi and Zha Jianguo [127–130, 134, 135] should be mentioned. We are preparing a survey "Generation in Chevalley groups", in which the available results will be presented in detail, an exhaustive bibliography will be compiled, and unsolved problems will be formulated. For the moment, therefore, we restrict ourselves to the essentials.

2. Classical groups. In the classical cases, a full range of results is available. To illustrate, we cite the following theorem, first proved in [141]. Recall that $H \sim F$ means that the subgroups H and F are conjugate in G.

MCLAUGHLIN'S THEOREM. *Let L be an irreducible subgroup of $\mathrm{GL}(n, K)$, $|K| \geq 3$, generated by root subgroups. Then either $L = \mathrm{SL}(n, K)$ or $n = 2l$ is even and $L \sim \mathrm{Sp}(2l, K)$.*

Of course, if F is a primitive irreducible subgroup of G containing X_α then, by Clifford's theorem (see, e.g., [69]), the normal subgroup L of F generated by all the subgroups of F conjugate to X_α is also irreducible, so that either $\mathrm{SL}(n, K) \leq F \leq \mathrm{GL}(n, K)$ or $n = 2l$ and $\mathrm{Sp}(2l, K) \leq F \leq \mathrm{GSp}(2l, K)$. However, it is easy to verify that $\mathrm{GSp}(2l, K)$ does not contain, up to conjugacy, the group $D(2l, K)$ of all diagonal matrices.

McLaughlin in fact proved a similar result for the field F_2, where some additional subgroups appear. Subsequently, Wagner, Zalesskiĭ and Serezhkin, Pollaczek, in a long series of papers, described all the finite irreducible subgroups generated by transvections (see references in [41, 42, 46]). Mclaughlin's theorem has been generalized and specialized in other directions. In particular, Brown and Humphries have "effectivized" it, that is, constructed an algorithm that makes it possible to establish, given a collection of root subgroups, which subgroup they generate.

Subgroups of orthogonal groups generated by root subgroups in the case of a finite base field have been studied by Stark (Saltzberg) in [158-160]. However, her result involved an inaccuracy, which was corrected by Kantor in the more general context of subgroups generated by root elements [114]. An attempt to generalize these results in part to the case of an infinite field was made by Andreassian [74]. The proofs of her results, however, were never published; and, the author has been informed by Kantor that the proof presented in Andreassian's thesis contains an error. The first proof of the following result was published by Li Shangzhi [128]. Independently, but later, this result was obtained by the author [24, 29].

THEOREM 19. *Let L be an irreducible subgroup of $G = \mathrm{SO}(n, K)$, $\mathrm{char} K \neq 2$, $|K| \geq 5$, generated by long root subgroups. Then one of the following possibilities holds*:
(1) $L = \Omega(n, K)$,
(2) $n = 4m$, $L \sim \mathrm{SU}(m, E)$, $[E:K] = 2$,
(3) $n = b$, $L \sim \mathrm{Spin}(7, K)$,
(4) $n = 7$, $L \sim \mathrm{G}_2(K)$.

In [128] — and elsewhere — Li Shangzhi has actually done more: he determined all the maximal subgroups of the classical groups that contain root subgroups. The following results, proved by Li Shangzhi in [128, 130] (see also [159]), complete the description of subgroups of the classical groups that are generated by root subgroups.

THEOREM 20. *Let $\mathrm{char} K \neq 2$. Then there are no irreducible subgroups in $G = \mathrm{SO}(m, K)$ that are generated by short root subgroups but do not contain long root subgroups.*

THEOREM 21. *Let $\mathrm{char} K \neq 2$ and let L be an irreducible subgroup of $G = \mathrm{Sp}(2l, K)$ generated by short root subgroups and containing no long root subgroups. Then $l = 2m$ and $L \sim \mathrm{Sp}(2m, E)$ for some extension E, $[E:K] = 2$.*

Thus, for the case of the classical groups everything is clear, although there are of course quite a lot of unsolved problems such as the "effectivization" of Theorems 19 and 21.

3. Exceptional groups. Here the situation is totally different. All we have is a series of papers by Cooperstein [89-91], in which he describes subgroups of the finite exceptional groups that are generated by long root subgroups but not contained in a proper parabolic subgroup. Here, for instance, one has such imbeddings as $G(F_4, K)$, $G(C_4, K) \leq G(E_6, K)$, but we shall not give the exact formulation of the result [54]. Later Cooperstein also described the "irreducible" subgroups of finite Chevalley groups generated by long root elements [92]. For an algebraically closed field, the maximal connected subgroups containing a long root subgroup were recently described by Seitz [156].

For the time being, the generalization of these results to arbitrary fields is an open problem. There are some partial results by the present author: the existence of two opposite root subgroups in any subgroup generated by long root subgroups but not contained in a proper parabolic subgroup; a description of orbits of a Chevalley group, acting by conjugacies on the sets of pairs and triples of long root subgroups; and so on (see [24, 29] and the references given there). Cooperstein and the author are presently working on this problem and hope to complete the solution in the near future.

As for subgroups generated by short root subgroups, the situation is even less clear. The problem has not been solved even for a finite base field. It is not even known exactly what subgroups can be generated by two short root subgroups of a Chevalley group of type F_4.

§10. Overgroups of nonsplit maximal tori

Thus, for almost any ground field we have a satisfactory description of subgroups of Chevalley groups that contain a split maximal torus. I do not believe there is any hope of obtaining an equally explicit description for overgroups of an arbitrary maximal torus over general fields. For some important classes of fields, such as local or global fields, there are nevertheless significant results, and for the finite case the results are almost definitive. A brief survey of these results follows.

1. Steinberg's theory. A remarkable paper of Seitz [150] gives an almost exhaustive description of subgroups of finite groups of Lie type that contain an arbitrary maximal torus. This description relies essentially on Steinberg's theory [162, 67, 85, 86]. It is convenient, therefore, to alter our notation slightly as compared to the rest of the paper. Thus, $\overline{G} = G(\Phi, \overline{K})$ will now denote not an extended Chevalley group but a Chevalley group of type Φ over the algebraic closure $\overline{K}$ of a finite field $K = F_q$ considered as an algebraic group. Similarly, $\overline{T} = T(\Phi, \overline{K})$, $\overline{B} = B(\Phi, \overline{K})$, etc.

Recall that a standard Frobenius endomorphism $\sigma = \sigma_q$ of $\mathrm{GL}(n, \overline{K})$ is defined by raising the matrix coefficients of σ to the qth power: $(x_{ij}) \to (x_{ij}^q)$. A standard Frobenius endomorphism of a Chevalley group G is the mapping $\sigma: \overline{G} \to \overline{G}$ induced by σ_q under a certain imbedding $i: \overline{G} \to \mathrm{GL}(n, \overline{K})$. Finally, a Frobenius endomorphism is a mapping $\sigma: \overline{G} \to \overline{G}$, some power of which is a standard Frobenius endomorphism. Steinberg showed that in this case the group $\overline{G}^\sigma$ of fixed points of σ is finite. Conversely, any finite group G of Lie type is such that $O^{p'}(\overline{G}^\sigma) \leq G \leq \overline{G}^\sigma$ for some σ, where p is the characteristic of K and $O^{p'}(G)$ denotes, as usual in the theory of finite groups, the subgroup of G generated by all p-elements. The major technical tool of the theory is the Lang-Steinberg theorem, according to which the mapping $g \mapsto g^{-1}\sigma(g)$ is surjective on $\overline{G}$.

By definition, a maximal torus T of G is of the form $G \cap \overline{T}$, where $\overline{T}$ is some σ-invariant algebraic maximal torus, that is, $\sigma(\overline{T}) = \overline{T}$. It is well known [67, 85, 86, 95, 162] that the conjugacy classes of maximal tori in G are in one-to-one correspondence with the σ-conjugacy classes of the Weyl group $W = W(\Phi)$. Recall that σ acts naturally on W and that two elements $x, y \in W$ are said to be σ-conjugate if there exists $w \in W$ such that $w^{-1}x\sigma(w) = y$. Since an untwisted Frobenius endomorphism acts trivially on W, it follows that, in Chevalley groups of normal types, maximal tori correspond to ordinary conjugacy classes of the Weyl group [44, 84]. Namely, if $\overline{T}$ is a σ-invariant torus contained in a σ-invariant Borel subgroup $\overline{B}$, then to each $w \in W$ there corresponds a torus T_w constructed in the following way: take $g \in \overline{G}$ such that $g^{-1}\sigma(g) = w$ and $T_w = G \cap g^{-1}Tg$. The order of the torus T_w, its index in the normalizer, etc., can easily be expressed in terms of w [44, 84, 85].

Maximal tori, especially minisotropic ones (that is, those not contained in proper parabolic subgroups), play a crucial role in the Deligne-Lusztig theory of complex representations of finite Chevalley groups [95, 139] (see also [85, 86]). Given a character of a torus T, one uses l-adic cohomology theory to construct a virtual character $R_{T,\theta}$ of a Chevalley group G, known as the Deligne-Lusztig character. If θ is in general position, then $R_{T,\theta}$ is, up to sign, a real character of G and, moreover, an irreducible one. In this procedure minisotropic tori lead to cuspidal characters. This explains the intense interest in maximal tori in Chevalley groups; they are also of extreme importance in describing conjugacy classes of elements. Besides the papers already cited, we also mention Veldcamp [171, 172].

2. Seitz's theory. We retain the notation of the previous subsection. Though Seitz [150] studies all finite groups of Lie type, we restrict ourselves to the normal types, i.e., the case in which σ is an untwisted Frobenius endomorphism.

Let $\overline{X}_\alpha$, $\alpha \in \Phi$, be a collection of elementary root subgroups of the group $\overline{G}$ with respect to $\overline{T}$. Since $\overline{T}$ is σ-invariant, these subgroups are permuted by the action of σ. Let $\Delta = \{\overline{X}_1, \ldots, \overline{X}_m\}$ be a σ-orbit of these root subgroups. Set $\overline{X} = \langle \overline{X}_1, \ldots, \overline{X}_m \rangle$. Then the groups $X = O^{p'}(\overline{X}^\sigma)$ will be called T-root subgroups of G [148, 150]. Every such group is either unipotent or a group of Lie type over a certain extension of the base field. If

T is split, all its root subgroups are unipotent; this is the case we considered previously. On the other hand, if T is minisotropic, i.e., its split rank is the least possible, then all the root subgroups are semisimple. In the general case one has subgroups of both types. Moreover, in the case of a nonsplit torus there may be nontrivial inclusions among the root subgroups.

Now let S be a set of the T-root subgroups that is closed in the sense that if X, Y, Z are three T-root subgroups such that X, $Y \in S$ and $Z \leq \langle X, Y \rangle$, then $Z \in S$. Let $G(T, S)$ be the subgroup generated by T and all the T-root subgroups $X \in S$. Let $N(T, S)$ be the normalizer of $G(T, S)$ in G. Then the main result of Seitz's paper can be stated as follows.

SEITZ'S THEOREM. *Let K be a finite field, $\operatorname{char} K \neq 2, 3$, $|K| \geq 13$, and T a maximal torus in a Chevalley group $G = G(\Phi, K)$ over K. Then, for any subgroup F of G that contains T, there exists a unique closed set S of T-root subgroups such that*

$$G(T, S) \leq F \leq N(T, S).$$

Seitz actually established more detailed and precise results in [150]. Among other important results, he showed that $G(T, S)$ is exactly the subgroup $\langle T^F \rangle$ generated by all the subgroups conjugate to T by elements of F; that if F contains two maximal tori, T_1 and T_2, then $\langle T_1^F \rangle = \langle T_2^F \rangle$; and that one can choose a maximal torus T, $T \leq F$, such that $F = G(T, S)N_F(T)$. The connection between Seitz's results and Lie algebras is discussed in [113]. The proofs in [150] rely essentially on the classification of finite simple groups.

The main results of [150] are rephrased in [149] and specialized to the classical cases in the geometric language. In particular, the maximal tori and their root subgroups are explicitly described for the classical groups in their natural representations. For these results, see also Kondrat'ev's survey [54].

Before Seitz's work, a description of overgroups of a nonsplit maximal torus was known only for the simplest case of a Singer cycle in the general linear group [115] (for the definition and properties of Singer cycles in the classical groups, see, e.g., [111, 112, 149, 80]). In this connection we mention also more general results of Hering [106] and Dempwolf [96] on finite linear groups containing an irreducible cyclic subgroup. Another proof for Singer cycles was recently given by Dye [98–102] in a wider context — the description of overgroups of stabilizers of spreads (analogous results were obtained by Li Shangzhi, but since he relied on the use of unipotent elements, he was forced to omit the case of tori [131–133]). Dye's proof is combinatorial-geometric; it does not depend on the assumption that $\operatorname{char} K \neq 2, 3$ and does not use the classification of finite simple groups.

3. Local fields. There is no other class of fields, except the finite fields, for which the description of subgroups containing nonsplit maximal tori has been accomplished to the same degree of completeness. The point is that a description of the overgroups of all maximal tori in Chevalley groups over a field K requires a good understanding of the structure of the finite extensions of K, the classification of K-forms of semisimple algebraic groups, etc. Only for certain very special classes of fields is this information readily available. It has in fact been shown [173] that many necessary facts from Steinberg's theory fail upon transition to a more general situation.

Borevich and Krupetskiĭ [5, 55, 56, 59, 60] described the subgroups of some classical groups over $\mathbb{R}$ that contain certain special tori. The problem for the compact Lie groups has been practically settled by Djoković [97], who has proved that overgroups of maximal tori are closed in the usual real topology. The situation is probably the same in other cases. Maximal tori were classified up to conjugacy by Kostant [126], Sugiura [163], and Rothschild [144]. For example, the conjugacy classes of maximal tori in a Chevalley group $G(\Phi, \mathbb{R})$ are in one-to-one correspondence with the involution classes of the Weyl group, so that numbers of classes are as follows: in E_6, 5; in E_7 and E_8, 10 each; in F_4, 8; in G_2, 4. For other $\mathbb{R}$-forms, only involutions that are products of reflections with respect to noncompact roots are admissible, so that the number of conjugacy classes of maximal tori certainly does not exceed the above values.

Krupetskiĭ [57–59, 61] considers overgroups of certain tori in some of the classical groups over non-Archimedean local fields. His work is enough to show that the answer cannot be as simple as for finite fields, since the ring of integral elements of the local field and its ideals come into play. The results are far from complete; in particular, the study of exceptional groups in this context has not yet begun. Problems of conjugacy for maximal tori have been studied by Kariyama and Morris [117, 143]. For example, it has been shown that, if the characteristic of K does not divide the order of the Weyl group $W(\Phi)$, then the number of conjugacy classes of maximal tori in a Chevalley group $G(\Phi, K)$ is finite. The same also holds without any assumption about the characteristic for tori that split over a weakly ramified extension.

4. Global fields. The case of global ground fields is, of course, even more difficult. Only two examples have been considered, and they clearly illustrate the problem. In [62], Krupetskiĭ considers subgroups containing the group of diagonal matrices in a unitary group over the division field of rational quaternions. In a recent paper [53] V. A. Koĭbaev described the overgroups of all maximal tori in $GL(2, \mathbb{Q})$. The description is phrased in arithmetical terms and differs in the most striking way from the description of overgroups of split tori. Thus, for example, every nonsplit torus has a continuum of different overgroups, depending on arithmetic parameters (the set of simple ideals, defined in terms of the Legendre symbol, in intermediate subrings R, $\mathbb{Z} \leq R \leq \mathbb{Q}$, etc.). Most recently, Koĭbaev has generalized this description to all global fields K and minisotropic tori in $GL(n, K)$ associated with cyclic Kummer extensions. This outstanding result indicates the probable nature of the answer in the general situation. Nevertheless, it is clear that a formidable amount of work will be needed to complete the description in the case of general groups and tori.

§11. Some unsolved problems

We will now state a number of unsolved problems that arise most naturally in connection with the topics discussed above. We begin with split tori.

1. Prove that, under the usual assumptions about the characteristic, the description of subgroups of $G = G(\Phi, K)$ containing $T = T(\Phi, K)$ is standard if $\Phi \neq C_l$ and $|K| \geq 7$.

We know that Seitz's result implies the bound $|K| \geq 13$. The answer is known to hold for the A_l and D_l series (see, in particular, [18, 24, 27, 30]). Problem 1 can be viewed as a specialization of the following problem of Seitz [147, Problem 1].

2. Describe the subgroups of $G = G(\Phi, K)$ containing $T = T(\Phi, K)$, on the assumption that $|K| \geq 4$.

For fields K with $|K| \leq 5$ the description is certainly nonstandard [45]. A complete answer to this problem has been obtained only for the series $\Phi = A_l$ by the author [18] and Koĭbaev [50–52] (see also the earlier [4, 46–49] where the case of the general linear group was considered). A "parastandard" description has been established for fields $\mathbb{F}_5$ and $\mathbb{F}_4$, although some new irreducible primitive subgroups containing T turn up, but all the other subgroups are constructed from them in the standard way. The situation for $\mathbb{F}_3$ is much more complex.

If the solution of problems 1 and 2 requires excessive efforts, it may be worth waiting for a complete classification of the subgroups of finite simple groups and other related groups. Vigorous efforts are underway to that effect (see, in particular, [75–78, 93, 94, 120–125, 132, 136–138, 142, 151–156, 166] and the references there). We already have a complete description of all maximal subgroups for all exceptional groups, except those of type E_7 and E_8 [76, 77, 93, 121, 122, 142, 155, 156, 166]. Solutions of problems 1 and 2 can clearly be derived from those results by brute force.

For the extended classical groups, i.e., GL_n, GSp_{2l}, SO_n, results analogous to Theorems 2 and 9 have been proved not only for fields but also for fairly large classes of rings, such as the semilocal rings (see, in particular, [3, 13, 15, 24, 27, 30, 33, 169]).

3. Prove analogs of the classification theorems for Chevalley groups over commutative semilocal rings.

This problem is still open even for SL_n (the only available result in this direction deals with discrete valuation rings [71]). There are other analogs of Theorems 2 and 9 for the classical groups, such as the classification theorems for subradical subgroups (see, in particular, [13, 15, 24, 27, 30, 33]), which are true for almost arbitrary rings. To be precise: let J be the Jacobson radical of a ring R. Let $B(\Phi, R, J)$ denote the subgroup of the Chevalley group $G = G(\Phi, R)$ generated by $B(\Phi, R)$ and all the root unipotents $x_\alpha(\xi)$, $\xi \in J$.

4. Describe the subgroups in $B(\Phi, R, J)$ that contain $T(\Phi, R)$.

Such a description would be a decisive step toward the solution of problem 3. In order to obtain a "standard" solution to problem 4, one must assume that R satisfies conditions of the following type: $R = \mathbb{Z}[R^*]$, and there exist $\varepsilon, \eta \in R^*$ such that $\varepsilon - 1, \eta - 1, \varepsilon - \eta, \varepsilon\eta - 1 \in R^*$; similar but slightly stronger conditions may also work.

It would be of great interest to obtain a uniform proof of Theorem 2, simultaneously including the cases of finite and infinite fields; as far as we know, there is no such proof even for SL_2. We point out two concrete possibilities in this direction; both may prove useful for the solution of problems 3 and 4.

5. Devise a proof of the classification theorems using not the Bruhat decomposition but the Birkhoff decomposition: $G = BWB^-$.

First, since it combines the Bruhat and Gauss decompositions, the Birkhoff decomposition bridges the gap between fields and subradical subgroups. Second, the coefficients are more closely related to the usual matrix coefficients than those of the Bruhat decomposition, thus enabling one to imitate the proofs in matrix language for the classical groups.

6. Prove the classification and conjugacy theorems using the matrix realizations of the exceptional Chevalley groups associated with modules of least dimension.

We are well acquainted with the dimensions of the minimal Weyl modules: 7 for G_2, 26 for F_4, 27 for E_6, 56 for E_7, and 248 for E_8. The exceptional groups are realized over these modules as the general isometry groups of certain multilinear forms, since the matrix coefficients of their elements satisfy especially simple equations. For G_2 and E_8 groups this was already known to Dickson in 1905; for further developments in connection with nonassociative algebras and geometry, see the references in [76–78, 88, 94, 105].

Formally speaking, the next problem is concerned with nonsplit tori, but it seems quite likely that it can be solved by the methods presented in this paper.

7. Generalize the results of this paper to subgroups of twisted Chevalley groups that contain a Cartan subgroup.

We now go on to questions relating to nonsplit tori. As already mentioned, a description of overgroups of maximal tori was obtained by Seitz [150], using the classification of finite simple groups.

8. Prove the results of [150] without using this classification and without the assumptions char $K \neq 2, 3$.

One expects that in many cases the condition $|K| \geq 13$ may also be weakened or eliminated. A good deal of computational work is still needed, apparently, to specialize the results of [150]: explicitly to calculate the lattice of overgroups of maximal tori, to establish the relationships between the lattices for different tori, and so on. We mention two concrete questions related to Seitz's paper.

9. Classify the maximal subgroups of $G(\Phi, F_q)$ that contain a maximal tori.

10. Classify the pronormal maximal tori in $G(\Phi, F_q)$.

One cannot expect Seitz's results to carry over completely to the case of infinite fields. However, the surprising parallel between the properties of solvable subgroups containing a maximal torus and Suprunenko's theory of maximally solvable linear groups [68] raises hopes for a fairly explicit solution of the following problem for any field.

11. Describe the solvable subgroups of $G(\Phi, K)$ that contain a maximal torus.

By removing the solvability condition, one may hope for such a description only for very special classes of fields.

12. Describe all the subgroups of $G(\Phi, K)$ that contain a maximal torus, when K is either local or global.

Another interesting problem is to describe the subgroups of G that contain only part of a maximal torus. Of course, the part should not be too small, if one is to expect a meaningful description. Several such problems for finite ground fields were stated in [54, 147, 150]. We point out one more arithmetical realization of this problem.

13. Let R be the ring of integral elements of a complete valued field K. Describe the subgroups of $G(\Phi, K)$ that contain $T(\Phi, R)$.

This problem was solved for GL_n by the author and Khamdan [36, 71]. A complete solution would be, in particular, a very broad generalization of results by Iwahori, Matsumoto, Bruhat, and Tits concerning the description of parahoric subgroups.

Another natural realization of the general problem is: what can be said about subgroups containing a not necessarily maximal torus? The following version of this problem was suggested by Berman in 1977.

14. Let S be a torus in G. Is it true that, knowing all the subgroups of $N_G(S)/S$, one can describe the subgroups of G that contain S in combinatorial terms (relative root system, etc.)?

This question is, of course, meaningful not only for Chevalley groups but also for other kinds of reductive groups. As always, when speaking about reductive groups, tori, etc., one means their groups of points over some not necessarily algebraically closed field K, and one is interested in all the subgroups, not merely the algebraic ones. In order to demonstrate the difficulty of problem 14, we mention the following very special "subquestion," which itself is an unsolved problem.

15. Let S be a maximal split subtorus (i.e., maximal with respect to all

split subtori) in G. Describe all subgroups of G that contain subtori $C_G(S)$.

This version of the problem eliminates all the difficulties due to the need to consider subgroups lying in the anisotropic kernel; hence this problem is closely related to our paper. Nevertheless, even problem 15 cannot be solved by directly applying our methods; it requires additional considerations. Some practical results of relevance for isotropic orthogonal groups have been obtained by Golubovsky.

During more than ten years of work on these problems, I have enjoyed the help and support of many of my colleagues; it would be difficult to thank all of them. Nevertheless, I must mention at least a few names. Z. I. Borevich, A. I. Kostrikin, and D. K. Faddeev actively encouraged me to work on the problem. Innumerable conversations and lengthy correspondence with B. B. Venkov, E. B. Vinberg, A. E. Zalesskiĭ, G. M. Seitz, V. A. Koĭbaev, and B. N. Cooperstein have been very useful. They and many others, including M. Aschbacher, R. G. Dye, O. King, P. Kleidman, M. U. Liebeck, and Li Shangzhi, have kept me abreast of developments, often sending me their papers long before publication. Without all this, it would have been impossible either to obtain the results presented here or to write this survey. I am deeply grateful to all of them.

References

1. Z. I. Borevich, *A description of subgroups of the general linear group containing the group of diagonal matrices*, Zap. Nauchn. Sem. Leningrad. Otdel. Mat. Inst. Steklov. (LOMI) **64** (1976), 12–29; English transl. in J. Soviet Math. **17** (1981), no. 2.
2. _____, *On the arrangement of subgroups*, Zap. Nauchn. Sem. Leningrad. Otdel. Mat. Inst. Steklov. (LOMI) **94** (1979), 5–12; English transl. in J. Soviet Math. **19** (1982), no. 1.
3. Z. I. Borevich and N. A. Vavilov, *Subgroups of the general linear groups over a semilocal ring containing the group of diagonal matrices*, Trudy Mat. Inst. Steklov **148** (1978), 43–57; English transl. in Proc. Steklov Inst. Math. **1980**, no. 4.
4. Z. I. Borevich and V. A. Koĭbaev, *Subgroups of the general linear groups over the field of five elements*, Algebra i Teoriya Chisel, Ordzhenikidze, 1978, pp. 9–37. (Russian)
5. Z. I. Borevich and S. L. Krupetskiĭ, *Subgroups of the unitary group containing the group of diagonal matrices*, Zap. Nauchn. Sem. Leningrad. Otdel. Mat. Inst. Steklov. (LOMI) **86** (1979), 19–29; English transl. in J. Soviet Math. **17** (1981), no. 4.
6. A. Borel, *Linear Algebraic Groups*, Springer-Verlag, Berlin and New York, 1969; 2nd. ed., 1991.
7. _____, *Properties and linear representations of Chevalley groups*, Lecture Notes in Math., vol. 131, Springer-Verlag, Berlin, 1970, pp. 1–55.
8. A. Borel and J. Tits, *Groupes réductifs*, Inst. Hautes Études Sci. Publ. Math. **27** (1965), 55–151.
9. _____, *Unipotent elements and parabolic subgroups of reductive groups*. I, Invent. Math. **12** (1971), no. 2, 95–104.
10. N. Bourbaki, *Groupes et algèbres de Lie*. Ch. IV–VI, Hermann, Paris, 1968; Ch. VII, VIII, Hermann, Paris, 1990.
11. N. A. Vavilov, *On parabolic subgroups of Chevalley groups over a semilocal ring*, Zap. Nauchn. Sem. Leningrad. Otdel. Mat. Inst. Steklov. (LOMI) **75** (1978), 43–58; English transl. in J. Soviet Math. **37** (1987), no. 2.
12. _____, *On conjugacy of subgroups of the general linear group containing the group of diagonal matrices*, Uspekhi Mat. Nauk **34** (1979), no. 5, 216–217. (Russian)

13. _____, *Bruhat decomposition for subgroups containing the group of diagonal matrices*, Zap. Nauchn. Sem. Leningrad. Otdel. Mat. Inst. Steklov. (LOMI) **103** (1980), 20–30; **114** (1983), 50–61; English transl. in J. Soviet Math. **24** (1984), no. 4; **27** (1984), no. 4.
14. _____, *On subgroups of extended Chevalley groups containing a maximal torus*, 16th All-Union Algebraic Conference, Abstracts of lectures, Part I, Leningrad, 1981, pp. 26–27. (Russian)
15. _____, *On subgroups of the general linear group over a semilocal ring, containing the group of diagonal matrices*, Vestnik Leningrad. Univ. Mat. Mekh. Astronom. (1981), no. 1, 10–15; English transl. in Vestnik Leningrad. Univ. Math. **14** (1981).
16. _____, *Parabolic subgroups of Chevalley groups over a commutative ring*, Zap. Nauchn. Sem. Leningrad. Otdel. Mat. Inst. Steklov. (LOMI) **116** (1982), 20–43; English transl. in J. Soviet Math. **26** (1984), no. 3.
17. _____, *Subgroups of Chevalley groups over a field, containing a maximal torus*, 17th All-Union Algebraic Conference, Abstracts of Lectures, Part I, Minsk, 1983, pp. 38–39. (Russian)
18. _____, *On subgroups of the special linear group, containing the group of diagonal matrices.* I, Vestnik Leningrad. Univ. Mat. Mekh. Astronom. (1985), no. 4, 3–7; II (1986), no. 1, 10–15; III (1987), no. 2, 3–8; IV (1988), no. 3, 10–15; English transl. in Vestnik Leningrad. Univ. Math. **18** (1985); **19** (1986); **20** (1987); **21** (1988).
19. _____, *Maximal subgroups of Chevalley groups containing a maximal split torus*, Visnik Kiïv Univ. Ser. Mat. Mekh. (1985), no. 27, 28–30. (Ukrainian)
20. _____, *On the structure of a Chevalley group of type E_6.* I, II, Manuscript No. 2962-B, deposited at VINITI, 1986; Manuscript No. 5228-B, deposited at VINITI, Leningrad, 1986. (Russian)
21. _____, *Maximal subgroups of Chevalley groups containing a maximal split torus*, Rings and Modules. Limit Theorems of Probability Theory. no. 1, Izdat. Leningrad. Univ., Leningrad, 1986, pp. 67–75. (Russian)
22. _____, *Bruhat decomposition of one-dimensional transformations*, Vestnik Leningrad. Univ. Mat. Mekh. Astronom. (1986), no. 3, 14–20; English transl. in Vestnik Leningrad. Univ. Math. **19** (1986).
23. _____, *On a problem of G. Seitz*, 10th All-Union Sympos. on Group Theory, Abstracts of Lectures, Gomel', 1986, p. 31. (Russian)
24. _____, *Subgroups of split classical groups*, Doctoral Dissertation, Leningrad, 1987. (Russian)
25. _____, *Weight elements of Chevalley groups*, Dokl. Akad. Nauk SSSR **298** (1988), no. 3, 524–527; English transl. in Soviet Math. Dokl. **37** (1988).
26. _____, *Conjugacy theorems for subgroups of extended Chevalley groups containing a split maximal torus*, Dokl. Akad. Nauk SSSR **299** (1988), no. 2, 269–272; English transl. in Soviet Math. Dokl. **37** (1988).
27. _____, *On subgroups of split orthogonal groups*, Sibirsk. Mat. Zh. **29** (1988), no. 3, 12–15; English transl. in Siberian Math. J. **29** (1988).
28. _____, *Bruhat decomposition of long root semisimple elements in Chevalley groups*, Rings and Modules. Limit Theorems of Probability Theory. no. 2, Izdat. Leningrad. Univ., Leningrad, 1988, pp. 18–39. (Russian)
29. _____, *Root semisimple elements and triples of root unipotent subgroups in Chevalley groups*, Problems of Algebra. no. 4, "Universitetskoe", Minsk, 1989, pp. 162–173. (Russian)
30. _____, *On subgroups of split classical groups*, Trudy Mat. Inst. Steklov. **183** (1989), 29–42; English transl. in Proc. Steklov Math. Inst. (1991), no. 4.
31. _____, *Bruhat decomposition of two-dimensional transformations*, Vestnik Leningrad. Univ. Mat. Mekh. Astronom. (1989), no. 3, 3–7; English transl. in Vestnik Leningrad. Univ. Math. **22** (1989).
32. _____, *Linear groups generated by one-parameter groups of one-dimensional transformations*, Uspekhi Mat. Nauk **44** (1989), no. 1, 189–190; English transl. in Russian Math. Surveys **44** (1989).

33. N. A. Vavilov and E. V. Dybkova, *Subgroups of the general symplectic group containing the group of diagonal matrices* I, II, Zap. Nauchn. Sem. Leningrad. Otdel. Mat. Inst. Steklov. (LOMI) **103** (1980), 31–47; **132** (1983), 44–56; English transl. in J. Soviet Math. **24** (1984), no. 4; **30** (1985), no. 1.
34. N. A. Vavilov and E. B. Plotkin, *Net subgroups of Chevalley groups*, Zap. Nauchn. Sem. Leningrad. Otdel. Mat. Inst. Steklov. (LOMI) **94** (1979), 40–49; **114** (1982), 62–76; English transl. in J. Soviet Math. **19** (1982), no. 1; **27** (1984), no. 4.
35. N. A. Vavilov and A. A. Semenov, *Bruhat decomposition of long root tori in Chevalley groups*, Zap. Nauchn. Sem. Leningrad. Otdel. Mat. Inst. Steklov. (LOMI) **175** (1989), 12–23; English transl. in J. Soviet Math. **57** (1991), no. 6.
36. N. A. Vavilov and I. Khamdan, *Subgroups of the general linear group over a local field*, Izv. Vyssh. Uchebn. Zaved. Mat. (1989), no. 12, 8–15; English transl. in Soviet Math. (Iz. VUZ) **32** (1989).
37. N. A. Vavilov and A. L. Kharebov, *On the lattice of subgroups of a Chevalley group containing a split maximal torus*, 19th All-Union Algebraic Conference, Abstracts of lectures. Part 2, L'vov, 1987, p. 47. (Russian)
38. E. B. Dybkova, *Subgroup structure in symplectic groups*, Candidate's Dissertation, Leningrad, 1986. (Russian)
39. E. B. Dynkin, *Semisimple subalgebras of semisimple Lie algebras*, Mat. Sb. **30** (1952), no. 2, 349–462. (Russian)
40. A. E. Zalesskiĭ, *Semisimple root elements of algebraic groups*, Preprint, AN BSSR Inst. Mat., Minsk, 1980. (Russian)
41. _____, *Linear groups*, Uspekhi Mat. Nauk **36** (1981), no. 5, 56–107; English transl. in Russian Math. Surveys **36** (1981).
42. _____, *Linear groups*, Itogi Nauki i Tekhniki: Sovremennye Problemy Matematiki: Algebra, Topologiya, Geometriya, vol. 21, VINITI, Moscow, 1983, pp. 135–182; English transl. in J. Soviet Math. **31** (1985), no. 3.
43. R. Carter, *Simple groups and simple Lie algebras*, J. London Math. Soc. (2) **40** (1965), 193–240.
44. _____, *Classes of conjugate elements in the Weyl group*, Lecture Notes in Math., vol. 131, Springer-Verlag, Berlin, 1970, pp. 297–318.
45. V. A. Koĭbaev, *Examples of nonmonomial linear groups without transvections*, Zap. Nauchn. Sem. Leningrad. Otdel. Mat. Inst. Steklov. (LOMI) **71** (1977), 153–154; English transl. in J. Soviet Math. **20** (1984), no. 6.
46. _____, *Subgroups of the general linear group over the four-element field*, Algebra i Teoriya Chisel. no. 4, Nal'chik, 1979, pp. 21–31. (Russian)
47. _____, *A description of D-complete subgroups in the general linear group over the three-element field*, Zap. Nauchn. Sem. Leningrad. Otdel. Mat. Inst. Steklov. (LOMI) **103** (1980), 76–78; English transl. in J. Soviet Math. **24** (1984), no. 4.
48. _____, *Subgroups of the general linear group over the three-element field*, Structural Properties of Algebraic Systems, Nal'chik, 1981, pp. 56–68. (Russian)
49. _____, *Subgroup structure in linear groups over finite fields*, Candidate's Dissertation, Leningrad, 1982. (Russian)
50. _____, *Subgroups of the special linear group over the five-element field, containing the group of diagonal matrices*, 9th All-Union Sympos. on Group Theory, Abstracts of Lectures, Moscow, 1984, pp. 210–211. (Russian)
51. _____, *Subgroups of the special linear group over the four-element field, containing the group of diagonal matrices*, 18th All-Union Algebraic Conference, Abstracts of Lectures. Part 1, Kishinev, 1985, p. 264. (Russian)
52. _____, *Intermediate subgroups of the special linear group of order 6 over the four-element field*, 10th All-Union Sympos. on Group Theory, Abstracts of Lectures, Minsk, 1986, p. 115. (Russian)
53. _____, *Subgroups of the group* $GL(2, \mathbb{Q})$ *containing a nonsplit maximal torus*, Dokl. Akad. Nauk SSSR **312** (1990), no. 1, 36–38; English transl. in Soviet Math. Dokl. **41** (1990).

54. A. S. Kondrat′ev, *Subgroups of finite Chevalley groups*, Uspekhi Mat. Nauk **41** (1986), no. 1, 57–96; English transl. in Russian Math. Surveys **41** (1986).
55. S. L. Krupetskiĭ, *On some subgroups of the unitary group over a quadratic extension of an ordered Euclidean field*, Algebra i Teoriya Chisel. no. 4, Nal′chik, 1979, pp. 39–48. (Russian)
56. _____, *Subgroups of an orthogonal group containing the group of block-diagonal matrices*, Zap. Nauchn. Sem. Leningrad. Otdel. Mat. Inst. Steklov. (LOMI) **94** (1979), 73–80; English transl. in J. Soviet Math. **19** (1982), no. 1.
57. _____, *On subgroups of the unitary group over a local field*, Zap. Nauchn. Sem. Leningrad. Otdel. Mat. Inst. Steklov. (LOMI) **94** (1979), 81–103; English transl. in J. Soviet Math. **19** (1982), no. 1.
58. _____, *On subgroups of the unitary group over a dyadic local field*, Zap. Nauchn. Sem. Leningrad. Otdel. Mat. Inst. Steklov. (LOMI) **103** (1980), 79–89; English transl. in J. Soviet Math. **24** (1984), no. 4.
59. _____, *Subgroup structure in unitary groups*, Candidate's Dissertation, Leningrad, 1980. (Russian)
60. _____, *Intermediate subgroups of the unitary group over the quaternion ring*, Zap. Nauchn. Sem. Leningrad. Otdel. Mat. Inst. Steklov. (LOMI) **116** (1982), 96–101; English transl. in J. Soviet Math. **26** (1984), no. 3.
61. _____, *Intermediate subgroups of the unitary group over the p-adic quaternion ring*, Rings and Matrix Groups., Ordzhonikidze, 1984, pp. 75–82. (Russian)
62. _____, *On subgroups of the unitary group over the quaternion ring, containing a maximal torus*, Rings and Modules. Limit Theorems of Probability Theory, no. 1, Izdat. Leningrad. Univ., Leningrad, 1986, pp. 103–115. (Russian)
63. E. B. Plotkin, *On net subgroups of twisted Chevalley groups*, Latv. Matem. Ezhegodnik (1984), no. 28, 179–193. (Russian)
64. _____, *On stabilization of the K_1-functor for Chevalley groups*, Manuscript No. 7648, deposited at VINITI, Riga, 1984. (Russian)
65. _____, *Net subgroups of Chevalley groups and problems of K_1-functor stabilization*, Candidate's Dissertation, Leningrad, 1985. (Russian)
66. A. A. Semenov, *Bruhat decomposition of root semisimple subgroups in the special linear group*, Zap. Nauchn. Sem. Leningrad. Otdel. Mat. Inst. Steklov. (LOMI) **160** (1987), 239–246; English transl. in J. Soviet Math. **52** (1990), no. 3.
67. T. A. Springer and R. Steinberg, *Classes of conjugate elements*, Lecture Notes in Math., vol. 131, Springer-Verlag, Berlin, 1970, pp. 167–266.
68. D. A. Suprunenko, *Matrix groups*, "Nauka", Moscow, 1972; English transl. in Translations of Mathematical Monographs, Vol. 45, Amer. Math. Soc., Providence, R.I., 1976.
69. R. Steinberg, *Lectures on Chevalley Groups*, Yale Univ., New Haven, Conn., 1967.
70. J. Tits, *Groupes semi-simple isotropes*, Colloq. Théorie des Groupes Algébriques (Bruxelles, 1962), Gauthier-Villars, Paris, 1962, pp. 137–147.
71. I. Khamdan, *Subgroups of the general linear group over the field of quotients of a semilocal ring*, Candidate's Dissertation, Leningrad, 1987. (Russian)
72. J. E. Humphreys, *Linear Algebraic Groups*, Springer-Verlag, Berlin and New York, 1975.
73. C. Chevalley, *Sur certains groupes simples*, Tôhoku Math. J. **7** (1959), 14–66.
74. A. Andreassian, *Irreducible subgroups of unitary and orthogonal groups, generated by long root subgroups*, Abstracts Amer. Math. Soc. **2** (1981), no. 1, 79.
75. M. Aschbacher, *On the maximal subgroups of the finite classical groups*, Invent. Math. **76** (1985), no. 3, 469–514.
76. _____, *Chevalley groups of type G_2 as the group of a trilinear form*, J. Algebra **108** (1987), no. 1, 193–259.
77. _____, *The 27-dimensional module for E_6*, Invent. Math. **89** (1987), no. 1, 159–195.
78. _____, *Some multilinear forms with large isometry groups*, Geom. Dedicata **25** (1988), no. 1–3, 417–465.
79. H. Azad, *Root groups*, J. Algebra **76** (1982), no. 1, 211–213.
80. B. Baumann, *Symmetrische Singer-Zyklen über Körpern der Charakteristik* 2, Mitt. Math. Sem. Giessen **163** (1984), 135–140.

81. S. Berman and R. Moody, *Extensions of Chevalley groups*, Israel J. Math. **22** (1975), no. 1, 42–51.
82. R. Brown and S. P. Humphries, *Orbits under symplectic transvections*, Proc. London Math. Soc. (3) **52** (1986), no. 3, 517–556.
83. R. W. Carter, *Simple groups of Lie type*, Wiley, New York and London, 1972.
84. _____, *Conjugacy classes in the Weyl group*, Compositio Math. **25** (1972), no. 1, 1–59.
85. _____, *Finite groups of Lie type: Conjugacy classes and complex characters*, Wiley, New York, 1985.
86. _____, *On the representation theory of the finite groups of Lie type over an algebraically closed field of characteristic* 0, Preprint, Univ. of Warwick, Warwick, 1987.
87. E. Cline, B. Parshall, and L. Scott, *Minimal elements of $N(H,P)$ and conjugacy of Levi complements in finite Chevalley groups*, J. Algebra **34** (1975), no. 3, 521–523.
88. A. M. Cohen and B. N. Cooperstein, *The 2-spaces of the standard $E_6(q)$-module*, Geom. Dedicata **25** (1988), 355–388.
89. B. N. Cooperstein, *Subgroups of the group $E_6(q)$ which are generated by root subgroups*, J. Algebra **46**, no. 2, 355–388.
90. _____, *The geometry of root subgroups in exceptional groups*, Geom. Dedicata **8** (1973), no. 3, 317–381; **15**, no. 1, 1–45.
91. _____, *Geometry of long root subgroups in groups of Lie type*, Proc. Sympos. Pure Math., vol. 37, Amer. Math. Soc., Providence, R.I., 1980, pp. 243–248.
92. _____, *Subgroups of exceptional groups of Lie type generated by long root elements*, J. Algebra **70** (1981), no. 1, 270–298.
93. _____, *Maximal subgroups of $G_2(2^n)$*, J. Algebra **70** (1981), no. 1, 23–36.
94. _____, *The fifty six dimensional module of groups of type E_6*, Preprint, Univ. California, Santa Cruz, 1988.
95. P. Deligne and G. Lusztig, *Representations of reductive groups over finite fields*, Ann. Math. **103** (1976), no. 1, 103–161.
96. U. Dempwolff, *Linear groups with large cyclic subgroups and translation planes*, Rend. Sem. Mat. Univ. Padova **77** (1987), no. 1, 69–113.
97. D. Ž. Djoković, *Subgroups of compact Lie groups containing a maximal torus are closed*, Proc. Amer. Math. Soc. **83** (1981), no. 1, 431–432.
98. R. H. Dye, *A maximal subgroup of $\text{PSp}_6(2^m)$ related to a spread*, J. Algebra **84** (1983), no. 1, 128–135.
99. _____, *Maximal subgroups of symplectic groups stabilizing spreads*, J. Algebra **87** (1984), no. 2, 493–509.
100. _____, *Maximal subgroups of $\text{PSp}_{6n}(q)$ stabilizing spreads of totally isotropic planes*, J. Algebra **99** (1986), no. 1, 191–209.
101. _____, *Maximal subgroups of finite orthogonal groups stabilizing spreads of lines*, J. London Math. Soc. (2) **33** (1986), no. 3, 279–293.
102. _____, *Spreads and classes of maximal subgroups of $\text{GL}_n(q)$, $\text{SL}_n(q)$, $\text{PGL}_n(q)$ and $\text{PSL}_n(q)$*, Preprint, Univ. Newcastle-upon-Tyne, 1987.
103. M. Golubitsky, *Primitive actions and maximal subgroups of Lie groups*, J. Differential Geom. **7** (1972), no. 1–2, 175–191.
104. M. Golubitsky and B. Rothschild, *Primitive subalgebras of exceptional Lie algebras*, Bull. Amer. Math. Soc. **77** (1971), no. 6, 983–986; Pacific J. Math. **39** (1971), no. 2, 371–393.
105. J. R. Faulkner and J. C. Ferrar, *Exceptional Lie algebras and related algebraic and geometric structures*, Bull. London Math. Soc. **9** (1977), no. 1, 1–35.
106. Ch. Hering, *Transitive linear groups and linear groups which contain irreducible subgroups of prime order*, Geom. Dedicata **2** (1974), no. 4, 425–460; J. Algebra **93** (1985), no. 1, 151–164.
107. J. E. Humphreys, *Introduction to Lie Algebras and Representation Theory*, Springer-Verlag, Berlin and New York, 1980.
108. S. P. Humphries, *Graphs and Nielsen transformations of symmetric, orthogonal and symplectic groups*, Quart. J. Math. Oxford Ser. 2 **36** (1985), no. 143, 297–313.
109. _____, *Generation of special linear groups by transvections*, J. Algebra **99** (1986), no. 2, 480–495.

110. _____, *Identification of subgroups of* $SL_n(F)$ *generated by transvections*, Preprint, Brigham Young Univ., Provo, Utah, 1987.
111. B. Huppert, *Endliche Gruppen*. I, Springer-Verlag, Berlin, 1967.
112. _____, *Singer-Zyklen in klassischen Gruppen*, Math. Z. **117** (1970), no. 1, 141–150.
113. M. Kaneda and G. M. Seitz, *On the Lie algebra of a finite group of Lie type*, J. Algebra **74** (1982), no. 2, 333–340.
114. W. M. Kantor, *Subgroups of classical groups generated by long root elements*, Trans. Amer. Math. Soc. **248** (1979), no. 2, 347–379.
115. _____, *Linear groups containing a Singer cycle*, J. Algebra **62** (1980), no. 1, 232–234.
116. _____, *Generation of linear groups*, The Geometric Vein: Coxeter Festschrift, Springer-Verlag, Berlin and New York, 1981, pp. 497–509.
117. K. Kariyama, *On the conjugacy classes of anisotropic maximal tori of a Chevalley group over a local field*, J. Algebra **99** (1986), no. 1, 22–49.
118. O. King, *On subgroups of the special linear group containing the special orthogonal group*, J. Algebra **96** (1985), no. 1, 178–193.
119. _____, *Subgroups of the special linear group containing the diagonal subgroup*, J. Algebra **123** (1990), no. 1, 198–204.
120. P. Kleidman, *The maximal subgroups of the finite 8-dimensional orthogonal groups* $P\Omega_8^+(q)$ *and of their automorphism groups*, J. Algebra **110** (1987), no. 1, 173–242.
121. _____, *The maximal subgroups of the Steinberg triality groups* $^3D_4(q)$ *and of their automorphism groups*, J. Algebra **115** (1988), no. 1, 182–199.
122. _____, *The maximal subgroups of the Chevalley groups* $G_2(q)$ *with q odd, the Ree groups* $^2G_2(q)$ *and their automorphism groups*, J. Algebra **117** (1988), no. 1, 30–71.
123. _____, *The low-dimensional finite classical groups and their subgroups*, London, 1989.
124. P. Kleidman and M. W. Liebeck, *A survey of the maximal subgroups of the finite simple groups*, Geom. Dedicata **25** (1988), no. 1-3, 375–389.
125. _____, *The subgroup structure of the finite classical groups*, London Math. Soc. Lecture Note Ser., vol. 129, Cambridge Univ. Press, Cambridge, 1988.
126. B. Kostant, *On the conjugacy classes of real Cartan subalgebras*, Proc. Nat. Acad. Sci. U.S.A. **41** (1955), no. 3, 967–970.
127. Li Shangzhi, *Maximal subgroups containing root subgroups in finite classical groups*, Kexue Tongbao **29** (1984), no. 1, 14–18.
128. _____, *Maximal subgroups in* $P\Omega(n, F, Q)$ *with root subgroups*, Sci. Sinica Ser. A **28** (1985), no. 8, 826–838.
129. _____, *Maximal subgroups in* $PSU(n, K, f)$ $(\nu(f) \geq 1)$ *containing root subgroups*, Acta Math. Sinica **29** (1986), no. 5, 232–244. (Chinese)
130. _____, *Maximal subgroups containing short root subgroups in* $PSp(2n, F)$, Acta Math. Sinica **3** (1987), no. 1, 82–91.
131. _____, *Overgroups in* $GL(nr, F)$ *of certain subgroups of* $SL(n, K)$, J. Algebra **125** (1989), no. 1, 215–235.
132. _____, *Maximal subgroups in classical groups over arbitrary fields*, Proc. Sympos. Pure Math., vol. 47, part 2, Amer. Math. Soc., Providence, R.I., 1987, pp. 487–493.
133. _____, *Overgroups of certain subgroups in the classical groups over division rings*, Classical groups and related topics (Beijing, 1987), Contemp. Math., vol. 82, Amer. Math. Soc., Providence, R.I., 1989, pp. 53–57.
134. Li Shangzhi and Zha Jianguo, *On certain classes of maximal subgroups in* $PSp(2n, F)$, Sci. Sinica Ser. A **25** (1982), no. 12, 1250–1257.
135. _____, *Certain classes of maximal subgroups in classical groups*, Proc. Internat. Group Theory Sympos., Beijing, 1984.
136. M. W. Liebeck, *On the orders of maximal subgroups of the finite classical groups*, Proc. London Math. Soc. (3) **50** (1985), no. 3, 426–446.
137. M. W. Liebeck and J. Sax, *On the orders of maximal subgroups of the finite exceptional groups of Lie type*, Proc. London Math. Soc. (3) **55** (1987), no. 2, 299–330.
138. M. W. Liebeck, J. Saxl, and G. M. Seitz, *On the overgroups of irreducible subgroups of the finite classical groups*, Proc. London Math. Soc. (3) **55** (1987), no. 3, 507–537.

139. G. Lusztig, *Characters of reductive groups over a finite field*, Princeton Univ. Press, Princeton, N.J., 1984.
140. H. Matsumoto, *Sur les sous-groupes arithmétiques des groupes semisimples déployés*, Ann. Sci. École Norm. Sup. (4) **2** (1969), 1–62.
141. J. McLaughlin, *Some groups generated by transvections*, Arch. Math. (Basel) **18** (1967), no. 4, 364–368.
142. E. T. Migliore, *The determination of the maximal subgroups of $G_2(q)$, q odd*, Thesis, Univ. California, Santa Cruz, 1982.
143. L. Morris, *Rational conjugacy classes of unipotent elements and maximal tori and some axioms of Shalika*, J. London Math. Soc. (2) **38** (1988), no. 1, 112–124.
144. L. P. Rothschild, *Invariant polynomials and conjugacy classes of real Cartan subalgebras*, Bull. Amer. Math. Soc. **77** (1971), no. 5, 762–764; Indiana Univ. Math. J. **21** (1971), no. 2, 115–120.
145. G. Seitz, *Small rank permutation representations of finite Chevalley groups*, J. Algebra **28** (1974), no. 3, 508–517.
146. _____, *Subgroups of finite groups of Lie type*, J. Algebra **61** (1979), no. 1, 16–27.
147. _____, *Properties of the known simple groups*, Proc. Sympos. Pure Math., vol. 37, Amer. Math. Soc., Providence, R.I., 1980, pp. 231–237.
148. _____, *The root subgroups of a maximal torus*, Proc. Sympos. Pure Math., vol. 37, Amer. Math. Soc., Providence, R.I., 1980, pp. 239–241.
149. _____, *On the subgroup structure of classical groups*, Comm. Algebra **10** (1982), no. 8, 875–885.
150. _____, *Root subgroups for maximal tori in finite groups of Lie type*, Pacific J. Math. **106** (1983), no. 1, 153–244.
151. _____, *Representations and maximal subgroups*, Proc. Sympos. Pure Math., vol. 47, Amer. Math. Soc., Providence, R.I., 1987, pp. 275–287.
152. _____, *The maximal subgroups of classical algebraic groups*, Mem. Amer. Math. Soc., vol. 67, no. 365, Amer. Math. Soc., Providence, RI, 1987.
153. _____, *Cross-characteristic embeddings of finite groups of Lie type*, Proc. London Math. Soc. (3) **60** (1990), no. 1, 166–200.
154. _____, *Representations and maximal subgroups of finite groups of Lie type*, Geom. Dedicata **25** (1988), 391–406.
155. _____, *Maximal subgroups of exceptional groups*, Classical Groups and Related Topics (Beijing, 1987), Contemp. Math., vol. 82, Amer. Math. Soc., Providence, R.I., 1989, pp. 143–157.
156. _____, *The maximal subgroups of exceptional algebraic groups*, Mem. Amer. Math. Soc., vol. 90, no. 441, Amer. Math. Soc., Providence, R.I., 1991.
157. G. B. Seligman, *Modular Lie Algebras*, Springer-Verlag, Berlin, 1967.
158. B. Stark (Saltzberg), *Some subgroups of $\Omega(V)$ generated by groups of root type. I*, Illinois J. Math. **17** (1973), no. 4, 584–607.
159. _____, *Some subgroups of $\Omega(V)$ generated by groups of root type*, J. Algebra **29** (1974), no. 1, 33–41.
160. _____, *Irreducible subgroups of orthogonal groups generated by groups of root type. I*, Pacific J. Math. **53** (1974), no. 2, 611–625.
161. M. R. Stein, *Stability theorems for K_1, K_2 and related functors modeled on Chevalley groups*, Japan. J. Math. **4** (1978), no. 1, 77–108.
162. R. Steinberg, *Endomorphisms of algebraic groups*, Mem. Amer. Math. Soc., no. 80, Amer. Math. Soc., Providence, R.I., 1968.
163. M. Sugiura, *Conjugate classes of Cartan subalgebras in real semi-simple Lie algebras*, J. Math. Soc. Japan **11** (1959), no. 2, 374–434.
164. K. Suzuki, *On parabolic subgroups of Chevalley groups over local rings*, Tôhoku Math. J. **28** (1976), no. 1, 57–66.
165. _____, *On parabolic subgroups of Chevalley groups over commutative rings*, Sci. Repts. Tokyo Kyoiku Daigaku Ser. A **13** (1977), no. 366–382, 225–232.
166. D. M. Testerman, *Irreducible subgroups of exceptional algebraic groups*, Mem. Amer. Math. Soc., vol. 75, no. 390, Amer. Math. Soc., Providence, R.I., 1988.

167. Ton Dao-rong, *A class of maximal subgroups in finite classical groups*, J. Algebra **106** (1987), no. 2, 536–542.
168. _____, *Second class of maximal subgroups in finite classical groups*, J. Algebra **108** (1987), no. 2, 578–588.
169. N. A. Vavilov, *On subgroups of split orthogonal groups in even dimensions*, Bull. Polish Acad. Sci. Math. **29** (1981), no. 9–10, 425–429.
170. _____, *A conjugacy theorem for subgroups of* GL_n *containing the group of diagonal matrices*, Colloq. Math. **54** (1987), no. 1, 9–14.
171. F. D. Veldkamp, *Roots and maximal tori in finite forms of semi-simple algebraic groups*, Math. Ann. **207** (1974), no. 2, 301–314.
172. _____, *Regular elements in anisotropic tori*, Contributions to Algebra (collection of papers dedicated to E. Kolchin), Academic Press, New York, 1977, pp. 389–424.
173. S. J. Gottlieb, *Algebraic automorphisms of algebraic groups with stable maximal tori*, Pacific J. Math. **72**, no. 2, 461–470.

Translated by N. ZOBIN

On the Boundary Integral Equation of the Neumann Problem in a Domain with a Peak

VLADIMIR MAZ'YA AND ALEXANDER SOLOV'EV

Abstract. The integral equations generated by the Neumann problem for the Laplace equation are considered on a plane contour. We assume that there is an inward or an outward cusp at this contour. Theorems on the solvability and the number of solutions to this integral equation are proved. Asymptotic formulae for solutions near the cusp are also obtained.
Bibliography: 9 titles.

In this article we study integral equations of the Neumann problem for the Laplace operator in a plane domain with inward or outward peaks at the boundary. We prove theorems on the unique solvability and asymptotics of solutions near peaks.

The classical method for solving boundary value problems is their reduction to boundary integral equations by using potentials. In the case of the Dirichlet and Neumann problems this procedure leads to equations which were studied traditionally by methods of the theory of Fredholm integral operators. However, as was shown by J. Radon in 1919 [1], if a plane domain has cusps at the boundary, then the Fredholm radius of the integral operator, generated by the double layer potential acting in the space of continuous functions, equals 1. Thus a direct application of the Fredholm theory is impossible if the contour contains cusps. In the present paper we use another approach, proposed by one of the authors several years ago [2, 3]. It enables one to get information about the inverse operators of boundary integral equations by applying theorems on inverse operators of auxiliary boundary value problems.

Let Ω be a plane simply-connected domain with piecewise smooth boundary Γ that has a unique peak at the origin. Suppose that near the point 0 either the domain Ω or the complementary domain Ω^c is given by the

---

1991 *Mathematics Subject Classification.* Primary 31A10, 31A35.

inequalities

$$\kappa_-(x) < y < \kappa_+(x), \qquad 0 < x < \delta,$$

in Cartesian coordinates x, y. Here $\kappa_\pm$ are C^∞-functions on $[0, \delta]$ satisfying the conditions

$$\kappa_\pm(0) = \kappa'_\pm(0) = 0, \qquad \kappa''_+(0) > \kappa''_-(0).$$

In the first case we shall speak of an external peak and in the second case of an internal one.

For $-1 < \mu < 1$ let $\mathfrak{N}_\mu$ be the class of functions on the contour Γ that can be represented as

$$\psi_\pm(x) = x^\mu \phi_\pm(x)$$

on the arcs $\Gamma_\pm = \{(x, \kappa_\pm(x)) \colon x \in [0, \delta]\}$, where $\phi_\pm$ are C^∞-functions on $[0, \delta]$ and $|\phi_+(0)| + |\phi_-(0)| \neq 0$.

By $\mathfrak{M}_\beta$ ($\beta > -1$) we denote the class of functions on Γ that satisfy

$$\sigma(x) = O(x^\beta), \qquad \sigma'(x) = O(x^{\beta-1}).$$

Further, we set $\mathfrak{M} = \bigcup \mathfrak{M}_\beta$.

We briefly describe the content of this paper.

We seek the solution v^e of the external Neumann problem with the boundary function ψ as a simple layer potential whose density γ is a solution of the integral equation

$$\gamma(p) - \frac{1}{\pi} \int_\Gamma \gamma(q) \frac{\partial}{\partial n_p} \log \frac{1}{r} \, ds_q = -\frac{1}{\pi} \psi(p), \qquad (1)$$

where $\partial/\partial n_p$ is the derivative with respect to the outward normal to Γ at the point $p \neq O$ and $r = |p - q|$.

The Neumann boundary value problem is discussed in detail in §3. We separately consider two cases: $\psi \in \mathfrak{N}_\mu$, $0 < \mu < 1$ and $\psi \in \mathfrak{N}_\mu$, $-1 < \mu \leq 0$.

The unique solvability of the integral equation (1) is studied in Theorems 1 and 2. We show that (1) is uniquely solvable in the class $\mathfrak{M}$ if Ω has an inward peak and the function $\psi \in \mathfrak{N}_\mu$, $0 < \mu < 1$, satisfies the condition

$$\int_\Gamma \psi \, ds = 0.$$

For Ω with external peak, the unique solvability in the class $\mathfrak{M}$ holds for all functions $\psi \in \mathfrak{N}_\mu$, $0 < \mu < 1$, that satisfy the conditions

$$\int_\Gamma \psi \, ds = 0, \qquad \int_\Gamma \psi \operatorname{Re}(1/\zeta) \, ds = 0,$$

where ζ is the conformal mapping of Ω^c onto the upper half-plane whose inverse mapping ζ^{-1} has the representation

$$\zeta^{-1}(\eta + i0) = a^2 \eta^2 + O(\eta^4 \log |\eta|), \qquad \eta \to 0.$$

The asymptotic behaviour of the solution near the peak is also studied in Theorems 1 and 2.

If Ω has an exterior peak, then the solution γ of (1) has the following representations on the arcs $\Gamma_\pm$:

$$\gamma(x) = \pm c_0 x^{\mu-1} + O(x^{-1/2}) \quad \text{for } 0 < \mu < 1/2,$$

$$\gamma(x) = \pm c_0 x^{-1/2} \log x \pm c_1 x^{-1/2} + O(1) \quad \text{for } \mu = 1/2$$

provided $\phi_+(0) \neq \phi_-(0)$. In other cases the solution $\gamma(x)$ admits the estimate $O(x^{-1/2})$ on $\Gamma_\pm$.

If $\psi \in \mathfrak{N}_\mu$, $-1 < \mu \leq 0$, then (1), generally speaking, has no solutions in the class $\mathfrak{M}$. In this case the solution of the external Neumann problem may be sought as the sum of a simple layer potential and a harmonic function in Ω^c which is chosen in such a way that one of the above mentioned theorems on unique solvability can be applied to the potential. The relating assertions are formulated as remarks to the theorems.

While studying integral equations of the Neumann problem, we use asymptotics of the conformal mapping of Ω and Ω^c to canonical domains. The construction of the asymptotics of the conformal mapping of a strip onto a domain with an exterior cusp and the justification of this expansion are carried out in §1. The asymptotics of the conformal mapping of the upper half-plane onto a domain with an interior peak and its justification is obtained in §2.

Following Hadamard [4], we denote by

$$\left| \int_{-\infty}^{\infty} f(t)\, dt \right.$$

the finite part of the divergent integral.

The results of this paper were partly announced in [5]. For a similar study of the integral equation of the Dirichlet problem see our article [9].

§1. Asymptotics of a conformal mapping of a domain with exterior peak onto a strip

1°. Let L_+ and L_- be two curves in the plane $w = u + iv$ given by the equations

$$v = g_\pm(u), \quad u \geq u_0, \quad g_+(u) > g_-(u).$$

Further, let L_1 be a Jordan curve lying in the half-plane $u \leq u_0$ and connecting the initial points of the curves L_+ and L_- (the curve L_1 may consist of one point). The contour L composed of L_+, L_-, and L_1 divides the plane w into two parts. The one that contains the domain $\{w: g_-(u) < v < g_+(u), u > u_0\}$ will be denoted by D and called the curvilinear semi-infinite strip.

First we shall seek the asymptotics of the imaginary part of the mapping $\zeta(u, v)$ of D onto the strip $T = \{\zeta = \tau + iv: |v| < \pi/2\}$ satisfying the condition $\operatorname{Re} \zeta(\infty) = \infty$. The function $\nu = \operatorname{Im} \zeta$ is a bounded harmonic function equal to $\pi/2$ on an arc containing the curve L_+ and to $-\pi/2$ on the complementary arc of the boundary curve containing L_-.

Suppose the functions $g_+(u)$, $g_-(u)$ have the asymptotic representations

$$g_\pm(u) = \sum_{k \geq 0} \alpha_\pm^{(k)} u^{-k}. \qquad (2)$$

We look for the asymptotic expansion of ν as a series in nonnegative powers of $1/u$:

$$\nu(u, v) = \sum_{k \geq 0} p_k(v) u^{-k}. \qquad (3)$$

In the kth step we construct the polynomial $p_k(v)$ in such a way that the coefficient of u^{-k} is compensated in the preceding discrepancy. As a result, the polynomials $p_k(v)$, $k = 0, 1, \ldots$, are uniquely defined by the relations

$$p_0(g_\pm(u)) = \pm\pi/2 + O(u^{-1}),$$
$$p_0(g_\pm(u)) + p_1(g_\pm(u))u^{-1} = \pm\pi/2 + O(u^{-2}),$$
$$p_0(g_\pm(u)) + p_1(g_\pm(u))u^{-1} + p_2(g_\pm(u))u^{-2} = \pm\pi/2 + O(u^{-3}),$$
$$\cdots\cdots\cdots\cdots\cdots\cdots\cdots\cdots\cdots\cdots$$
$$p_k''(v) = 0, \qquad k = 0, 1, 2,$$
$$(u^{-k} p_k(v))_{vv}'' + (u^{-k+2} p_{k-2}(v))_{uu}'' = 0, \qquad k = 3, 4, \ldots.$$

Here p_0, p_1, p_2 are polynomials of the first degree. In what follows we shall use the following explicit expression for p_0:

$$p_0(v) = a + bv = -\frac{\pi}{2}\frac{\alpha_+ + \alpha_-}{\alpha_+ - \alpha_-} + \frac{\pi}{\alpha_+ - \alpha_-} v.$$

We set $p_1(v) = \alpha + \beta v$ and $p_2(v) = \omega + \sigma v$.

LEMMA 1. *Suppose the asymptotic equality* (2) *can be differentiated three times. Then $\nu(u, v)$ has the following representation for $u \to +\infty$:*

$$\nu(u, v) = p_0(v) + p_1(v)u^{-1} + p_2(v)u^{-2} + O(u^{-3}), \qquad (4)$$

which can be differentiated once, i.e.,

$$\nu_u'(u, v) = -p_1(v)u^{-2} - 2p_2(v)u^{-3} + O(u^{-4}),$$
$$\nu_v'(u, v) = p_0'(v) + p_1'(v)u^{-1} + p_2'(v)u^{-2} + O(u^{-3}).$$

Lemma 1 is proved in supplement 1° of §4.

Under the assumption that (2) can be differentiated four times one can prove that the relation (4) can be differentiated twice.

The real part τ can be restored from the imaginary part of the conformal mapping ζ of the curvilinear semi-infinite strip D onto the strip T. We have
$$\tau(u, v) = \lambda + bu + \beta \log u - \sigma u^{-1} + O(u^{-2}), \qquad (5)$$
where b, β, and σ are coefficients of the polynomials p_0, p_1, and p_2; and λ is an arbitrary constant. We set $\lambda = \beta \log b$. Unifying (4) and (5) we find
$$\zeta(w) = bw + \beta \log w + (\beta \log b + ia) - (\sigma - i\alpha)w^{-1} + O(w^{-2}).$$
The inversion of this expansion gives the conformal mapping of the strip T onto the semi-infinite strip D,
$$w = w_*(\zeta) = b^{-1}[\zeta - \beta \log \zeta - ia + \beta^2 \zeta^{-1} \log \zeta$$
$$+ (\sigma b + i(a\beta - \alpha b))\zeta^{-1} + o(\zeta^{-1})], \quad \operatorname{Re} \zeta \to +\infty. \quad (6)$$
We note that application of the Warschawski asymptotic formula [6] enables one to get only three principal terms in the expansion of w_*.

$2°$. The inversion $u + iv = (x + iy)^{-1}$ maps the domain Ω with an exterior cusp described in the introduction onto the curvilinear semi-infinite strip D, bounded for large u by the curves L_+ and L_- defined by the functions $g_\pm(u) = \alpha_\pm + \alpha_\pm^{(1)} u^{-1} + \cdots$, where
$$\alpha_\pm = -\kappa''_\mp(0)/2, \qquad \alpha_+ > \alpha_-, \quad \text{and} \quad \alpha_\pm^{(1)} = -\kappa'''_\mp(0)/6.$$
Using the expansion (6) and making the inversion we find the following representation of the conformal mapping w of the strip T onto Ω:
$$w(\zeta) = b\zeta^{-1} + b\beta\zeta^{-2} \log \zeta + iab\zeta^{-2} + \beta^2 b \zeta^{-3}(\log \zeta)^2$$
$$- \beta b(\beta - 2ia)\zeta^{-3} \log \zeta - b[(a^2 + b\sigma) + i(\beta a - \alpha b)]\zeta^{-3} + o(\zeta^{-3}). \qquad (7)$$
Separating the real and imaginary parts in (7) we find
$$x = \operatorname{Re} w(\tau \pm i\pi/2)$$
$$= b[\tau^{-1} + \beta \tau^{-2} \log \tau + \beta^2 \tau^{-3}(\log \tau)^2 - \beta^2 \tau^{-3} \log \tau$$
$$- b(\sigma - b\alpha_\pm^2)\tau^{-3} + o(\tau^{-3})], \qquad (8)$$
$$y = \operatorname{Im} w(\tau \pm i\pi/2)$$
$$= -b^2[\alpha_\pm \tau^{-2} + 2\beta\alpha_\pm \tau^{-3} \log \tau + b\alpha_\pm^{(1)} \tau^{-3} + o(\tau^{-3})].$$
Taking the inverse of (8), we obtain
$$\tau = \tau_\pm(x) = bx^{-1} + \beta \log \frac{1}{x} + \beta \log b - (\sigma - b\alpha_\pm^2)x + o(x). \qquad (9)$$
Finally, straightforward calculations yield
$$|w'(\tau \pm i\pi/2)| = b\tau^{-2} + 2b\beta\tau^{-3} \log \tau - b\beta\tau^{-3} + 3b\beta^2 \tau^{-4}(\log \tau)^2$$
$$- 5b\beta^2 \tau^{-4} \log \tau - b(3b\sigma + b^2\alpha_\pm^2 - \beta^2)\tau^{-4} + o(\tau^{-4}),$$
$$\tau \to +\infty. \qquad (10)$$

§2. Asymptotics of a conformal mapping of a domain with inward peak onto the upper half-plane

Let Ω be the domain with inward peak described in the introduction. By the transformation $u+iv = (x+iy)^{1/2}$, we map the plane, with the slit along the positive part of the real axis, onto the upper half-plane. Then the image of Ω is the domain $\widetilde{\Omega}$ bounded in the neighbourhood of the origin by the graph of the function $v = v(u)$ with the expansion

$$v(u) = \beta_\pm u^3 + \beta_\pm^{(1)} u^4 + \beta_\pm^{(2)} u^5 + o(u^5), \qquad u \to \pm 0, \qquad (11)$$

where $\beta_\pm = \kappa_\pm''(0)/4$, $\beta_\pm^{(1)} = -\kappa_\pm'''(0)/12$.

We denote by $\widetilde{\theta}$ the conformal mapping of the upper half-plane $\zeta = \xi + i\eta$ onto $\widetilde{\Omega}$ normalized by the condition $\widetilde{\theta}(0) = 0$. From the Kellogg theorem [7, p. 426] it follows that the function $u(\eta) = \operatorname{Re} \widetilde{\theta}(\eta + i0)$ has the representation

$$u(\eta) = a_\theta \eta + b_\theta \eta^2 + O(|\eta|^{2+\varepsilon}), \qquad 0 < \varepsilon < 1, \qquad (12)$$

for $\eta \to 0$. Substituting (12) into (11), we obtain

$$v(\eta) = \operatorname{Im} \widetilde{\theta}(\eta + i0) = a_\theta^3 \beta_\pm \eta^3 + (3a_\theta^2 b_\theta \beta_\pm + a_\theta^4 \beta_\pm^{(1)}) \eta^4 + O(|\eta|^{4+\varepsilon}), \qquad (13)$$
$$0 < \varepsilon < 1.$$

The derivative $\widetilde{\theta}'$ belongs to the Hardy space H_1 in the upper half-plane; therefore $u' = -H(v') + \text{const}$, where H is the Hilbert transform.

LEMMA 2. *Suppose ϕ has the expansion*

$$\phi(t) = a_\pm t^3 + b_\pm t^4 + O(|t|^{4+\varepsilon}), \qquad 0 < \varepsilon < 1, \ t \to \pm 0,$$

which can be differentiated once, and let ϕ' be summable on the axis $(-\infty, \infty)$. Then the function $H(\phi')$ admits the representation

$$H(\phi')(\eta) = -\frac{1}{\pi} \int_{-\infty}^{\infty} \phi'(t) \frac{dt}{t} - \frac{\eta}{\pi} \int_{-\infty}^{\infty} \phi'(t) \frac{dt}{t^2} + \frac{3(a_+ - a_-)}{\pi} \eta^2 \log |\eta|$$
$$- \frac{\eta^2}{\pi} \left| \int_{-\infty}^{\infty} \phi'(t) \frac{dt}{t^3} \right| + \frac{4}{\pi}(b_+ - b_-)\eta^3 \log |\eta| - \frac{\eta^3}{\pi} \left| \int_{-\infty}^{\infty} \phi'(t) \frac{dt}{t^4} \right| + O(|\eta|^{3+\varepsilon}).$$

The proof of this lemma is given in supplement 2° of §4.

We set $\phi = v$ in the lemma. By (12) and (13) we have

$$\widetilde{\theta}(\eta + i0) = a_\theta \eta + b_\theta \eta^2 - a_\theta^3(\beta_+ - \beta_-)\pi^{-1} \eta^3 \log |\eta| + (d + ia_\theta^3 \beta_\pm)\eta^3$$
$$- f\eta^4 \log |\eta| + (h + ia_\theta^4 \beta_\pm^{(1)})\eta^4 + O(|\eta|^{4+\varepsilon}).$$

We make the coefficient b_θ vanish by a conformal mapping of the upper half-plane onto itself. After simple transformations we obtain the following expansion of the conformal mapping of the upper half-plane onto Ω:

$$\theta(\eta + i0) = a_\theta^2 \eta^2 - 2\pi^{-1} a_\theta^4 (\beta_+ - \beta_-)\eta^4 \log |\eta| + 2a_\theta(d + ia_\theta^3 \beta_\pm)\eta^4$$
$$- 2a_\theta f \eta^5 \log |\eta| + 2a_\theta(h + ia_\theta^4 \beta_\pm^{(1)})\eta^5 + O(|\eta|^{5+\varepsilon}),$$

which can be differentiated twice. In particular,
$$\begin{aligned}x = \operatorname{Re}\theta(\eta + i0) \\ = a_\theta^2\eta^2 - 2\pi^{-1}a_\theta^4(\beta_+ - \beta_-)\eta^4\log|\eta| + 2a_\theta d\eta^4 \\ - 2a_\theta f\eta^5\log|\eta| + 2a_\theta h\eta^5 + O(|\eta|^{5+\varepsilon}), \quad 0 < \varepsilon < 1.\end{aligned} \quad (14)$$

Taking the inverse of (14), we find
$$\begin{aligned}\eta = \eta_\pm(x) = \pm(a_\theta^{-1}x^{1/2} - \pi^{-1}a_\theta^{-1}(\beta_+ - \beta_-)x^{3/2}\log(a_\theta x^{-1/2}) - da_\theta^{-4}x^{3/2}) \\ \mp fa_\theta^{-5}x^2\log(a_\theta x^{-1/2}) \mp ha_\theta^{-5}x^2 + O(x^{2+\varepsilon}), \quad 0 < \varepsilon < 1/2.\end{aligned} \quad (15)$$

Finally, by straightforward calculations, we find the asymptotics of the derivative of the modulus of the conformal mapping
$$\begin{aligned}|\theta'(\eta + i0)| = 2a_\theta^2|\eta| - 8\pi^{-1}a_\theta^4(\beta_+ - \beta_-)|\eta|^3\log|\eta| \\ + (8a_\theta d - 2\pi^{-1}a_\theta^4(\beta_+ - \beta_-))|\eta|^3 + O(|\eta|^{3+\varepsilon}), \quad 0 < \varepsilon < 1.\end{aligned} \quad (16)$$

§3. Unique solvability and asymptotics of solution of the integral equation of the Neumann problem

1°. Let Ω have an outward or an inward peak.

LEMMA 3. *The homogeneous integral equation of the exterior Neumann problem has only the trivial solution in the class $\mathfrak{M}$.*

PROOF. Let $\gamma \in \mathfrak{M}$ be a solution of the homogeneous integral equation (1). Then the function
$$V(z) = \int_\Gamma \gamma(q)\log\frac{1}{r}\,ds_q, \quad r = |z - q|,$$
is a solution of the interior Neumann problem with the boundary function $2\pi\gamma$. Using the integral representation for harmonic functions, we obtain
$$\begin{aligned}V(z) &= \frac{1}{2\pi}\int_\Gamma\left(\log\frac{1}{\gamma}\frac{\partial}{\partial n_q}V(q) - V(q)\frac{\partial}{\partial n_q}\log\frac{1}{r}\right)ds_q \\ &= V(z) - \frac{1}{2\pi}\int_\Gamma V(q)\frac{\partial}{\partial n_q}\log\frac{1}{r}\,ds_q, \quad z \in \Omega.\end{aligned}$$

This implies that $V(p)$, $p \in \Gamma$, is a solution of the homogeneous integral equation of the internal Dirichlet problem. This equation has only the trivial solution in the class of bounded functions. So V vanishes on the boundary of the domain. Since the potential $V(z)$ is continuous in $\overline{\Omega}$ and harmonic in Ω, $V = 0$ in Ω. Therefore, the limit values of the normal derivative of V are equal to zero. From the equality $\partial V/\partial n = 2\pi\gamma$ on Γ it follows that $\gamma = 0$ on Γ. The lemma is proved.

2°. Asymptotics of the solution (outward peak). Let Ω be a domain with an outward peak and let ω be the conformal mapping of the upper half-plane onto Ω^c normalized by the condition

$$\omega(\eta + i0) = a_\omega^2 \eta^2 + O(\eta^4 \log|\eta|).$$

The inverse mapping of ω will be denoted by ζ.

THEOREM 1. I. *If ψ belongs to the class $\mathfrak{N}_\mu$, $0 < \mu < 1$, and*

$$\int_\Gamma \psi(q)\, ds_q = 0, \quad \int_\Gamma \psi(q) \operatorname{Re}(1/\zeta(q))\, ds_q = 0, \quad \phi_+(0) - \phi_-(0) \neq 0,$$

then the integral equation (1) is uniquely solvable in the class $\mathfrak{M}$ and its solution can be represented in the form:

(a) *for $0 < \mu < 1/2$,*

$$\gamma(x) = \mp \frac{1}{\pi} \frac{\phi_+(0) - \phi_-(0)}{\kappa_+''(0) - \kappa_-''(0)} \frac{\tan \pi\mu}{\mu + 1} x^{\mu - 1} + O(x^{-1/2});$$

(b) *for $\mu = 1/2$,*

$$\gamma(x) = \mp \frac{4}{3\pi^2} \frac{\phi_+(0) - \phi_-(0)}{\kappa_+''(0) - \kappa_-''(0)} \frac{\log(a_\omega x^{-1/2}) + 1/3}{x^{1/2}}$$

$$\mp \frac{2}{3\pi^2} \frac{a_\omega^{-3} x^{-1/2}}{\kappa_+''(0) - \kappa_-''(0)} \left| \int_\Gamma \psi(q) \operatorname{Re}\left(\frac{1}{\zeta(q)}\right)^3 ds_q + O(x^{1/2} (\log x)^2) \right|;$$

(c) *for $1/2 < \mu < 1$,*

$$\gamma(x) = \mp \frac{2}{3\pi^2} \frac{a_\omega^{-3} x^{-1/2}}{\kappa_+''(0) - \kappa_-''(0)} \int_\Gamma \psi(q) \operatorname{Re}\left(\frac{1}{\zeta(q)}\right)^3 ds_q$$

$$\mp \frac{1}{\pi} \frac{\phi_+(0) - \phi_-(0)}{\kappa_+''(0) - \kappa_-''(0)} \frac{\tan \pi\mu}{\mu + 1} x^{\mu - 1} + O(x^{1/2} \log x).$$

II. *If ψ belongs to the class $\mathfrak{N}_\mu$, $0 \leq \mu < 1$, and*

$$\int_\Gamma \psi(q)\, ds_q = 0, \quad \int_\Gamma \psi(q) \operatorname{Re}(1/\zeta(q))\, ds_q = 0, \quad \phi_+(0) - \phi_-(0) = 0,$$

then the integral equation (1) is uniquely solvable in the class $\mathfrak{M}$ and its solution can be represented in the form:

(a) *for $\mu = 0$,*

$$\gamma(x) = \mp \frac{2}{3\pi^2 a_\omega^3} \frac{x^{-1/2}}{\kappa_+''(0) - \kappa_-''(0)} \int_\Gamma \psi(q) \operatorname{Re}\left(\frac{1}{\zeta(q)}\right)^3 ds_q$$

$$- \frac{\phi_+(0) + \phi_-(0)}{4\pi} + O(x^\varepsilon), \quad 0 < \varepsilon < 1/2;$$

(b) *for* $0 < \mu < 1/2$,

$$\gamma(x) = \mp \frac{2}{3\pi^2 a_\omega^3} \frac{x^{-1/2}}{\kappa_+''(0) - \kappa_-''(0)} \int_\Gamma \psi(q) \operatorname{Re}\left(\frac{1}{\zeta(q)}\right)^3 ds_q$$
$$- \frac{1}{\pi}\left(\frac{\phi_+(0) + \phi_-(0)}{4} \pm \frac{\phi_+'(0) - \phi_-'(0)}{\kappa_+''(0) - \kappa_-''(0)} \frac{\tan \pi\mu}{\mu + 2}\right) x^\mu + O(x^{1/2} \log x);$$

(c) *for* $\mu = 1/2$,

$$\gamma(x) = \mp \frac{2}{3\pi^2 a_\omega^3} \frac{x^{-1/2}}{\kappa_+''(0) - \kappa_-''(0)} \int_\Gamma \psi(q) \operatorname{Re}\left(\frac{1}{\zeta(q)}\right)^3 ds_q + O(x^{1/2} \log x),$$

and

$$\gamma(x) = \pm \frac{E x^{1/2} \log(a_\omega x^{-1/2})}{\kappa_+''(0) - \kappa_-''(0)}$$
$$- \left(\frac{\phi_+(0) + \phi_-(0)}{4\pi} \mp \frac{F}{\kappa_+''(0) - \kappa_-''(0)}\right) x^{1/2} + O(x^{1/2+\varepsilon}),$$
$$0 < \varepsilon < 1/2,$$

if

$$\int_\Gamma \psi(q) \operatorname{Re}(1/\zeta(q))^3 ds_q = 0,$$

where

$$E = \frac{2f}{\pi^2 a_\omega^6} \int_\Gamma \psi(q) \operatorname{Re}\left(\frac{1}{\zeta(q)}\right)^2 ds_q - \frac{4}{5\pi^2}(\phi_+'(0) - \phi_-'(0)),$$

$$F = \frac{2h}{\pi^2 a_\omega^6} \int_\Gamma \psi(q) \operatorname{Re}\left(\frac{1}{\zeta(q)}\right)^2 ds_q - \frac{4}{25\pi^2}(\phi_+'(0) - \phi_-'(0))$$
$$- \frac{2}{5\pi^2 a_\omega^5} \overline{\int_\Gamma \psi(q) R\left(\frac{1}{\zeta(q)}\right)^5 ds_q};$$

(d) *for* $1/2 < \mu < 1$,

$$\psi(x) = \mp \frac{2}{3\pi^2 a_\omega^3} \frac{x^{-1/2}}{\kappa_+''(0) - \kappa_-''(0)} \int_\Gamma \psi(q) \operatorname{Re}\left(\frac{1}{\zeta(q)}\right)^3 ds_q + O(x^{1/2} \log x),$$

and

$$\gamma(x) = \pm \frac{E x^{1/2} \log(a_\omega x^{-1/2})}{\kappa_+''(0) - \kappa_-''(0)} \pm \frac{F x^{1/2}}{\kappa_+''(0) - \kappa_-''(0)}$$
$$- \left(\frac{\phi_+(0) + \phi_-(0)}{4\pi} \pm \frac{1}{\pi}\frac{\phi_+'(0) - \phi_-'(0)}{\kappa_+''(0) - \kappa_-''(0)} \frac{\tan \pi\mu}{\mu + 2}\right) x^\mu + O(x),$$

if

$$\int_\Gamma \psi(q) \operatorname{Re}(1/\zeta(q))^3 ds_q = 0,$$

where

$$E = \frac{2f}{\pi^2 a_\omega^6} \int_\Gamma \psi(q) \operatorname{Re}\left(\frac{1}{\zeta(q)}\right)^2 ds_q,$$

$$F = \frac{2h}{\pi^2 a_\omega^6} \int_\Gamma \psi(q) R\left(\frac{1}{\zeta(q)}\right)^2 ds_q - \frac{2a_\omega^{-5}}{5\pi^2} \int_\Gamma \psi(q) \operatorname{Re}\left(\frac{1}{\zeta(q)}\right)^5 ds_q.$$

Here f and h are coefficients in the asymptotic expansion of the conformal mapping ω.

PROOF. 1. We introduce the solution v^e of the external Neumann problem

$$\begin{cases} \Delta v^e = 0 & \text{in } \Omega^c, \\ \partial v^e/\partial n = \psi & \text{on } \Gamma = \partial\Omega, \end{cases}$$

which tends to zero at infinity. Let u^i be a harmonic extension of v^e from the boundary Γ to Ω. From the asymptotic representation of the normal derivative $\partial u^i/\partial n$ (see subsections 2–4 of the proof), it follows that, in all the cases considered in the theorem, $\partial u^i/\partial n$ belongs to $\mathfrak{M}$.

It is well known that

$$v^e(p) = \frac{1}{2\pi} \int_\Gamma \left(\frac{\partial u^i}{\partial n} - \psi\right) \log \frac{1}{r} ds_q, \qquad r = |p-q|, \quad p \in \Gamma.$$

This and Lemma 3 imply that the density of the potential

$$\gamma = (2\pi)^{-1}(\partial u^i/\partial n - \psi) \tag{17}$$

is the only solution of equation (1).

2. Since the proof of all the cases of the theorem follows the same pattern we consider only one of them in detail. Let $\psi \in \mathfrak{N}_{1/2}$ and let

$$\psi_\pm(x) = x^{1/2}(A + B_\pm x + C_\pm x^2 + \cdots)$$

(the case IIc). We set

$$\tilde{\psi}(\eta) = \psi(\omega(\eta + i0))|\omega'(\eta + i0)|,$$
$$\tilde{\psi}_0(\eta) = (\tilde{\psi}(\eta) - \tilde{\psi}(-\eta))/2,$$
$$\tilde{\psi}_e(\eta) = (\tilde{\psi}(\eta) + \tilde{\psi}(-\eta))/2.$$

By (14) and (16), we have

$$\Psi(\eta) = -\int_\eta^\infty \tilde{\psi}_e(t)\,dt = \pm\overline{A}|\eta|^3 \pm \overline{B}|\eta|^5 \log\frac{1}{|\eta|}$$
$$\pm \overline{C}|\eta|^5 + O(|\eta|^{5+\varepsilon}), \qquad 0 < \varepsilon < 1.$$

The solution of the Neumann problem in the upper half-plane $\{(\eta, \xi): \xi > 0\}$ with the given boundary function $\overline{\psi}_e$ is given by the formula

$$\tilde{v}_e(\eta + i0) = \frac{1}{\pi} \int_{-\infty}^\infty \tilde{\psi}_e(t) \log|\eta - t|\,dt$$
$$= \frac{1}{\pi} \int_{-\infty}^\infty \tilde{\psi}_e(t) \log|t|\,dt + \frac{\eta}{\pi} \int_{-\infty}^\infty \Psi(t) \frac{t^{-1}\,dt}{\eta - t}.$$

After calculating the asymptotics of the last integral, we obtain

$$\tilde{v}_e(\eta) - \tilde{v}_e(0) = -\frac{\eta^2}{2\pi}\int_{-\infty}^{\infty}\tilde{\psi}(u)u^{-2}\,du + \overline{A}_1|\eta|^3 + \overline{B}_1|\eta|^4$$
$$+ \overline{C}_1|\eta|^5\log\frac{1}{|\eta|} + \overline{D}_1|\eta|^5 + O(|\eta|^{5+\varepsilon}).$$

In a similar way, we find the asymptotic expansion of the solution of the problem

$$\begin{cases} \Delta\tilde{v}_0 = 0 & \text{in the upper half-plane,} \\ \partial\tilde{v}_0/\partial n = \tilde{\psi}_0 & \text{on the boundary of the upper half-plane.} \end{cases}$$

As a result we have

$$\tilde{v}_0(\eta) = \pm\frac{|\eta|}{\pi}\int_{-\infty}^{\infty}\tilde{\psi}(u)\frac{du}{u} \mp \frac{|\eta|^3}{3\pi}\int_{-\infty}^{\infty}\tilde{\psi}(u)\frac{du}{u^3} \pm \frac{2}{5\pi}(B_+ - B_-)a_\omega^5|\eta|^5\log|\eta|$$
$$\mp \frac{|\eta|^5}{5\pi}\left|\int_{-\infty}^{\infty}\tilde{\psi}(u)\frac{du}{u}\right. + O(\eta^6\log|\eta|).$$

For the solution γ constructed in subsection 1 to belong to the class $\mathfrak{M}$, we must assume that

$$\int_{-\infty}^{\infty}\tilde{\psi}(u)\frac{du}{u} = \int_{\Gamma}\psi(q)\operatorname{Re}\left(\frac{1}{\zeta(q)}\right)ds_q = 0.$$

Making the inverse substitution (see (15)), we obtain

$$\tilde{v}_0(\eta_\pm(x)) = \mp\frac{x^{3/2}}{3\pi a_\omega^3}\int_{-\infty}^{\infty}\tilde{\psi}(u)\frac{du}{u^3} + O(x^{5/2}\log x),$$

if

$$\int_{-\infty}^{\infty}\tilde{\psi}(u)u^{-3}\,du \neq 0,$$

and

$$\tilde{v}_0(\eta_\pm(x)) = \frac{2}{5\pi}(B_+ - B_-)x^{5/2}\left[\log(a_\omega x^{-1/2}) + \frac{1}{5}\right]$$
$$\mp \frac{x^{5/2}}{5\pi a_\omega^5}\left|\int_{-\infty}^{\infty}\tilde{\psi}(u)\frac{du}{u^5}\right. + O(x^{5/2+\varepsilon}), \qquad 0 < \varepsilon < 1/2,$$

if

$$\int_{-\infty}^{\infty}\tilde{\psi}(u)u^{-3}\,du = 0.$$

In a similar way, we find

$$\tilde{v}_e(\eta_\pm(x)) = \tilde{v}_e(\eta(0)) - \frac{x}{2\pi a_\omega^2}\int_{-\infty}^{\infty}\tilde{\psi}(u)\frac{du}{u^2}$$

$$= A^{\#}x^{3/2} + B^{\#}x^2\log\frac{1}{x} + C^{\#}x^2 + D^{\#}x^{5/2}\log\frac{1}{x}$$

$$\pm \frac{f}{\pi a_\omega^6}\int_{-\infty}^{\infty}\tilde{\psi}(u)\frac{du}{u^2}x^{5/2}\log(a_\omega x^{-1/2})$$

$$\pm \frac{h}{\pi a_\omega^6}\int_{-\infty}^{\infty}\tilde{\psi}(u)\frac{du}{u^2}x^{5/2} + O(x^{5/2+\varepsilon}), \qquad 0 < \varepsilon < 1/2,$$

where f and h are coefficients in the asymptotic expansion of the conformal mapping ω (see §2).

The asymptotic expansion of the function v_e can be derived from the equality

$$v^e(x) = \tilde{v}_e(\eta_\pm(x)) + \tilde{v}_0(\eta_\pm(x)).$$

3. Now we find the asymptotics of the normal derivative $\partial u^i/\partial n$ of the harmonic extension u^i of the function v^e from the boundary Γ to Ω. The function $w(\log(i\zeta^{-1}))$, where w is the conformal mapping of the strip T onto Ω (see (7)), maps the upper half-plane $\zeta = \eta + i\xi$ onto Ω. The normal derivative of $u^i(w(\log i\zeta^{-1}))$, which equals

$$(\partial u^i/\partial n)(w(\log i\zeta^{-1}))|(w(\log i\zeta^{-1}))'|,$$

is the Hilbert transform of the function $\partial v^e(w(\log i\eta^{-1}))/\partial \eta$. After calculating the asymptotics of the Hilbert transform and making the inverse substitution (see (9), (10)), we obtain

$$\frac{\partial u^i}{\partial \eta}(x) = \mp\frac{2bx^{-1/2}}{3\pi^2 a_\omega^3}\int_{-\infty}^{\infty}\tilde{\psi}(u)\frac{du}{u^3} + O(x^{1/2}\log x),$$

if

$$\int_{-\infty}^{+\infty}\tilde{\psi}(u)u^{-3}\,du \neq 0$$

and

$$\frac{\partial u^i}{\partial n}(x) = \pm bEx^{1/2}\log(a_\omega x^{-1/2}) \pm bFx^{1/2} + O(x^{1/2+\varepsilon}), \qquad 0 < \varepsilon < 1/2,$$

if

$$\int_{-\infty}^{\infty}\tilde{\psi}(u)u^{-3}\,du = 0,$$

where

$$E = \frac{2f}{\pi^2 a_\omega^6}\int_{-\infty}^{\infty}\tilde{\psi}(u)\frac{du}{u^2} - \frac{4}{5\pi^2}(B_+ - B_-);$$

$$F = \frac{2h}{\pi^2 a_\omega^6}\int_{-\infty}^{\infty}\tilde{\psi}(u)\frac{du}{u^2} - \frac{4}{25\pi^2}(B_+ - B_-) - \frac{2}{5\pi^2 a_\omega^5}\left|\int_{-\infty}^{\infty}\tilde{\psi}(u)\frac{du}{u^5}\right..$$

Substituting the asymptotic expansions of the functions $\partial u^i/\partial n$ and ψ into (17) we find the required representation for the solution of the equation (1) in the case II(c).

4. In the remaining cases of the theorem the solution of the exterior Neumann problem has the representations:

in the case I(a),

$$v^e(x) - v^e(0) = B_1 x \pm \frac{(A_+ - A_-)x^{\mu+1}}{\pi(\mu+1)} \int_0^\infty \frac{u^{2\mu}}{1-u^2} du + B_2 x^{\mu+1} + O(x^{3/2});$$

in the case I(b),

$$v^e(x) - v^e(0) = B_1 x \mp \frac{2(A_+ - A_-)}{3\pi}(\log(a_\omega x^{-1/2}) + 1/3)x^{3/2}$$

$$\mp \frac{x^{3/2}}{3\pi a_\omega^3} \overline{\left| \int_\Gamma \psi(q) \operatorname{Re}\left(\frac{1}{\zeta(q)}\right)^3 ds_q \right|} + B_2 x^{3/2} + O(x^2 \log x);$$

in the case I(c),

$$v^e(x) - v^e(0) = B_1 x \mp \frac{x^{3/2}}{3\pi a_\omega^3} \int_\Gamma \psi(q) \operatorname{Re}\left(\frac{1}{\zeta(q)}\right)^3 ds_q + B_2 x^{3/2}$$

$$\pm \frac{x^{\mu+1}}{\pi(\mu+1)}(A_+ - A_-) \int_0^\infty \frac{u^{2\mu-2}}{1-u^2} du + O(x^2 \log x);$$

in the case II(a),

$$v^e(x) - v^e(0) = B_1 x \log x + B_2 x \mp \frac{x^{3/2}}{3\pi a_\omega^3} \int_\Gamma \psi(q) \operatorname{Re}\left(\frac{1}{\zeta(q)}\right)^3 ds_q$$

$$+ B_3 x^2 + O(x^{2+\varepsilon}), \quad 0 < \varepsilon < 1/2;$$

in the case II(b),

$$v^e(x) - v^e(0) = B_1 x + B_2 x^{\mu+1} \mp \frac{x^{3/2}}{3\pi a_\omega^3} \int_\Gamma \psi(q) \operatorname{Re}\left(\frac{1}{\zeta(q)}\right)^3 ds_q$$

$$+ B_3 x^2 \log \frac{1}{x} + B_4 x^2 + B_5 x^{\mu+2} \log \frac{1}{x} + B_6 x^{\mu+2}$$

$$\pm \frac{B_+ - B_-}{\pi(\mu+2)} x^{\mu+2} \int_0^\infty \frac{u^{2\mu}}{1-u^2} du + O(x^{5/2} \log x);$$

in the case II(d),

$$v^e(x) - v^e(0) = B_1 x \mp \frac{x^{3/2}}{3\pi a_\omega^3} \int_\Gamma \psi(q) \operatorname{Re}\left(\frac{1}{\zeta(q)}\right)^3 ds_q$$

$$+ B_2 x^{\mu+1} + B_3 x^2 \log \frac{1}{x} + B_4 x^2 + O(x^{5/2} \log x),$$

if

$$\int_\Gamma \psi(q) \operatorname{Re}\left(\frac{1}{\zeta(q)}\right)^3 ds_q \neq 0,$$

and

$$v^e(x) - v^e(0) = B_1 x + B_2 x^{\mu+1} + B_3 x^2 \log \frac{1}{x} + B_4 x^2 \pm E x^{5/2} \log(a_\omega x^{-1/2})$$
$$\pm F x^{5/2} \pm \frac{B_+ - B_-}{\pi(\mu+2)} x^{\mu+2} \int_0^\infty \frac{u^{2\mu-2}}{1-u^2} du + B_5 x^{\mu+2} \log \frac{1}{x}$$
$$+ B_6 x^{\mu+2} + O(x^{5/2+\varepsilon}), \qquad \mu - 1/2 < \varepsilon < 1/2,$$

if

$$\int_\Gamma \psi(q) \operatorname{Re} \left(\frac{1}{\zeta(q)} \right)^3 ds_q = 0,$$

where

$$E = \frac{f}{\pi a_\omega^6} \int_\Gamma \psi(q) \operatorname{Re} \left(\frac{1}{\zeta(q)} \right)^2 ds_q;$$

$$F = \frac{h}{\pi a_\omega^6} \int_\Gamma \psi(q) \operatorname{Re} \left(\frac{1}{\zeta(q)} \right)^2 ds_q - \frac{1}{5\pi a_\omega^5} \int_\Gamma \psi(q) \operatorname{Re} \left(\frac{1}{\zeta(q)} \right)^5 ds_q.$$

In the same way as in subsection 2, we find the asymptotics of the normal derivative $\partial u^i / \partial n$ of the harmonic extension u^i of the function v^e from Γ to Ω. We have:

in the case I(a),

$$\frac{\partial u^i}{\partial n}(x) = \mp 2 \frac{\phi_+(0) - \phi_-(0)}{\kappa''_+(0) - \kappa''_-(0)} \frac{\tan \pi \mu}{\mu + 1} x^{\mu-1} + O(x^{-1/2}),$$

in the case I(b),

$$\frac{\partial u^i}{\partial n}(x) = \mp \frac{4}{3\pi} \frac{\phi_+(0) - \phi_-(0)}{\kappa''_+(0) - \kappa''_-(0)} x^{-1/2} (\log(a_\omega x^{-1/2}) + 1/3)$$
$$\mp \frac{4}{3\pi} \frac{a_\omega^{-3} x^{-1/2}}{\kappa''_+(0) - \kappa''_-(0)} \left| \int_\Gamma \psi(q) \operatorname{Re} \left(\frac{1}{\zeta(q)} \right)^3 ds_q \right| + O(x^{1/2}(\log x)^2),$$

in the case I(c),

$$\frac{\partial u^i}{\partial n}(x) = \mp \frac{4}{3\pi a_\omega^3} \frac{x^{-1/2}}{\kappa''_+(0) - \kappa''_-(0)} \int_\Gamma \psi(q) \operatorname{Re} \left(\frac{1}{\zeta(q)} \right)^3 ds_q$$
$$\mp 2 \frac{\phi_+(0) - \phi_-(0)}{\kappa''_+(0) - \kappa''_-(0)} \frac{\tan \pi \mu}{\mu + 1} x^{\mu-1} + O(x^{1/2} \log x),$$

in the case II(a),

$$\frac{\partial u^i}{\partial n}(x) = \mp \frac{4}{3\pi a_\omega^3} \frac{x^{-1/2}}{\kappa''_+(0) - \kappa''_-(0)} \int_\Gamma \psi(q) \operatorname{Re} \left(\frac{1}{\zeta(q)} \right)^3 ds_q + O(x^\varepsilon),$$
$$0 < \varepsilon < 1/2;$$

in the case II(b),

$$\frac{\partial u^i}{\partial n}(x) = \mp \frac{4}{3\pi a_\omega^3} \frac{x^{-1/2}}{\kappa_+''(0) - \kappa_-''(0)} \int_\Gamma \psi(q) \operatorname{Re}\left(\frac{1}{\zeta(q)}\right)^3 ds_q$$
$$\mp 2 \frac{\phi_+'(0) - \phi_-'(0)}{\kappa_+''(0) - \kappa_-''(0)} \frac{\tan \pi\mu}{\mu+2} x^\mu + O(x^{1/2} \log x);$$

in the case II(d),

$$\frac{\partial u^i}{\partial n}(x) = \mp \frac{4}{3\pi a_\omega^3} \frac{x^{-1/2}}{\kappa_+''(0) - \kappa_-''(0)} \int_\Gamma \psi(q) \operatorname{Re}\left(\frac{1}{\zeta(q)}\right)^3 ds_q + O(x^{1/2} \log x),$$

if

$$\int_\Gamma \psi(q) \operatorname{Re}\left(\frac{1}{\zeta(q)}\right)^3 ds_q \neq 0$$

and

$$\frac{\partial u^i}{\partial n}(x) = \pm \frac{2\pi(E \log(a_\omega x^{-1/2}) + F)}{\kappa_+''(0) - \kappa_-''(0)} x^{1/2} \mp 2 \frac{\phi_+'(0) - \phi_-'(0)}{\kappa_+''(0) - \kappa_-''(0)} \frac{\tan \pi\mu}{\mu+2} x^\mu + O(x),$$

if

$$\int_\Gamma \psi(q) \operatorname{Re}\left(\frac{1}{\zeta(q)}\right)^3 ds_q = 0,$$

where

$$E = \frac{2f}{\pi^2 a_\omega^6} \int_\Gamma \psi(q) \operatorname{Re}\left(\frac{1}{\zeta(q)}\right)^2 ds_q;$$

$$F = \frac{2h}{\pi^2 a_\omega^6} \int_\Gamma \psi(q) \operatorname{Re}\left(\frac{1}{\zeta(q)}\right)^2 ds_q - \frac{2a_\omega^{-5}}{5\pi^2} \int_\Gamma \psi(q) \operatorname{Re}\left(\frac{1}{\zeta(q)}\right)^5 ds_q.$$

By substituting the asymptotic expansions of the functions $\partial u/\partial n$ and ψ into (17), we obtain the required representations for the solution of the equation (1). The theorem is proved.

REMARK. In the case $\mu = 0$, $\phi_+(0) \neq \phi_-(0)$, one can seek the solution of the exterior Neumann problem as the sum of the harmonic function

$$u_1(x, y) = \frac{1}{2}(\phi_+(0) - \phi_-(0)) \operatorname{Im} \frac{zz_0}{z_0 - z},$$
$$z = x + iy \in \Omega^e, \qquad z_0 \in \Omega,$$

and the simple layer potential with density γ which is the solution of integral equation (1) with the right-hand side

$$-\pi^{-1} \psi_1(z) = -\pi^{-1} \left(\psi(z) - \left(\frac{\partial u_1}{\partial n}\right)(z) \right), \qquad z \in \Gamma.$$

Since

$$(\partial u_1/\partial n)(z) \sim \pm(\phi_+(0) - \phi_-(0))/2, \qquad z \in \Gamma_\pm,$$

the function ψ_1 has the following representation as $x \to 0$:

$$\psi_1(x) = \frac{1}{2}(\phi_+(0) + \phi_-(0)) + B_\pm^\# x + \cdots .$$

By applying the Green function and taking into account that $\operatorname{Re}(1/\zeta(z))$ is the solution of the exterior Neumann problem with zero boundary condition on $\Gamma\setminus\{0\}$, we obtain

$$\int_\Gamma \frac{\partial u_1}{\partial n}(q) \operatorname{Re}\left(\frac{1}{\zeta(q)}\right) ds_q = \int_\Gamma u_1(q) \frac{\partial}{\partial n} \operatorname{Re}\left(\frac{1}{\zeta(q)}\right) ds_q = 0.$$

Therefore, according to II(a) of Theorem 1, the integral equation for the density of the simple layer potential is uniquely solvable in the class $\mathfrak{M}$ and the solution γ can be represented in the form

$$\gamma(x) = \mp \frac{2x^{-1/2}}{3\pi^2 a_\omega^3(\kappa_+''(0) - \kappa_-''(0))} \int_\Gamma \psi_1(q) \operatorname{Re}\left(\frac{1}{\zeta(q)}\right)^3 ds_q$$
$$- \frac{\phi_+(0) + \phi_-(0)}{4\pi} + O(x^\varepsilon), \qquad 0 < \varepsilon < 1/2.$$

Consider the case $-1 < \mu < 0$ and $\mu \neq -1/2$. We are looking for the solution of the exterior Neumann problem in the form of the sum of the simple layer potential

$$u_2(x, y) = \int_\Gamma \gamma(q) \log\frac{1}{r} ds_q$$

and the harmonic function

$$u_1(x, y) = \frac{\phi_+(0)}{\mu + 1} \operatorname{Im}\left(\frac{zz_0}{z_0 - z}\right)^{1+\mu} + \frac{\phi_-(0) + \phi_+(0)\cos 2\pi\mu}{(1+\mu)\sin 2\pi\mu} \operatorname{Re}\left(\frac{zz_0}{z_0 - z}\right)^{1+\mu}$$

(where $z = x + iy \in \Omega^c$ and z_0 is a fixed point of Ω), for which

$$(\partial u_1/\partial n)(z) = \phi_\pm(0)x^\mu + A_\pm^\# x^{\mu+1} + O(x^{\mu+2}), \qquad z \in \Gamma_\pm, \; x \to 0.$$

We choose γ in such a way that the limit values of the normal derivative of u_2 on Γ equal $\psi - \partial u_1/\partial n$. Then one can apply Theorem 1 to Eq. (1) with this right-hand side.

In the case $\mu = -1/2$ one should take $u_1(x, y)$ to be

$$u_1(x, y) = -\frac{2}{\pi}(\phi_+(0) - \phi_-(0)) \operatorname{Re}\left(\frac{zz_0}{z_0 - z}\right)^{1/2} \log i\sqrt{\frac{z_0 - z}{zz_0}}$$
$$+ (\phi_+(0) + \phi_-(0)) \operatorname{Im}\left(\frac{zz_0}{z_0 - z}\right)^{1/2},$$

with the normal derivative

$$(\partial u_1/\partial n)(z) = \phi_\pm(0)x^{-1/2} + A_\pm^\# x^{1/2} \log x$$
$$+ B_\pm^\# x^{1/2} + O(x^{3/2} \log x), \qquad z \in \Gamma_\pm.$$

In addition to Theorem 1 we note that if ψ in the right-hand side of (1) has the expansion
$$\psi(x) = A_\pm^\# x^{1/2} \log x + B_\pm^\# x^{1/2} + O(x^{3/2} \log x), \qquad x \to +0,$$
and satisfies the conditions
$$\int_\Gamma \psi(q) \, ds_q = 0, \qquad \int_\Gamma \psi(q) \operatorname{Re}(1/\zeta(q)) \, ds_q = 0,$$
then the integral equation (1) is uniquely solvable in $\mathfrak{M}$ and its solution can be represented in the form
$$\gamma(x) = \pm \frac{A_+^\# - A_-^\#}{3\pi^2} \frac{x^{-1/2}(\log x)^2}{\kappa_+''(0) - \kappa_-''(0)} + O(x^{-1/2} \log x).$$

3°. Asymptotics of the solution (inward peak). Suppose the domain has an inward peak. We choose the conformal mapping θ of the upper half-plane $\{(\eta, \xi) : \xi > 0\}$ onto Ω in such a way that
$$\theta(\eta + i0) = a_\theta^2 \eta^2 + O(\eta^4 \log|\eta|).$$

THEOREM 2. I. *If ψ belongs to the class $\mathfrak{N}_\mu$, $0 < \mu < 1$, and*
$$\int_\Gamma \psi(q) \, ds_q = 0, \qquad \phi_+(0) + \phi_-(0) \neq 0,$$

then the integral equation (1) is uniquely solvable in the class $\mathfrak{M}$ and its solution can be represented in the form:

(a) *for $0 < \mu < 1/2$,*
$$\gamma(x) = -\frac{1}{\pi} \frac{\phi_+(0) + \phi_-(0)}{\kappa_+''(0) - \kappa_-''(0)} \frac{\tan \pi\mu}{\mu + 1} x^{\mu-1} + O(x^{-1/2});$$

(b) *for $\mu = 1/2$,*
$$\gamma(x) = -\frac{4}{3\pi^2} \frac{\phi_+(0) + \phi_-(0)}{\kappa_+''(0) - \kappa_-''(0)} x^{-1/2} \log(a_\theta x^{-1/2})$$
$$- \frac{x^{-1/2}}{4\pi^2 a_\theta} \left| \int_{-\infty}^\infty (v^e \circ \theta)_\eta'(\eta + i0) \frac{d\eta}{\eta} \right| + O(1);$$

(c) *for $1/2 < \mu < 1$,*
$$\gamma(x) = -\frac{x^{-1/2}}{4\pi^2 a_\theta} \int_{-\infty}^\infty (v^e \circ \theta)_\eta'(\eta + i0) \frac{d\eta}{\eta}$$
$$- \frac{1}{\pi} \frac{\phi_+(0) + \phi_-(0)}{\kappa_+''(0) - \kappa_-''(0)} \frac{\tan \pi\mu}{\mu + 1} x^{\mu-1} + O(1).$$

II. *If ψ belongs to the class $\mathfrak{N}_\mu$, $0 \leq \mu < 1$, and*
$$\int_\Gamma \psi(q) \, ds_q = 0, \qquad \phi_+(0) + \phi_-(0) = 0,$$

then the integral equation (1) *is uniquely solvable in the class* $\mathfrak{M}$ *and its solution can be represented in the form:*

(a) for $0 \le \mu < 1/2$,

$$\gamma(x) = \frac{-x^{-1/2}}{4\pi^2 a_\theta} \int_{-\infty}^{\infty} (v^e \circ \theta)'_\eta(\eta + i0) \frac{d\eta}{\eta} \mp \frac{1}{4\pi^2 a_\theta^2} \int_{-\infty}^{\infty} (v^e \circ \theta)'_\eta(\eta + i0) \frac{d\eta}{\eta^2}$$
$$- \frac{1}{\pi} \left(\frac{\phi'_+(0) + \phi'_-(0)}{\kappa''_+(0) - \kappa''_-(0)} \frac{\tan \pi\mu}{\mu + 2} \pm \frac{\phi_+(0) - \phi_-(0)}{4} \right) x^\mu + O(x^{1/2} \log x),$$

(b) for $\mu = 1/2$,

$$\gamma(x) = -\frac{x^{-1/2}}{4\pi^2 a_\theta} \int_{-\infty}^{\infty} (v^e \circ \theta)'_\eta(\eta + i0) \frac{d\eta}{\eta} + O(1),$$

and

$$\gamma(x) = \mp \frac{1}{4\pi^2 a_\theta^2} \int_{-\infty}^{\infty} (v^e \circ \theta)'_\eta(\eta + i0) \frac{d\eta}{\eta^2} - \frac{4}{5\pi^2} \frac{\phi'_+(0) + \phi'_-(0)}{\kappa''_+(0) - \kappa''_-(0)} x^{1/2} \log(a_\theta x^{-1/2})$$
$$+ \left(\frac{-1}{4\pi^2 a_\theta^3} \left| \int_{-\infty}^{\infty} (v^e \circ \theta)'_\eta(\eta + i0) \frac{d\eta}{\eta^3} \mp \frac{\phi_+(0) - \phi_-(0)}{4\pi} \right) x^{1/2}$$
$$+ O(x(\log x)^2),$$

if

$$\int_{-\infty}^{\infty} (v^e \circ \theta)'_\eta(\eta + i0) \eta^{-1} d\eta = 0;$$

(c) for $1/2 < \mu < 1$,

$$\gamma(x) = -\frac{x^{-1/2}}{4\pi^2 a_\theta} \int_{-\infty}^{\infty} (v^e \circ \theta)'_\eta(\eta + i0) \frac{d\eta}{\eta} + O(1),$$

and

$$\gamma(x) = \mp \frac{1}{4\pi^2 a_\theta^2} \int_{-\infty}^{\infty} (v^e \circ \theta)'_\eta(\eta + i0) \frac{d\eta}{\eta^2} - \frac{x^{1/2}}{4\pi^2 a_\theta^3} \int_{-\infty}^{\infty} (v^e \circ \theta)'_\eta(\eta + i0) \frac{d\eta}{\eta^3}$$
$$- \frac{1}{\pi} \left(\frac{\phi'_+(0) + \phi'_-(0)}{\kappa''_+(0) - \kappa''_-(0)} \frac{\tan \pi\mu}{\mu + 1} \pm \frac{\phi_+(0) - \phi_-(0)}{4} \right) x^\mu + O(x \log x),$$

if

$$\int_{-\infty}^{\infty} (v^e \circ \theta)'_\eta(\eta + i0) \eta^{-1} d\eta = 0.$$

PROOF. 1. We introduce the solution v^e of the exterior Neumann problem with a boundary function ψ that tends to zero at infinity. By u^i we denote a harmonic extension of v^e from Γ to Ω. The argument in the proof below shows that, under the conditions of the theorem, $\partial u^i/\partial n$ belongs to $\mathfrak{M}$ in all the cases under consideration.

The equality

$$v^e(p) = \frac{1}{2\pi} \int_\Gamma \left(\frac{\partial u^i}{\partial n} - \psi\right) \log \frac{1}{r} ds_q, \qquad r = |p-q|, \ p \in \Gamma,$$

and Lemma 3 imply that the density

$$\gamma = \frac{1}{2\pi}\left(\frac{\partial u^i}{\partial n} - \psi\right) \tag{18}$$

of the potential is the only solution of the equation (1).

2. We consider one of the assertions of the theorem in detail. Suppose

$$\psi(x) = x^\mu(A_\pm + B_\pm x + \cdots), \qquad A_\pm = \phi_\pm(0),$$
$$\phi_+(0) + \phi_-(0) \neq 0, \qquad 1/2 < \mu < 1$$

(this is the case I(c)). Let $\tilde{v}^e$ denote the function $v^e \circ \omega$, where ω is the conformal mapping of the strip T onto Ω^c (see (7)). By (8) and (10) we have

$$(\partial \tilde{v}^e/\partial n)(\tau \pm i\pi/2) = \tilde{\psi}(\tau \pm i\pi/2)$$
$$= b^{\mu+1} A_\mp \tau^{-\mu-2} + \beta b^{\mu+1}(\mu+2)A_\mp \tau^{-\mu-3} \log \tau$$
$$+ b^{\mu+1}(B_\mp b - A_\mp \beta)\tau^{-\mu-3} + O(\tau^{-\mu-4}(\log \tau)^2).$$

We consider the following problems:

$$\Delta v = 0, \tag{19}$$

$$\frac{\partial v}{\partial n}\left(\tau \pm \frac{i\pi}{2}\right) = \pm \varepsilon_1 \tau^{-\mu-2} \pm \varepsilon_2 \tau^{-\mu-3} \log \tau \pm \varepsilon_3 \tau^{-\mu-3} + O(\tau^{-\mu-4}(\log \tau)^2);$$

$$\Delta v = 0, \tag{20}$$

$$\frac{\partial v}{\partial n}\left(\tau \pm \frac{i\pi}{2}\right) = \delta_1 \tau^{-\mu-2} + \delta_2 \tau^{-\mu-3} \log \tau + \delta_3 \tau^{-\mu-3} + O(\tau^{-\mu-4}(\log \tau)^2),$$

where n is the inward normal to the boundary of the strip T. For the solutions of the problems (19), (20), we have

$$v\left(\tau \pm \frac{i\pi}{2}\right) = \mp \frac{\pi}{2}\left(\varepsilon_1 \tau^{-\mu-2} + \varepsilon_2 \tau^{-\mu-3} \log \tau + \varepsilon_3 \tau^{-\mu-3}\right) + O\left(\tau^{-\mu-4}(\log \tau)^2\right),$$

$$v\left(\tau \pm \frac{i\pi}{2}\right) = \frac{2\delta_1}{\pi\mu(\mu+1)} \tau^{-\mu} + \frac{2\delta_2}{\pi(\mu+1)(\mu+2)} \tau^{-\mu-1} \log \tau + O(\tau^{-\mu-1}),$$

respectively. Hence $\tilde{v}^e$ admits the representation

$$\tilde{v}^e\left(\tau \pm \frac{i\pi}{2}\right) = \frac{b^{\mu+1}}{\pi\mu(\mu+1)}(A_+ + A_-)\tau^{-\mu} + \frac{\beta b^{\mu+1}}{\pi(\mu+1)}(A_+ + A_-)\tau^{-\mu-1} \log \tau$$
$$+ F\tau^{-\mu-1} + O(\tau^{-\mu-2}(\log \tau)^2).$$

By making the inverse substitution (see (9)), we obtain the following asymptotic formula for the solution of the exterior Neumann problem:

$$v^e(x) = \frac{b}{\pi\mu(\mu+1)}(A_+ + A_-)x^\mu + Fb^{-\mu-1}x^{\mu+1} + O(x^{\mu+2}(\log x)^2).$$

3. Now we find the normal derivative of the harmonic function u^i. According to (14) we have

$$v^e(x(\eta)) = \frac{b}{\pi\mu(\mu+1)}(A_+ + A_-)\eta^{2\mu} + E^{\#}\eta^{2\mu+2}\log\frac{1}{|\eta|} + F^{\#}\eta^{2\mu+2} + O(|\eta|^4).$$

We put $\tilde{u}^i = u^i \circ \theta$. Then in the same way as in §2 we find

$$\frac{\partial \tilde{u}^i}{\partial n}(\eta) = \frac{-1}{\pi}\int_{-\infty}^{\infty}(v^e \circ \theta)'_t(t+i0)\frac{dt}{t}$$

$$+ \frac{4ba_\theta^{2\mu}(A_+ + A_-)}{\pi^2(\mu+1)}\int_0^\infty \frac{t^{2\mu-2}}{1-t^2}dt|\eta|^{2\mu-1} + O(|\eta|).$$

By making the inverse substitution (see (15)), we obtain

$$\frac{\partial u^i}{\partial n}(x) = \frac{-x^{-1/2}}{2\pi a_\theta}\int_{-\infty}^{\infty}(v^e \circ \theta)'_\eta(\eta+i0)\frac{d\eta}{\eta}$$

$$- 2\frac{\phi_+(0) + \phi_-(0)}{\kappa''_+(0) - \kappa''_-(0)}\frac{\tan\pi\mu}{\mu+1}x^{\mu-1} + O(1).$$

4. By duplicating the arguments used in subsection 2 we find that in the cases I(a), (b), as in I(c), the solution of the exterior Neumann problem has the following representation:

$$v^e(x) - v^e(0) = \frac{b}{\pi\mu(\mu+1)}(A_+ + A_-)x^\mu + Fx^{\mu+1} + O(x^{\mu+2}(\log x)^2).$$

Analogously, in the cases II(a)–II(c), we have

$$v^e(x) - v^e(0) = \frac{b}{\pi(\mu+1)(\mu+2)}(B_+ + B_-)x^{\mu+1} + O(x^{\mu+2}(\log x)^2).$$

Similarly to subsection 3, let $\tilde{u}^i$ be the harmonic extension of the function $v^e(\theta(\eta+i0))$, $\eta \in \mathbb{R}$, onto the upper half-plane. The normal derivative of $\tilde{u}^i$ is

$$\frac{\partial \tilde{u}^i}{\partial n}(\eta) = \frac{1}{\pi}\int_{-\infty}^{\infty}\frac{\partial \tilde{u}^i}{\partial \eta}(t)\frac{dt}{\eta-t}.$$

By calculating the asymptotics of the last integral and making the inverse substitution, we obtain:

in the case I(a),

$$\left(\frac{\partial u^i}{\partial n}\right)_{\pm}(x) = -2\frac{\phi_+(0) + \phi_-(0)}{\kappa''_+(0) - \kappa''_-(0)}\frac{\tan\pi\mu}{\mu+1}x^{\mu-1} + O(x^{-1/2});$$

in the case I(b),

$$\left(\frac{\partial u^i}{\partial n}\right)_{\pm}(x) = -\frac{8}{3\pi}\frac{\phi_+(0)+\phi_-(0)}{\kappa''_+(0)-\kappa''_-(0)}x^{-1/2}\log(a_\theta x^{-1/2})$$
$$-\frac{x^{-1/2}}{2\pi a_\theta}\left|\int_{-\infty}^{\infty}(v^e\circ\theta)'_\eta(\eta+i0)\frac{d\eta}{\eta}\right|+O(1);$$

in the case II(a),

$$\left(\frac{\partial u^i}{\partial n}\right)_{\pm}(x) = -\frac{x^{-1/2}}{2\pi a_\theta}\int_{-\infty}^{\infty}(v^e\circ\theta)'_\eta(\eta+i0)\frac{d\eta}{\eta}$$
$$\mp\frac{1}{2\pi a_\theta^2}\int_{-\infty}^{\infty}(v^e\circ\theta)'_\eta(\eta+i0)\frac{d\eta}{\eta^2}$$
$$-2\frac{\phi'_+(0)+\phi'_-(0)}{\kappa''_+(0)-\kappa''_-(0)}\frac{\tan\pi\mu}{\mu+2}x^\mu+O(x^{1/2}\log x);$$

in the case II(b),

$$\left(\frac{\partial u^i}{\partial n}\right)_{\pm}(x) = -\frac{x^{-1/2}}{2\pi a_\theta}\int_{-\infty}^{\infty}(v^e\circ\theta)'_\eta(\eta+i0)\frac{d\eta}{\eta}+O(1),$$

and

$$\left(\frac{\partial u^i}{\partial n}\right)_{\pm}(x) = \frac{\mp 1}{2\pi a_\theta^2}\int_{-\infty}^{\infty}(v^e\circ\theta)'_\eta(\eta+i0)\frac{d\eta}{\eta^2}$$
$$-\frac{8}{5\pi}\frac{\phi'_+(0)+\phi'_-(0)}{\kappa''_+(0)-\kappa''_-(0)}x^{1/2}\log(a_\theta x^{-1/2})$$
$$-\frac{x^{1/2}}{2\pi a_\theta^3}\left|\int_{-\infty}^{\infty}(v^e\circ\theta)'_\eta(\eta+i0)\frac{d\eta}{\eta^3}\right|+O(x(\log x)^2),$$

if

$$\int_{-\infty}^{\infty}(v^e\circ\theta)'_\eta(\eta+i0)\eta^{-1}\,d\eta=0;$$

in the case II(c),

$$\left(\frac{\partial u^i}{\partial n}\right)_{\pm}(x) = -\frac{x^{-1/2}}{2\pi a_\theta}\int_{-\infty}^{\infty}(v^e\circ\theta)'_\eta(\eta+i0)\frac{d\eta}{\eta}+O(1),$$

and

$$\left(\frac{\partial u^i}{\partial n}\right)_{\pm}(x) = \frac{\mp 1}{2\pi a_\theta^2}\int_{-\infty}^{\infty}(v^e\circ\theta)'_\eta(\eta+i0)\frac{d\eta}{\eta^2}-\frac{x^{1/2}}{2\pi a_\theta^3}\int_{-\infty}^{\infty}(v^e\circ\theta)'_\eta(\eta+i0)\frac{d\eta}{\eta^3}$$
$$-2\frac{\phi'_+(0)+\phi'_-(0)}{\kappa''_+(0)-\kappa''_-(0)}\frac{\tan\pi\mu}{\mu+1}x^\mu+O(x\log x),$$

if
$$\int_{-\infty}^{\infty} (v^e \circ \theta)'_\eta (\eta + i0) \eta^{-1} \, d\eta = 0.$$

By substituting decompositions for the functions $\partial u^i / \partial n$ and ψ into (18), we obtain the required representations in all cases under consideration. The theorem is proved.

REMARK. In the case $\phi_+(0) + \phi_-(0) \neq 0$, $\mu = 0$, the solution of the exterior Neumann problem is sought as the sum of the harmonic function

$$u_1(x, y) = 2 \frac{\phi_+(0) + \phi_-(1)}{\kappa_+''(0) - \kappa_-''(0)} \log \left| \frac{zz_0}{z_0 - z} \right|, \qquad z = x + iy \in \Omega^c, \qquad z_0 \in \Omega,$$

and the simple layer potential

$$u_2(x, y) = \int_\Gamma \gamma(q) \log \frac{1}{r} \, ds_q, \qquad r = |z - q|,$$

whose derivative along the exterior normal is $\psi_1 = \psi - \partial u_1 / \partial n$. The density γ of the potential is the solution of the integral equation of the exterior Neumann problem with the right-hand side $-\pi^{-1} \psi_1(z)$, $z \in \Gamma$.

Since

$$\psi_1(z) = \mp \frac{\kappa_+''(0) \phi_-(0) + \kappa_-''(0) \phi_+(0)}{\kappa_+''(0) - \kappa_-''(0)} + O(x), \qquad z \in \Gamma, \ x \to +0;$$

then, according to II(a) of Theorem 2, the integral equation is uniquely solvable in the class $\mathfrak{M}$ and its solution γ has the asymptotic representation

$$\gamma(x) = \frac{-x^{-1/2}}{4\pi^2 a_\theta} \int_{-\infty}^{\infty} (u_2 \circ \theta)'_\eta (\eta + i0) \frac{d\eta}{\eta} \mp \frac{1}{4\pi^2 a_\theta^2} \int_{-\infty}^{\infty} (u_2 \circ \theta)'_\eta (\eta + i0) \frac{d\eta}{\eta^2}$$
$$\pm \frac{1}{2\pi} \frac{\kappa_+''(0) \phi_-(0) + \kappa_-''(0) \phi_+(0)}{\kappa_+''(0) - \kappa_-''(0)} + O(x^{1/2} \log x).$$

If $-1 < \mu < 0$, then we can take

$$u_1(x, y) = \frac{2}{\mu(\mu+1)} \frac{\phi_+(0) + \phi_-(0)}{\kappa_+''(0) - \kappa_-''(0)} \operatorname{Re} \left(\frac{zz_0}{z_0 - z} \right)^\mu$$
$$+ \frac{\phi_+(0) \kappa_-''(0) + \phi_-(0) \kappa_+''(0)}{(\mu+1)(\kappa_+''(0) - \kappa_-''(0))} \operatorname{Im} \left(\frac{zz_0}{z_0 - z} \right)^{1+\mu},$$

where $z = x + iy \in \Omega^c$ and z_0 is a fixed point of Ω. Since

$$\left(\frac{\partial u_1}{\partial n} \right)_\pm (z) = \phi_\pm(0) x^\mu \pm \frac{\mu+2}{\mu+1} \kappa_\pm'''(0) \frac{\phi_+(0) + \phi_-(0)}{3(\kappa_+''(0) - \kappa_-''(0))} x^{\mu+1} + O(x^{\mu+2}),$$

the function $\psi_1 = \psi - \partial u_1 / \partial n$ has the representation

$$\psi_1(z) = A_\pm^\# x^{\mu+1} + B_\pm^\# x^{\mu+2} + \cdots, \qquad z \in \Gamma.$$

Now Theorem 2 is applicable to the integral equation (1) with the right-hand side $-\pi^{-1} \psi_1$.

4°. The internal Neumann problem. The solution of the integral equation of the internal Neumann problem

$$\gamma(p) + \frac{1}{\pi} \int_\Gamma \gamma(q) \frac{\partial}{\partial n_p} \log \frac{1}{r} ds_q = \frac{1}{\pi} \psi(p), \qquad p \in \Gamma, \tag{21}$$

is sought by the same scheme as the solution of Eq. (1).

We only note that the harmonic function v^i, whose derivative along the outward normal is equal to ψ on Γ, should be normalized in such a way that the harmonic extension u^e of the function v^i from the boundary Γ to Ω tends to zero at infinity.

By changing signs in the right-hand sides of the asymptotic formulas in Theorem 1 (Theorem 2), we obtain representations for the solutions of the integral equation (21) for the domain Ω with inward (outward) peak.

§4. Supplements

1°. Proof of Lemma 1. According to (3)

$$\Delta\left[\nu(u, v) - \sum_{k=0}^{5} p_k(v) u^{-k}\right] = O(u^{-6}),$$

$$\nu(u, g_\pm(u)) - \sum_{k=0}^{5} p_k(g_\pm(u)) u^{-k} = O(u^{-6}).$$

We set

$$W(u, v) = u^4 \left(\nu(u, v) - \sum_{k=0}^{5} p_k(v) u^{-k}\right) \tag{22}$$

and note that

$$u^4 \Delta(u^{-4} W) = \Delta W - 8 u^{-1} \partial W / \partial u + 20 u^{-2} W.$$

Let $\eta \in C_0^\infty(\mathbb{R})$ be a cut-off function vanishing for $u < 1$ and equal to unity for $u > 2$. We set $\eta_R(u) = \eta(u/R)$. By Λ_0 we denote the restriction of the Laplace operator to the space $W_2^1(D)$ and by Λ_R we mean the operator defined on $W_2^1(D)$ by the differential expression

$$-8 \eta_R(u) u^{-1} \partial / \partial u + 20 \eta_R(u) u^{-2}.$$

Since the operator $\Lambda_0(\overset{\circ}{W}{}_2^1(D) \to W_2^{-1}(D))$ is invertible, the operator $(\Lambda_0 + \Lambda_R)(\overset{\circ}{W}{}_2^1(D) \to W_2^{-1}(D))$ is also invertible for large R. Hence for such R the boundary value problem,

$$(\Lambda_0 + \Lambda_R) \widetilde{W}_1 = F \quad \text{in } D, \qquad \widetilde{W}_1 = 0 \quad \text{on } \partial D, \tag{23}$$

is uniquely solvable in $W_2^1(D)$ for any F in $L^2(D)$.

Let χ be a cut-off function from $C_0^\infty(-1, 1)$, equal to unity for $-1/2 < u < 1/2$ and let $\chi^\tau(u) = \chi(u - \tau)$, $u \in \mathbb{R}$.

From the well-known local a priori estimate for solutions of elliptic boundary value problems (see [8]) and from the Sobolev imbedding theorem, it follows that the function $\widetilde{W}_1$ and its gradient $\nabla \widetilde{W}_1$ are bounded in D.

We set $Q(u, v) = u^4 \Delta(u^{-4} W(u, v))$ and $q_\pm(u) = W(u, g_\pm(u))$. By $\widetilde{W}_2$ we denote the function

$$\eta_R(u)(q_-(u)(g_+(u) - v) + q_+(u)(v - g_-(u)))/(g_+(u) - g_-(u));$$

it belongs to $W_2^3(D)$ and is bounded on the strip D together with its gradient. We take the function $Q(u, v) - (\Lambda_0 + \Lambda_R)\widetilde{W}_2$ belonging to $W_2^1(D)$ as $F(u, v)$ in (23). Then the function $\widetilde{W} = \widetilde{W}_1 + \widetilde{W}_2$ is the solution of the boundary value problem

$$\Delta\widetilde{W} - \eta_R(u)\frac{8}{u}\frac{\partial}{\partial u}\widetilde{W} + \eta_R(u)\frac{20}{u^2}\widetilde{W} = Q \quad \text{in } D,$$

$$\widetilde{W} = \eta_R q_\pm \quad \text{on } L_\pm, \qquad \widetilde{W} = 0 \quad \text{on } L_1,$$

in $W_2^1(D)$ for sufficiently large R. This and (22) imply that the harmonic function

$$\nu_1(u, v) = u^{-4}\widetilde{W}(u, v) + \sum_{k=0}^{5} p_k(v) u^{-k}$$

satisfies the relations $\nu_1(g_\pm(u)) = \pm \pi/2$ for sufficiently large u.

We show that $\nu_2 = \nu - \nu_1$ decays exponentially in D as $u \to +\infty$. Let $w = \omega(\zeta)$ denote the conformal mapping of the right half-plane $\{\zeta = \eta + i\xi, \eta > 0\}$ onto the curvilinear semi-infinite strip D, normalized by the condition $\omega(0) = \infty$. The function $\nu_2 \circ \omega(\zeta)$ is harmonic and bounded in the right half-plane, vanishes on some interval of the imaginary axis containing the origin, say at $\{|\xi| < \xi_0\}$. By representing this function as the Poisson integral, we find that

$$|\nu_2 \circ \omega(\zeta)| \leq \text{const}|\zeta| \quad \text{for } |\xi| < \xi_0/2.$$

According to the Warschawski theorem [6] on the asymptotics of the conformal mapping of the curvilinear semi-infinite strip, the mapping $\zeta = \omega^{-1}(w)$ inverse to $w = \omega(\zeta)$ has the representation

$$\log \omega^{-1}(w) = -bw - \beta \log w + O(1)$$

for $u = \operatorname{Re} w \to +\infty$. This implies

$$|\nu_2(w)| \leq \text{const} \exp(-bu/2). \tag{24}$$

So

$$\nu(u, v) = \nu_1(u, v) + O(\exp(-bu/2)).$$

Hence

$$\nu(u, v) = p_0(v) + p_1(v) u^{-1} + p_2(v) u^{-2} + O(u^{-3}), \qquad u \to +\infty.$$

We now show that the gradient of the function ν_2 decays exponentially as $u \to +\infty$. The function $\eta_R \nu_2$ belongs to $\overset{\circ}{W}{}^1_2(D)$ if R is large enough. After integration by parts and an application of the Schwarz inequality, we obtain

$$\int_D |\nabla(\eta_R \nu_2)|^2 \, du\,dv \leq \frac{\mathrm{const}}{R} \left(\int_{R<u<2R} |\nu_2|^2 \, du\,dv + \int_D |\nabla(\eta_R \nu_2)|^2 \, du\,dv \right).$$

Therefore, for sufficiently large R, we have

$$\int_D |\nabla(\eta_R \nu_2)|^2 \, du\,dv \leq \frac{\mathrm{const}}{R} \int_{R<u<2R} |\nu_2|^2 \, du\,dv. \tag{25}$$

Consider the function $\chi^\tau \nu_2$ which is the solution of the boundary value problem

$$\Delta u = 2\nabla \chi^\tau \cdot \nabla \nu_2 + (\nabla \chi^\tau)\nu_2 \text{ in } D, \quad u = 0 \text{ on } \partial D,$$

for large positive τ. The local a priori estimate (see [8]) and inequalities (25), (24) yield

$$\|\chi^\tau \nu_2\|^2_{W^3_2(D)} \leq \mathrm{const} \int_{u \geq \tau/2} |\nu_2|^2 \, du\,dv \leq \mathrm{const}\, \exp(-b\tau/2).$$

This and the Sobolev imbedding theorem imply

$$\max_{|u-\tau|<1/2} |\nabla \nu_2(u,v)| \leq \mathrm{const}\, \exp(-b\tau/2).$$

Therefore,

$$|\nabla \nu_2(u,v)| \leq \mathrm{const}\, \exp(-bu/4)$$

for sufficiently large u. This proves the possibility of differentiation of the asymptotic decomposition of ν. The lemma is proved.

$2°$. **Proof of Lemma 2.** For positive η we have

$$\begin{aligned}H(\phi')(\eta) &= \frac{1}{\pi} \int_{-\infty}^{\infty} \phi'(t) \frac{dt}{\eta - t} \\ &= \frac{1}{\pi} \int_0^\eta [\phi'(\eta - t) - \phi'(\eta + t)] \frac{dt}{t} + \frac{1}{\pi} \int_0^{2\eta} \phi'(-t) \frac{dt}{\eta + t} \\ &\quad + \frac{2}{\pi} \int_{2\eta}^\infty (\phi')_o(t) \frac{t\,dt}{\eta^2 - t^2} + \frac{2}{\pi} \eta \int_{2\eta}^\infty (\phi')_e(t) \frac{dt}{\eta^2 - t^2}, \end{aligned} \tag{26}$$

where $(\phi')_e$ and $(\phi')_o$ are the even and odd components of ϕ', respectively. Straightforward calculations yield

$$\frac{1}{\pi} \int_0^\eta [\phi'(\eta - t) - \phi'(\eta + t)] \frac{dt}{t} = -\frac{12}{\pi} a_+ \eta^2 - \frac{80}{3\pi} b_+ \eta^3 + O(\eta^{3+\varepsilon}),$$

$$\frac{1}{\pi} \int_0^{2\eta} \phi'(-t) \frac{dt}{\eta + t} = \frac{3}{\pi} a_- \eta^2 \log 3 - \frac{4}{\pi} b_- \left(\frac{8}{3} - \log 3 \right) \eta^3 + O(\eta^{3+\varepsilon}). \tag{27}$$

Further, using the equality

$$(1 - t^2)^{-1} = 1 + t^2 + \frac{t^4}{1 - t^2},$$

we obtain

$$\frac{2}{\pi} \int_{2\eta}^{\infty} (\phi')_0(t) \frac{t\,dt}{\eta^2 - t^2} = -\frac{1}{\pi} \int_{-\infty}^{\infty} \phi'(t) \frac{dt}{t} + \frac{3(a_+ - a_-)}{\pi} \eta^2 \log \eta$$

$$+ \frac{1}{\pi} \left[\left(6 + \frac{3}{2} \log 3 \right)(a_+ - a_-) - \left| \int_{-\infty}^{\infty} \phi'(t) \frac{dt}{t^3} \right| \right] \eta^2$$

$$+ \frac{1}{\pi} \left(\frac{56}{3} - 2 \log 3 \right)(b_+ + b_-)\eta^3 + O(\eta^{3+\varepsilon}). \quad (28)$$

Finally,

$$\frac{2}{\pi} \eta \int_{2\eta}^{\infty} (\phi')_e(t) \frac{dt}{\eta^2 - t^2} = -\frac{\eta}{\pi} \int_{-\infty}^{\infty} \phi'(t) \frac{dt}{t^2}$$

$$+ \frac{1}{\pi} \left(6 - \frac{3}{2} \log 3 \right)(a_+ + a_-)\eta^2 + \frac{4}{\pi}(b_+ - b_-)\eta^3 \log \eta$$

$$+ \frac{1}{\pi}(8 + 2\log 3)(b_+ - b_-)\eta^3 - \frac{1}{\pi}\eta^3 \left| \int_{-\infty}^{\infty} \phi'(t) \frac{dt}{t^4} \right|$$

$$+ O(\eta^{3+\varepsilon}). \quad (29)$$

By substituting (27)–(29) into (26) and taking into account that

$$\frac{1}{\pi} \int_{-\infty}^{\infty} \phi'(t) \frac{dt}{\eta - t} = -\frac{1}{\pi} \int_{-\infty}^{\infty} \phi'(-t) \frac{dt}{|\eta| - t},$$

for $\eta < 0$, we obtain the required asymptotic expansion of the function $H(\phi')$. The lemma is proved.

References

1. J. Radon, *Über die Randwertaufgaben beim logarithmischen Potential*, Sitzungsber. Akad. Wiss. Abt. 2a Wien (7) **128** (1919), 1123–1167.
2. V. G. Maz'ya, *The integral equations of potential theory in domains with piecewise smooth boundary*, Uspekhi. Mat. Nauk (4) **36** (1981), 229–230.
3. ＿＿, *Boundary integral equations*, Sovremennye problemy matematiki, Fundamental'nye napravleniya, vol. 27, Analyz. 4, VINITI, Moscow, 1988, pp. 131–228; English transl. in Encyclopaedia of Mathematical Sciences, vol. 27, Analysis 4, Springer-Verlag, Berlin and New York, 1991.
4. J. Hadamard, *Lectures on Cauchy's problem in linear partial differential equations*, Yale University Press, New Haven, 1922.
5. V. G. Maz'ya and A. A. Solov'ev, *Asymptotics of the solution of the integral equation of the Neumann problem in a plane domain with cusps on the boundary*, Soobshch. Akad. Nauk Gruzin. SSR (1) **30** (1988), 17–20.
6. S. E. Warschawski, *On conformal mapping of infinite strips*, Trans. Amer. Math. Soc. **51** (1942), 280–335.
7. G. M. Goluzin, *Geometric theory of functions of complex variable*, "Nauka", Moscow, 1966.

8. S. Agmon, A. Douglis, and L. Nirenberg, *Estimates near the boundary for solutions of elliptic partial differential equations satisfying general boundary conditions*, Comm. Pure Appl. Math. **12** (1959), 623–727.
9. V. G. Maz'ya, and A. A. Solov'ev, *On the integral equation of the Dirichlet problem in a plane domain with cusps on the boundary*, Mat. Sb. (9) **180** (1989), 1211–1233.

Translated by T. SHAPOSHNIKOVA

Singular Stationary Nonhomogeneous Linear-Quadratic Optimal Control

A. V. MEGRETSKIĬ AND V. A. YAKUBOVICH

Linear-quadratic optimization, also known as analytical regulator design, has been studied quite thoroughly in the nonsingular case. The first results were apparently obtained by Kalman [1] and Letov [2]; we also mention Lur'e [3] and Willems [4]. For stationary systems and an infinite time interval, the problem is formulated as follows.

Given a controlled object and an initial state

$$dx/dt = Ax + bu, \qquad (0.1)$$

$$x(0) = a, \qquad (0.2)$$

where $x(t) \in \mathbb{R}^n$ is the state, $u(t) \in \mathbb{R}^m$ is the control, and A and b are constant matrices. Measurable controls $u(\cdot)$ are admissible if they satisfy the "stability" condition

$$|x(\cdot)| \in \mathfrak{L}_2(0, \infty), \qquad |u(\cdot)| \in \mathfrak{L}_2(0, \infty), \qquad (0.3)$$

where $x(\cdot)$ is defined in terms of $u(\cdot)$ by Equations (0.1) and (0.2). We must find an optimal control from the condition

$$\Phi(x(\cdot), u(\cdot)) = \int_0^\infty \mathfrak{G}(x(t), u(t))\, dt \to \inf, \qquad (0.4)$$

where $\mathfrak{G}(x, u)$ is a given quadratic form with constant coefficients

$$\mathfrak{G}(x, u) = x^* G x + 2 x^* g u + u^* \Gamma u. \qquad (0.5)$$

Here $G = G^*$, $\Gamma = \Gamma^*$, and the asterisk denotes the transpose of a matrix (or, for complex matrices, the Hermitian conjugate). The matrices A, b, G, g, and Γ are real.

For all $a \in \mathbb{R}^n$, we assume that the set of admissible controls is not empty. This is equivalent to the condition that the pair (A, b) be stabilizable, i.e.,

---

1991 *Mathematics Subject Classification.* Primary 49J15.

there exists an $n \times m$ matrix r such that $A + br^*$ is a Hurwitz matrix; in other words, $\operatorname{Re} \lambda_j < 0$, where the λ_j are the eigenvalues of $A + br^*$.

The first complete analysis of the linear-quadratic problem was probably [4]. There, and elsewhere, another formulation without the "stability" conditions (0.3) imposed on the controls, was also considered. In that case some special clarifications are necessary regarding the formula (0.4) for the cost functional, since the integrand (0.4) may be nonintegrable and indefinite. The easiest case is when $\mathfrak{G}(x, u) = |Lx + Du|^2$ is positive semidefinite (L and D are constant matrices). It is then assumed that $\Phi(x(\cdot), u(\cdot)) = \infty$ if $|L(x(\cdot) + D(u(\cdot))| \notin \mathfrak{L}_2(0, \infty)$; otherwise, Φ is found by formula (0.4).

Notice that problem (0.1)–(0.4) is more complicated and richer in content than problem (0.1), (0.2), (0.4) without the stability condition. It is also more interesting for applications. As a rule, the solutions of these two problems are different. Consider, for example, the problem

$$dx/dt = x + u, \qquad x(0) = a, \qquad \Phi = \int_0^\infty u(t)^2 dt \to \min,$$

where $x(t) \in \mathbb{R}^1$ and $u(t) \in \mathbb{R}^1$. Without the stability condition, $\inf \Phi = 0$ and $x(t) = a \exp(t)$, $u(t) \equiv 0$ is an optimal process. With the stability condition, it follows, for example, from Theorem 1 (see below) that $\inf \Phi = 2a^2$ and an optimal process $x(t) = a \exp(-t)$, $u(t) = -2a \exp(-t)$ is produced by the regulator $u = -2x$.

We will consider problem (0.1)–(0.4) with the stability condition (0.3), as well as generalizations of the problem with condition (0.3) replaced by a similar "quasistability" condition.

Early studies of problem (0.1)–(0.4) assumed that $\mathfrak{G}(x, u) \geq 0$ or even $\mathfrak{G}(x, u) > 0$ if $|x| + |u| \neq 0$. The last condition, and some of its generalizations to the case $\mathfrak{G} \geq 0$, clearly guarantees the existence of an optimal control. It was under these assumptions that problem (0.1)–(0.4) was considered in [1–3] and elsewhere. In later publications the case of an indefinite form $\mathfrak{G}$ was also considered (some applications reduce to this case). In that context the question arose of criteria for the existence of an optimal control. Problem (0.1)–(0.4) is said to be *nonsingular* if it has exactly one solution for any $x(0) = a$. In what follows, effective criteria for nonsingularity will be given.

A proof of the existence of an "optimal regulator" solving the optimization problem (0.1)–(0.4) was a remarkable result of [1]–[4] and others. Namely, a continuum of problems (0.1)–(0.4) with different initial states $x(0) = a$ have a feedback-implementable solution

$$u(t) = h^* x(t). \tag{0.6}$$

The $n \times m$ matrix h does not depend on $x(0) = a$; so the solution $(x(\cdot), u(\cdot))$ of system (0.1), (0.2), and (0.6), which depends on a, guarantees an optimal control $u(\cdot)$ in problem (0.1)–(0.4) for any $a \in \mathbb{R}^n$. The matrix h may be found from the (nonlinear) equations introduced, by Lur'e in connection with stability of nonlinear control systems, a few years before the formulation of problem (0.1)–(0.4) (for more details see [5]).

Engineers were using linear regulators of the type (0.6) (PD- and PID-regulators) long before the mathematical theory was developed. The coefficients were chosen empirically to ensure that the processes in the system satisfied the necessary requirements.

The optimization problem (0.1)–(0.4) is closely related to the absolute stability of nonlinear systems. Various properties of many such systems make it necessary to find a quadratic form (Lyapunov function) $V_0(x) = x^* H_0 x$, where $H_0 = H_0^* = $ const, such that $\dot{V}_0 + \mathfrak{G}$ is positive definite. Here, $\dot{V}_0 = 2x^* H_0(Ax + bu)$ is the derivative along the flow of system (0.1), and $\mathfrak{G}$ is a given quadratic form (0.5). Since $\mathfrak{G}$ is indefinite, one is interested more in the existence rather than the construction of the form V_0.

The final results of numerous studies of the nonsingular problem may be summarized in the following theorem (see [4], [6], [7] etc.).

THEOREM 1. $1°$. *The following statements are equivalent*:
  (I) *Problem (0.1)–(0.4) is nonsingular, i.e., for any $a \in \mathbb{R}^n$ there is a unique optimal admissible control.*
  (II) *The problem of the Lyapunov function is solvable: there is a quadratic form $V_0 = x^* H_0 x$, $H_0 = H_0^*$ such that the form*
  $$\dot{V}_0 + \mathfrak{G} = 2x^* H_0(Ax + bu) + \mathfrak{G}(x, u)$$
  in the variables x and u is positive definite.
  (III) *The frequency condition*
  $$\exists \delta > 0 : \mathfrak{G}(\tilde{x}, \tilde{u}) \geq \delta(|\tilde{x}|^2 + |\tilde{u}|^2)$$
  holds for all $\omega \in \mathbb{R}^1$ and all pairs $(\tilde{x}, \tilde{u}) \in \mathbb{C}^n \times \mathbb{C}^m$ in the set $\mathfrak{N}[\omega]$ defined by the relation $i\omega \tilde{x} = A\tilde{x} + b\tilde{u}$. Here $\mathfrak{G}(\tilde{x}, \tilde{u})$ is the Hermitian extension of the form (0.5):
  $$\mathfrak{G}(\tilde{x}, \tilde{u}) = \tilde{x}^* G \tilde{x} + 2\operatorname{Re}(\tilde{x}^* g \tilde{u}) + \tilde{u}^* \Gamma \tilde{u}.$$
  (IV) *The matrix Γ in (0.5) is positive definite ($\Gamma > 0$) and there exist real $n \times n$ and $n \times m$ matrices $H = H^*$ and h such that Lur'e's equations*
  $$HA + A^* H + G = h\Gamma h^*, \qquad Hb + g = -h\Gamma \qquad (0.8)$$
  hold, and the matrix $A + bh^*$ is Hurwitz.
  (V) $\Gamma > 0$, *and the frequency condition*
  $$\det(K - i\omega J) \neq 0, \quad \forall \omega \in \mathbb{R}^1$$
  holds, where
  $$J = \begin{bmatrix} 0 & -I_n \\ I_n & 0 \end{bmatrix}, \qquad K = \begin{bmatrix} g\Gamma^{-1}g^* - G & A^* - g\Gamma^{-1}b^* \\ A - b\Gamma^{-1}g^* & b\Gamma^{-1}b^* \end{bmatrix}.$$

(VI) $\Gamma > 0$, and there exist a quadratic form $V = x^*Hx$, $H = H^*$, and an $n \times m$ matrix h such that
$$\dot{V} + \mathfrak{G} = |\Gamma^{1/2}(u - h^*x)|^2$$
identically, where $\dot{V} = 2x^*H(Ax + bu)$ and $A + bh^*$ is a Hurwitz matrix.

(VII) The functional Φ in (0.4) is positive definite in the subspace $\mathfrak{M}_0$ of pairs of functions $(x(\cdot), u(\cdot))$ satisfying conditions (0.1)–(0.3) with $a = 0$, that is,
$$\exists \delta > 0 : \int_0^\infty \mathfrak{G}(x(t), u(t))\,dt \geq \delta \int_0^\infty (|x(t)|^2 + |u(t)|^2)\,dt$$
for all $(x(\cdot), u(\cdot)) \in \mathfrak{M}_0$.

$2°$. *Let any one of the conditions* (I)–(VII) *hold. Then the matrix H in conditions* (IV) *and* (VI) *is uniquely defined by the formula*
$$H = -\Psi_0 X_0^{-1},$$
where the columns of the $2n \times n$ matrix $\text{col}(X_0, \Psi_0)$ *form an arbitrary real basis in the sum of those root subspaces of the matrix $J^{-1}K$ that correspond to eigenvalues in the half-plane* $\text{Re}\,\lambda < 0$. *For any* $x(0) = a$, *the optimal control is given by* (0.6), *where h is determined in terms of H by the second relation of* (0.8).

The condition in Theorem 1 that is the most convenient for the verification of nonsingularity is (III). Define an $m \times m$ matrix $\Pi(i\omega) = \Pi(i\omega)^*$ by
$$\mathfrak{G}[(i\omega I_n - A)^{-1}b\tilde{u}, \tilde{u}] = \tilde{u}^*\Pi(i\omega)\tilde{u}. \qquad (0.9)$$

If A has no eigenvalues on the imaginary axis, we can rewrite condition (III) as
$$\exists \delta > 0 : \Pi(i\omega) \geq \delta I_m > 0, \quad \forall \omega \in \mathbb{R}^1. \qquad (0.10)$$

If $m = 1$ (one regulating device only), $\Pi(i\omega)$ is a real function.

It is easy to see (see, for example, [7, Lemma 2.5]) that if condition (0.10) is badly violated, i.e.,
$$\exists \tilde{u}_0 \in \mathbb{C}^m \; \exists \omega_0 \in \mathbb{R}^1 : \tilde{u}_0^* \Pi(i\omega_0)\tilde{u}_0 < 0,$$
where $\det(A - i\omega_0 I_n) \neq 0$, then $\inf \Phi = -\infty$ (where the inf is taken over all admissible controls). Therefore, below we will always assume that
$$\Pi(i\omega) \geq 0, \quad \omega \in \mathbb{R}^1, \quad \det(i\omega I_n - A) \neq 0. \qquad (0.11)$$

We will also assume that
$$\exists \omega_0 \in \mathbb{R}^1 : \Pi(i\omega_0) > 0, \quad \det(i\omega_0 I_n - A) \neq 0. \qquad (0.12)$$

It can be shown that the general case with $\inf \Phi \neq -\infty$ reduces to the case in which (0.12) holds.

To complete the definition of the set $\mathfrak{N}[\omega]$ in Theorem 1 (III) we define $\mathfrak{N}[\infty] = \{(\tilde{x}, \tilde{u}) \in \mathbb{C}^n \times \mathbb{C}^m : \tilde{x} = 0\}$. If any of the conditions (I)–(VII) does not hold, then for some $\omega_* \in \mathbb{R}^1 \cup \{\infty\}$ the form $\mathfrak{G}$ is not positive definite on the subspace $\mathfrak{N}[\omega_*]$ of $\mathbb{C}^n \times \mathbb{C}^m$. In that case problem (0.1)–(0.4) is called *singular*, and one says that *the frequency inequality* (0.7) *degenerates* at $\omega = \omega_*$. If $\det(i\omega I_n - A) \neq 0$ for all $\omega \in \mathbb{R}^1$, degeneracy at $\omega = \omega_*$ means that $\det \Pi(i\omega_*) = 0$. In particular, degeneracy for $\omega = \infty$ means that $\det \Gamma = 0$. It is this case of degeneracy that occurs most often in applications. Note that (0.11) implies $\Gamma \geq 0$.

We have already considered the simplest linear-quadratic optimization problem under various assumptions. Similar problems were studied for the nonhomogeneous equation

$$p\xi = A\xi + b\eta + \zeta_1, \qquad p = d/dt, \tag{0.13}$$

where $\zeta_1 = a\zeta_0$, $\zeta_0 \in \mathbb{R}^q$, is a perturbation, and a is a constant $n \times q$ matrix. In this case, the cost functional (0.4) and class (0.3) of admissible processes are different but the optimization problem is again said to be nonsingular if conditions (II)–(VI) of Theorem 1 hold.

The class of admissible processes (ξ, η) for the object (0.13) is often defined by a set of stabilizing regulators

$$\alpha(p)\eta = \beta(p)\chi, \qquad p = d/dt \tag{0.14}$$

($\alpha(\lambda)$ and $\beta(\lambda)$ are matrix polynomials; $\alpha(\lambda)$ is a square matrix that is invertible almost everywhere; the function $W(\lambda) = \alpha(\lambda)^{-1}\beta(\lambda)$ is bounded as $|\lambda| \to \infty$; system (0.13), (0.14) is stable when $\zeta_0 \equiv 0$). In (0.14) χ is the given (measurable) output of the system

$$\chi = c^*\xi + d^*\eta + \zeta_2, \tag{0.15}$$

where c and d are constant real matrices of the appropriate orders.

If $\zeta = (\zeta_1, \zeta_2)$ is a stationary random process (possibly generalized), the cost functional will be

$$P\Phi = \overline{\lim_{t \to \infty}} M\mathfrak{G}(\xi(t), \eta(t)) \to \inf, \tag{0.16}$$

and a further constraint must be imposed on $\zeta(\cdot)$ to guarantee the existence of the mean value in (0.16). A special case of this problem will be considered in §5. In particular, if (0.15) has the form $\chi = \xi$ and ζ_0 is white noise (a generalized stationary stochastic process with constant spectral density), the optimal nonsingular regulator in class (0.14) is $\eta = h^*\xi$, where h is the matrix of Theorem 1.

Let us consider now singular problems. To fix ideas, we will consider problem (0.1)–(0.4). Since here $\inf \Phi$ is not attained for all $a = x(0) \in \mathbb{R}^n$, it is natural to extend the class of admissible processes without changing $\inf \Phi$, so that an optimal process will exist in a wider class. This can be done in different ways, depending on additional conditions adopted, but the generalized admissible processes will always be the distributions (in the sense of Schwartz and others) and we must then give the integral (0.4) a reasonable meaning. The fundamental goal is to construct a minimizing sequence of admissible regulators for the problem. In this research area there are many different approaches and results. We mention only papers [8]–[13], where the problem is treated in very similar settings (stationary case, infinite time interval). However, here we will be concerned with a more general formulation: linear terms will be admitted in the cost functionals, and degeneracy on the imaginary axis will be permitted, hence, a different technique will be necessary. We will also study the relations between the initial problem and generalizations, obtain explicit formulas for the generalized optimal regulator and develop formulas for a "good" optimizing sequence of regulators in the initial problem. The results will then be applied to the solution of stochastic problems.

In the context of singular problems, we cannot but touch on their history, which, quite unexpectedly, is connected with the names of N. N. Luzin, G. V. Shchipanov, and many others.

Nikolaï Nikolaevich Luzin originally had no intention of working in singular optimal control. As far as we know, he never formulated the problems or mentioned them. Recently, however, it became known that in the last years of his life his scientific activity was closely related to the field. Everything began in 1936, with the appereance of two articles in *Pravda*: "An answer to Academician Luzin" (No. 180, July 2) and "On enemies in a Soviet disguise" (No. 181, July 3). We think the reader can guess their content; if necessary, they are not inaccessible. A clear impression of the events that were then unfolding may be gained from the following extracts from resolutions adopted at general meetings of the staff of the Mathematical Institute of the Soviet Academy of Sciences (director, academician I. M. Vinogradov) and the Institute of Mathematics, Mechanics and Astronomy at Moscow University [14]:
" ... the scientific community did not discern in these facts the enemy's face, covered by the disguise of a Soviet academician ... ," " ... we must openly admit that the position regarding N. Luzin was one of rotten liberalism, favouring the foul anti-Soviet activity of N. Luzin and facilitating it ... ," " ... the meeting considers the Luzin's behavior to be incompatible with his position as a member of the scientific councils of the University and the Institute of Mechanics, and queries the presidium of the Academy of Sciences as to Luzin's further tenure as a full member of the Academy ... ,"
" ... denunciation of the wrecking activity of Luzin and his ilk will further contribute to the future successful development of Soviet mathematics and the flourishing of the creative abilities of the youth, uniting the mass of scientific workers even more closely around the party and our ardently beloved comrade Stalin, leader of the nations ... ."

The journal *Uspekhi Matematicheskikh Nauk* also echoed the events, with the editorial "Ridding the scientific community of *Luzinism*".

Some time later Nikolaï Nikolaevich lost his job and his means of substinence. He was not expelled from the Academy, though the presidium censured him; but in those days academicians did not receive pecuniary aid. Of course, this was not the worst thing that could have happened in those days.

Nikolaï Nikolaevich was helped by academician V. S. Kulebiakin who, probably, showing outstanding courage, secured him a job at the Institute of Automation and Telemechanics. Luzin was later drawn into the notorious "Shchipanov discussion" which was, as we now know, closely related to singular optimization.

In 1938 Professor G. V. Shchipanov posed the problem of designing controls with an invariance property and proposed ways to solve it. In mathematical terms the problem was to design the regulator (0.14) in such a way that a given component $z = q^*\xi$ of the state of the object (0.13) would not depend on the input ζ_0. (The voltage in the circuit should not depend on the load, and so on.) Very often, however, Shchipanov's mathematically obvious solution did not satisfy various practical requirements (stability of the closed-loop system, and so on) and was, therefore, subjected to criticism. The debate that developed was overly sharp and, for those times (1939), dangerous and unnecessary, as it threatened the participants with "organizational conclusions" whose nature was very far from being scientific.

For simplicity, let χ, z, and ζ_0 be scalars. In a modern context, Shchipanov's idea is quite reasonable. All we need is to discard the unattainable ideal of absolute invariance, i.e., complete independence of z of the deterministic perturbation ζ_0, requiring instead only the minimal possible invariance. A measure of dependence is defined by the functional

$$R = \overline{\lim_{t \to \infty}} \sup_{\zeta_0 \in \mathfrak{L}_2(0,\infty)} \frac{|z(t)|^2}{\|\zeta_0(\cdot)\|_{\mathfrak{L}_2(0,\infty)}}.$$

In the problem of the regulator optimization (0.14), $R = R(\alpha, \beta)$. It turns out [16] that R coincides with $P\Phi$ in (0.16) for $\mathfrak{G} = |z|^2 = |q^*\xi|^2$, if (0.16) is calculated under the assumption that ζ_0 is the normalized white noise ($M\zeta_0 = 0$, $M\zeta_0(\tau)\zeta_0(t) = \delta(t-\tau)$). Since $\mathfrak{G}$ is such that $\Gamma = 0$, the problem is singular. Hence an optimal regulator may not exist. If a regulator does exist and $\min R = 0$, then absolute invariance in Shchipanov's sense is possible. If $\inf R > 0$ or $\inf R = 0$, but the infimum is not attainable (this can indeed happen), then absolute invariance is impossible. In this case we can design an optimizing sequence of "good" (in particular, stabilizing) regulators (0.14). Thus, the solution to the problem of properly formulating Shchipanov's problem is an optimal regulator, if it exists, or a sequence (for the details, see [16]).

Of course, such a solution was impossible in the prewar years, if only because the mathematical technique of distributions did not exist; moreover, the analogous nonsingular problems had not been formulated, let alone solved. In Luzin's study [17] and the papers he wrote in cooperation with his student and colleague P. I. Kuznetsov, the problem of invariance in automation, properly formulated, was solved to a certain extent. From the modern point of view the results seem somewhat naive. Nevertheless, Luzin's study [17] marked the beginning of modern control theory.

For development of the engineering theory of invariance, whose origins may also be traced to Luzin [17], the reader may consult Kuchtenko's survey [18].

§1. Basic notation and terminology

Throughout this paper, $\mathfrak{D}_+^n$ is the space of infinitely differentiable functions $\varphi : [0, \infty) \to \mathbb{R}^n$ with compact support, endowed with the standard topology of the Schwartz test functions [19], $\mathfrak{L}_2^n \langle s, k \rangle$, $s, k \in \{0, 1, 2, \dots\}$, is the completion of $\mathfrak{D}_+^n$ in the norm $\|\cdot\|_{s,k}$, where

$$\|\varphi\|_{s,k} = k \int_0^\infty t^{k-\frac{1}{2}} |\varphi(t)| \, dt + \left(\frac{1}{2} \int_0^\infty (|\varphi(t)|^2 + |(p^s \varphi)(t)|^2) \, dt \right)^{1/2}, \quad (1.1)$$

$p = d/dt$; $(\mathfrak{D}_+^n)'$ is the space of distributions (in Schwartz's sense) with support on the ray $[0, \infty)$, i.e., the space of continuous functionals on $(\mathfrak{D}_+^n)'$ with the pointwise convergence topology; $W_2^n \langle s, k \rangle$ is the space dual to $\mathfrak{L}_2^n \langle s, k \rangle$. Elements of $\mathfrak{D}_+^n$, $\mathfrak{L}_2^n \langle s, k \rangle$, and $W_2^n \langle s, k \rangle$ are naturally identified with the distributions in $(\mathfrak{D}_+^n)'$, so that

$$\mathfrak{D}_+^n \subset \mathfrak{L}_2^n \langle s, k \rangle \subset W_2^n \langle s, k \rangle \subset (\mathfrak{D}_+^n)',$$
$$\mathfrak{L}_2^n \langle 0, 0 \rangle = W_2^n \langle 0, 0 \rangle \stackrel{\text{def}}{=} \mathfrak{L}_2^n.$$

Let $\delta^{(q)}(t)$ be the qth generalized derivative of the generalized Dirac delta function $\delta(t)$ ($\delta^{(q)} \in W_2^1 \langle q+1, k \rangle$ for $k \geq 0$). The norm (1.1) with $s = 0$, $k = 0$ is denoted by $\|\cdot\|$.

Let A and b be real matrices of the orders $n \times n$ and $n \times m$ respectively. Let $\mathfrak{G}$ be the Hermitian form of (0.5) and let $\tau : \mathfrak{L}_2^n \times \mathfrak{L}_2^m \to \mathbb{R}^1$ be a linear functional such that

$$\tau(x(\cdot), u(\cdot)) = \int_0^\infty \{\tau_x(t)^* x(t) + \tau_u(t)^* u(t)\} \, dt,$$

where $\tau_x(\cdot) \in \mathfrak{L}_2^n$, $\tau_u(\cdot) \in \mathfrak{L}_2^m$. For arbitrary $a \in \mathbb{R}^n$ we define the affine space $\mathfrak{M}[a]$ of *admissible processes* as the set of pairs $(x(\cdot), u(\cdot)) \in \mathfrak{L}_2^n \times \mathfrak{L}_2^m$ of functions satisfying the equation

$$px = Ax + bu + a\delta(t), \qquad p = d/dt. \quad (1.2)$$

We also define a functional $\Phi : \mathfrak{L}_2^n \times \mathfrak{L}_2^m \to \mathbb{R}^1$ by

$$\Phi(x(\cdot), u(\cdot)) = 2\tau(x(\cdot), u(\cdot)) + \int_0^\infty \mathfrak{G}(x(t), u(t)) \, dt. \quad (1.3)$$

It is readily seen that the admissible processes, i.e., elements of $\mathfrak{M}[a]$, satisfy (0.1) for $t > 0$ with initial condition $x(t) \to a$ as $t \to +0$, and the "stability" condition $x(t) \to 0$ as $t \to \infty$ (up to a set of measure zero). Functionals of (1.3) occur in problems in which one has to minimize the deviation of a process from a given "programmed process" $(x_0(\cdot), u_0(\cdot)) \in \mathfrak{L}_2^n \times \mathfrak{L}_2^m$, with

$$\Phi(x(\cdot), u(\cdot)) = \text{const} + \int_0^\infty \mathfrak{G}(x(t) - x_0(t), u(t) - u_0(t)) \, dt.$$

In this case

$$\tau_x = -Gx_0 - gu_0, \qquad \tau_u = -g^* x_0 - \Gamma u_0.$$

The problem of minimizing Φ on the set $\mathfrak{M}[a]$ of admissible processes is called *nongeneralized stationary linear-quadratic optimal control* ($LQ(\mathfrak{M}[a], \Phi)$ for short). Matrices A, b, G, g, Γ in (1.2) and (0.5) are called the *coefficients* of $LQ(\mathfrak{M}[a], \Phi)$, the vector $a \in \mathbb{R}^n$ in (1.2) is called the *initial state*, and the functions $\tau_x(\cdot), \tau_u(\cdot)$ form the *input*. An element $v \in \mathfrak{M}[a]$ for which Φ attains its minimum is called an *optimal process*. If $LQ(\mathfrak{M}[a], \Phi)$ is a singular problem, then an optimal process may not exist. Notice that being singular is a property of the coefficients rather than of either the initial state or the input $\tau_x(\cdot), \tau_u(\cdot)$.

In the singular case we can enlarge the space $\mathfrak{M}[a]$ to a space $\mathfrak{M}_*[a] \supset \mathfrak{M}[a]$ and extend the functional Φ to $\mathfrak{M}_*[a]$ so that the minimum of Φ on $\mathfrak{M}_*[a]$ is attainable. The elements of the extension are called *generalized processes* and $\mathfrak{M}_*[a]$ is called the *space of generalized processes*. In the natural definition of $\mathfrak{M}_*[a]$, proposed in [20], generalized processes are classes of optimizing sequences in $LQ(\mathfrak{M}[a], \Phi)$ for all possible inputs $\tau_x(\cdot), \tau_u(\cdot)$. In §2 we will consider another definition of generalized processes, as pairs $(x, u) \in (\mathfrak{D}_+^n)' \times (\mathfrak{D}_+^m)'$ of distributions. The two approaches in fact lead to the same result. An element $v \in \mathfrak{M}_*[a]$ at which the minimum of Φ is attained, is called a *generalized optimal process*.

§2. Generalized processes

If condition (VI) of Theorem 1 holds, it is easy to see that the cost functional (0.4) can be represented in the space $\mathfrak{M}[a]$ of admissible processes as

$$\Phi(x(\cdot), u(\cdot)) = a^* H a + \int_0^\infty |\Gamma^{1/2}(u(t) - h^* x(t))|^2 dt. \qquad (2.1)$$

Since the matrix $A + bh^*$ is Hurwitz, $a^* H A = \min\{\Phi(v) : v \in \mathfrak{M}[a]\}$ and (0.6) is an optimal regulator in the problem $LQ(\mathfrak{M}[a], \Phi)$ for all $a \in \mathbb{R}^n$. Condition (VI) is equivalent to the frequency inequality (0.7). For a general linear-quadratic problem (when inequality (0.7) degenerates for some $\omega \in \mathbb{R}^1 \cup \{\infty\}$ and, possibly, $\tau \neq 0$), an analog of (2.1) exists.

DEFINITION 2.1. Let the functional Φ of (1.3) admit the following representation in the spaces $\mathfrak{M}[a]$ of admissible processes:

$$\Phi(x(\cdot), u(\cdot)) = \Phi_*(a) + \|l^*x + \kappa u - \varphi\|^2, \qquad (2.2)$$
$$\forall a \in R^n, \quad \forall(x, u) \in \mathfrak{M}[a],$$

where l and κ are constant real matrices and $\varphi(\cdot) \in \mathfrak{L}_2^m$. If $\Phi_*(a)$ is the infimum of Φ on $\mathfrak{M}[a]$ and the matrix

$$N(\lambda) = \begin{bmatrix} \lambda I_n - A & -b \\ l^* & \kappa \end{bmatrix} \qquad (2.3)$$

is nonsingular for $\operatorname{Re}\lambda > 0$, then (2.2) is called the *canonical representation* of Φ.

Any representation (2.1) is clearly canonical if $\tau = 0$, $\det \Gamma \neq 0$, provided that $A + bh^*$ is a Hurwitz matrix.

THEOREM 2.1. *If the functional (1.3) is lower semibounded on the set $\mathfrak{M}[a]$ of admissible processes, it admits a canonical representation (2.2), where l^*, κ, and φ are uniquely determined up to multiplication on the left by a constant unitary matrix. The matrices l and κ in (2.2) are dependent only on the coefficients of $LQ(\mathfrak{M}[a], \Phi)$.*

In what follows we assume that a canonical representation exists. The problem of evaluating the matrices l, κ and the functions $\varphi(\cdot)$ and Φ_* in (2.2) will be considered in §3. For the moment, we remark that formula (2.2) also determines Φ when $x(\cdot)$ and $u(\cdot)$ are distributions. Indeed, to ensure that the functional in (2.2) is well defined, we must demand that

$$l^*x + \kappa u \in \mathfrak{L}_2^m. \qquad (2.4)$$

Let (2.2) be a canonical representation. We define the space $\mathfrak{M}_*[a]$ of *generalized processes* as the set of pairs $(x, u) \in (\mathfrak{D}_+^n)' \times (\mathfrak{D}_+^m)'$ of distributions that satisfy equation (1.2) and condition (2.4). The functional Φ is extended from $\mathfrak{M}[a]$ to $\mathfrak{M}_*[a]$ by formula (2.2). The problem of minimizing Φ on $\mathfrak{M}_*[a]$ is referred to as the *generalized stationary linear-quadratic optimal control* ($LQ(\mathfrak{M}_*[a], \Phi)$ for short).

THEOREM 2.2. *Let (2.2) be the canonical representation of Φ. Then the exact lower bounds of Φ on $\mathfrak{M}_*[a]$ and of Φ on $\mathfrak{M}[a]$ are equal and the problem $LQ(\mathfrak{M}_*[a], \Phi)$ has a unique solution: the generalized optimal process (x_*, u_*) defined by (1.2) and the generalized optimal regulator*

$$u = Rx + \alpha(p)a\delta(t) + \beta(p)\varphi, \qquad p = d/dt. \qquad (2.5)$$

Here the polynomials α and β and the matrix R are given by

$$R = U_{-1} + R_0(I_n - X_{-1}),$$
$$\alpha(\lambda) = U(\lambda) - RX(\lambda),$$
$$\beta(\lambda) = (I_m - R(\lambda I_n - A)^{-1}b)(l^*(\lambda I_n - A)^{-1}b + \kappa)^{-1},$$

where R_0 is an arbitrary real $m \times n$ matrix and $X(\lambda)$, $U(\lambda)$, X_{-1}, and U_{-1} are defined by the Laurent expansion of

$$N(\lambda)^{-1} \begin{bmatrix} I_n \\ 0 \end{bmatrix} = \begin{bmatrix} X(\lambda) \\ U(\lambda) \end{bmatrix} = \sum_{j=-\infty}^{k} \begin{bmatrix} X_j \\ U_j \end{bmatrix} \lambda^j$$

around $\lambda = \infty$; $N(\lambda)$ is the matrix (2.3).

REMARK. In the special case where $\tau \equiv 0$, $\mathfrak{G} \geq 0$, and the frequency inequality (0.7) does not degenerate for $\omega \in \mathbb{R}^1$, the expression (2.5) for a generalized optimal process was obtained in [8], [9], [11], but without explicit formulas for α, β, and R.

As we have already noted, there are other definitions of spaces of generalized processes. For example, in [8], [9] Φ is assumed from the start to have the form $\Phi = \|Lx + Du\|^2$, where L and D are constant real matrices. The same authors define generalized processes as pairs (x, u) of distributions that satisfy (1.2) and are regular for $t > 0$, such that $Lx + Du \in \mathfrak{L}_2^k$ and $|x(t)| \to 0$ as $t \to \infty$. In this context an optimal process need not exist. Replacing the regularity of x and u for $t > 0$ and the "stability" condition $|x(t)| \to 0$ as $t \to \infty$ by a weaker condition of quasistability of the process (x, u) (see below), we can define generalized processes without assuming that Φ admits a canonical representation (2.2). The rigorous assertion will now be given.

DEFINITION 2.2. A distribution $f \in (\mathfrak{D}_+^k)'$ is *quasistable* if its Laplace transform $\tilde{f}$ [16] exists and is analytic in the open right half-plane $\operatorname{Re} \lambda > 0$.

LEMMA 2.3. *Let* Φ *be a functional*

$$\Phi(x(\cdot), u(\cdot)) = \|Lx + Du - y_0\|^2 + \mathrm{const}, \qquad (2.6)$$

let (2.2) *be the canonical representation of* Φ, *and let* $\mathfrak{M}_*[a]$ *be the space of generalized processes. Then* $\mathfrak{M}_*[a]$ *is the set of pairs* $(x, u) \in (\mathfrak{D}_+^n)' \times (\mathfrak{D}_+^m)'$ *of quasistable processes satisfying* (1.2) *and the condition* $Lx + Du \in \mathfrak{L}_2^k$, *and the functionals* (2.2) *and* (2.6) *coincide on* $\mathfrak{M}_*[a]$.

REMARK. The authors of [8] consider the space $\mathfrak{M}_c[a]$ of generalized processes without the stability property. The elements of $\mathfrak{M}_c[a]$ are pairs $(x, u) \in (\mathfrak{D}_+^n)' \times (\mathfrak{D}_+^m)'$ of distributions satisfying (1.2) and the condition $Lx + Du \in \mathfrak{L}_2^k$. The functional Φ is defined on $\mathfrak{M}_c[a]$ by formula (2.6). It can be shown that if $\ker D = \{0\}$, then $\mathfrak{M}_c[a]$ is the set of admissible processes in the linear-quadratic problem "without stability", (0.1), (0.2), (0.4). But if $\ker D \neq \{0\}$, then $\mathfrak{M}_c[a]$ is the "natural" space [20] of generalized processes for the linear-quadratic problem "without stability".

Lemma 2.3 implies, in particular, that the generalized processes in the class $\mathfrak{M}_*[a]$ are distributions with analytic Laplace transforms in the open right half-plane. Actually, the generalized processes lie in a much narrower class of distributions. The definition of $\mathfrak{M}_*[a]$ depends essentially on "frequencies" $\omega \in \mathbb{R}^1 \cup \{\infty\}$ at which the frequency inequality (0.7) degenerates, as well as on the degree of that degeneracy.

DEFINITION 2.3. Let A be a matrix having no eigenvalues on the imaginary axis and let $\Pi(i\omega) = \Pi(i\omega)^* \geq 0$ be the matrix function defined in (0.9). Then the *degree of degeneracy* of inequality (0.7) at a point $\omega = \omega_0 \in \mathbb{R}^1 \cup \{\infty\}$ is the multiplicity of the pole of $\Pi(i\omega)^{-1}$ at $\omega = \omega_0$.

Using inequality (0.11), it can be shown that the degree of degeneracy is always even; it differs from zero only at points of degeneracy. The degree of degeneracy depends only on A, b, and $\mathfrak{G}$ and is invariant under a feedback transformation $u = u_1 + r^*x$, i.e., under the substitutions $A \to A_1$, $b \to b_1$, $\mathfrak{G} \to \mathfrak{G}_1$, where $A_1 = A + br^*$, $b_1 = b$, $\mathfrak{G}_1(x, u) = \mathfrak{G}(x, u + r^*x)$, and r is an arbitrary constant matrix such that $\det(i\omega I_n - A_1) \neq 0$ for $\omega \in \mathbb{R}^1$. Hence, if A has eigenvalues on the imaginary axis, the degree of degeneracy of the triple $(A, b, \mathfrak{G})$ is defined as that of the triple $(A_1, b_1, \mathfrak{G}_1)$, where $b_1 = b$, $\mathfrak{G}_1(x, u) = \mathfrak{G}(x, u + r^*x)$, $A_1 = A + br^*$, and $\det(i\omega I_n - A_1) \neq 0$ for $\omega \in \mathbb{R}^1$.

THEOREM 2.4. *Let $2s$ be the degree of degeneracy of the inequality* (0.7) *at $\omega = \infty$ and let $2k$ be its maximal degree of degeneracy for $\omega \in \mathbb{R}^1$. Let $W_* = W_2^n\langle s, k\rangle \times W_2^m\langle s, k\rangle$. Then*

1°. $\mathfrak{M}_*[a] \subset W_*$.
2°. *If $v_* \in \mathfrak{M}_*[a]$ is an optimal generalized process and $\{v_j\}$ an optimizing sequence of generalized processes, i.e., $v_j \in \mathfrak{M}_*[a]$ and $\Phi(v_j) \to \Phi(v_*)$, then $v_j \to v_*$ in W_* (a fortiori $v_j \to v_*$ in $(\mathfrak{D}_+^n)' \times (\mathfrak{D}_+^m)'$).*

REMARK. The convergence of optimizing sequences of a special type was proved in [10] for $k = 0$ and the space $(\mathfrak{D}_+^n)' \times (\mathfrak{D}_+^m)'$ instead of W_* (i.e., (0.7) degenerates only at $\omega = \infty$).

Theorem 2.4 yields a qualitative description of the space of generalized processes (hence also of generalized optimal processes). In nonsingular problems $\mathfrak{M}[a] = \mathfrak{M}_*[a]$ and so an optimal process always exists. If the frequency inequality (0.7) degenerates at some $\omega \in \mathbb{R}^1$, the generalized processes may increase at a polynomial rate as $t \to \infty$. If degeneracy occurs at $\omega = \infty$, then generalized processes are δ-functions or their derivatives.

Let the coefficients of the problem and the initial state be fixed. Then it is easy to see that every generalized process $v \in \mathfrak{M}_*[a]$ is a generalized optimal process for a suitable choice of τ in (1.3). Thus, Theorem 2.4 yields a "natural" definition of generalized processes as the generalized limits of optimizing sequences of admissible processes for all possible inputs (see also [20]).

§3. Canonical representation of the cost functional

In this section the following important "technical" questions are considered: how can we ascertain whether the functional (1.3) is lower semibounded on $\mathfrak{M}[a]$? How can we find the "coefficients" l, κ, $\varphi(\cdot)$, and Φ_* of the canonical representation (2.2) of Φ? For $\tau_x \equiv 0$, $\tau_u \equiv 0$, the nonsingular problem was partially solved by Theorem 1. A complete solution for the nonsingular problem is given in [21]. In singular problems representation of the functional in canonical form (2.2) is more complicated except in the case $\tau \equiv 0$, when the solution follows readily from [22].

First, let us formulate some properties of l, κ, $\varphi(\cdot)$, Φ_* on which numerical methods for their determination are based.

LEMMA 3.1. *Let* (2.2) *be the canonical representation of the functional* (1.3). *Then there exist a unique symmetric* $n \times n$ *matrix* H, *a function* $\psi(\cdot) \in \mathfrak{L}_2^n$, *and a vector* $\psi_0 \in \mathbb{R}^n$ *such that*

$$G + HA + A^*H = ll^*, \qquad Hb + g = l\kappa, \qquad \kappa^*\kappa = \Gamma, \qquad (3.1)$$

$$p\psi + A^*\psi - l\varphi = \tau_x + \psi_0 \delta(t), \qquad b^*\psi - \kappa^*\varphi = \tau_u, \qquad (3.2)$$

$$\Phi_* = a^*HA - \|\varphi(\cdot)\|^2 - 2\psi_0^* a. \qquad (3.3)$$

Conversely, if $H = H^*$, $\psi \in \mathfrak{L}_2^n$, $\varphi \in \mathfrak{L}_2^m$, *conditions* (3.1), (3.2), *and* (3.3) *hold, and the matrix* (2.3) *is invertible for* $\operatorname{Re} \lambda > 0$, *then* (2.2) *is the canonical representation of* (1.3).

Equations (3.1), also known as the Lure$'$ equations, are a natural generalization of system (0.8). An effective algorithm to determine matrices $H = H^*$, l, κ that satisfy (3.1), such that the matrix (2.3) is invertible for $\operatorname{Re} \lambda > 0$, may be found in [22]. Once H, l, and κ have been determined, one must find a (unique!) vector $\psi_0 \in \mathbb{R}^n$ such that the solution $(\psi, \varphi) \in (\mathfrak{D}_+^n)' \times (\mathfrak{D}_+^m)'$ of (3.2) is in $\mathfrak{L}_2^n \times \mathfrak{L}_2^m$. Taking Laplace transforms in (3.2), we obtain

$$\tilde{\theta}(\lambda) = E(\lambda)^{-1}(\tilde{\tau}(\lambda) + E_0 \psi_0), \qquad (3.4)$$

where

$$\tilde{\tau} = \begin{bmatrix} \tilde{\tau}_x \\ \tilde{\tau}_u \end{bmatrix}; \quad \tilde{\theta} = \begin{bmatrix} \tilde{\psi} \\ \tilde{\varphi} \end{bmatrix}; \quad E_0 = \begin{bmatrix} I^n \\ 0 \end{bmatrix}; \quad E(\lambda) = \begin{bmatrix} \lambda I_n + A^* & -l \\ b^* & -\kappa^* \end{bmatrix}. \qquad (3.5)$$

Therefore, we can find ψ_0 from the condition $\tilde{\theta} \in \mathfrak{H}_2$, where $\mathfrak{H}_2$ is the class of vector-valued functions with components in the Hardy space in the right half-plane. If there is no such ψ_0, then $\inf \Phi = -\infty$ on $\mathfrak{M}[a]$ for all $a \in \mathbb{R}^n$; if ψ_0 is known we can calculate $\psi(\cdot)$ and $\varphi(\cdot)$ once their transforms $\tilde{\psi}$ and $\tilde{\varphi}$ have been obtained from (3.4) and (3.5).

Suppose the functional $\tilde{\tau}$ in (3.4) and (3.5) is rational. Then, as shown in [23], $\inf \Phi \neq -\infty$ on $\mathfrak{M}[a]$, and hence the required $\psi_0 \in \mathbb{R}^n$ exists. It is obvious that in this case ψ_0 is uniquely determined by the condition that

$\tilde{\theta}(\lambda) \to 0$ as $\lambda \to \infty$ and that $\tilde{\theta}(\lambda)$ and $E(\lambda)^{-1}$ have no common poles in the half-plane $\operatorname{Re}\lambda \geq 0$. It can be shown that these conditions are equivalent to a system of n independent linear equations in ψ_0.

If $\tilde{\tau}$ is not rational, the following lemma reduces everything to the previous case.

LEMMA 3.2. *Let* $H = H^*$, l, κ *be matrices satisfying* (3.1) *and such that the matrix* (2.3) *is nonsingular for* $\operatorname{Re}\lambda > 0$; *let* $E(\lambda)$, E_0, *and* $\tilde{\tau}(\lambda)$ *be defined by* (3.5). *Then*

(1) *system* (3.4) *has a solution* $\tilde{\theta} \in \mathfrak{H}_2$, $\psi_0 \in \mathbb{R}^n$ *if and only if there exists a rational vector-valued function* $\tilde{\theta}_0(\lambda)$ *such that* $E(\lambda)^{-1}\tilde{\tau}(\lambda) - \tilde{\theta}_0(\lambda)$ *is square integrable over the imaginary axis*.

(2) *If* (3.4) *has a solution* $\tilde{\theta} \in \mathfrak{H}_2$, $\psi_0 \in \mathbb{R}^n$, *then the function* $\tilde{\theta}_0$ *in* (1) *can be chosen so that* $E(\lambda)^{-1}\tilde{\tau}(\lambda) - \tilde{\theta}_0(\lambda)$ *is in class* $\mathfrak{H}_2$, *and* $\tilde{\tau}_0 = E\tilde{\theta}_0 \in \mathfrak{H}_2$. *In that case* $\psi_0 \in \mathbb{R}^n$ *is uniquely determined by the condition*

$$E(\lambda)^{-1}(\tilde{\tau}_0(\lambda) + E_0\psi_0) \in \mathfrak{H}_2.$$

§4. Construction of optimizing sequences

The generalized regulator (2.5) effectively determines a generalized optimal process. In singular nongeneralized optimal control, equation (2.5) is a limiting relation satisfied by optimizing sequences; it gives no information as to how to construct such sequences. In principle, this can be done by regularization.

Consider a "perturbed" optimal control problem $LQ(\mathfrak{M}[a], \Phi_\varepsilon)$ with a parameter ε, obtained from $LQ(\mathfrak{M}[a], \Phi)$ by replacing the form $\mathfrak{G}$ with a form $\mathfrak{G}_\varepsilon$ that is positive definite on $\mathfrak{N}[\omega]$ for all $\omega \in \mathbb{R}^1 \cup \{\infty\}$. In particular, we can take $\mathfrak{G}_\varepsilon$ to be

$$\mathfrak{G}_\varepsilon(x, u) = \mathfrak{G}(x, u) + \varepsilon(|x|^2 + |u|^2), \qquad \varepsilon > 0. \tag{4.1}$$

The "perturbed" problem has an optimal solution — the process defined by (0.1), (0.2), and the linear regulator

$$u(t) = h_\varepsilon^*(t) + \tau_\varepsilon(t), \tag{4.2}$$

where the matrix h_ε and the function $\tau_\varepsilon(\cdot)$ do not depend on the initial state $a \in \mathbb{R}^n$ and such that $A + bh_\varepsilon^*$ is Hurwitz and $\tau_\varepsilon(\cdot) \in \mathfrak{L}_2^m$. It is well known (see, for example, [16]) that if $\mathfrak{G}_\varepsilon \to \mathfrak{G}$ (for $\mathfrak{G}_\varepsilon$ defined by (4.1), $\varepsilon \to 0$), then (4.2) is an optimizing sequence for the problem $LQ(\mathfrak{M}[a], \Phi)$.

The construction of optimizing sequences by regularization (4.1) has serious disadvantages. First, using (4.2) to evaluate h_ε and $\tau_\varepsilon(\cdot)$ for each new ε, we must solve the problem $LQ(\mathfrak{M}[a], \Phi_\varepsilon)$ anew each time, which involves considerable computational difficulties. Second, regulators (4.2) may determine a nonoptimal process even for those initial states $a \in \mathbb{R}^n$ for which the optimal process in the problem $LQ(\mathfrak{M}[a], \Phi)$ exists. The disadvantages can be avoided by a suitable choice of $\mathfrak{G}_\varepsilon$. Namely, if the matrices l and κ of Lemma 3.1 and the "frequencies" $\omega \in \mathbb{R}^1 \cup \{\infty\}$ at which the inequality (0.7) becomes degenerate are known, we can construct an optimizing sequence of regulators (4.2) using only rational operations over matrices, so that each regulator produces an optimal process for any initial state $a \in \mathbb{R}^n$ for which such a process exists (such sequences will be called *strong optimizing sequences*). A corresponding algorithm is given in [13], [24] for the general case. Below we will describe a result that can be used to construct a strong optimizing sequence in the important special case $\tau \equiv 0$.

LEMMA 4.1. *Let A, b, l, κ be real matrices of orders $n \times n$, $n \times m$, $n \times m$, and $m \times m$, respectively, let $N(\lambda)$ be the matrix (2.3), and let $\alpha(\lambda) = \det N(\lambda)$.*

1°. *Let $\lambda_0 = i\omega$, $\omega \in \mathbb{R}^1$, be a root of the polynomial α, and let $v \in \mathbb{C}^{n+m}$ be a vector such that $v^* N(i\omega) = 0$ ($v = \mathrm{col}(p, q)$, where $p \in \mathbb{C}^n$, $q \in \mathbb{C}^m$, $|q| = 1$). Set*

$$l_1 = l + \varepsilon \operatorname{Re}(pq^*), \qquad \kappa_1 = \kappa, \qquad (4.3)$$

where $\varepsilon > 0$ and Re denotes the componentwise real part. Let $\alpha_1(\lambda) = \det N_1(\lambda)$, where

$$N_1(\lambda) = \begin{bmatrix} \lambda I_n - A & -b \\ l_1^* & \kappa_1 \end{bmatrix}. \qquad (4.4)$$

Then for some $\theta \in [0, 1/4]$ ($\theta = 0$ for $\omega = 0$, $p \in \mathbb{R}^n$, $q \in \mathbb{R}^m$), we have

$$\alpha_1(\lambda) = (\lambda^2 + \varepsilon\lambda + \omega^2 + \varepsilon^2\theta)(\lambda^2 + \omega^2)^{-1}\alpha(\lambda),$$

and for all $(x, u) \in \mathfrak{M}[a]$, $a \in \mathbb{R}^n$,

$$\|l^*x + \kappa u\|^2 \leq \varepsilon |p^*a|^2 + \|l_1^*x + \kappa_1 u\|^2, \qquad (4.5)$$

where we set $p^*a = 0$ and $l_1^*x + \kappa_1 u \equiv 0$ for $l^*x + \kappa u \equiv 0$.

2°. *Let $\det \kappa = 0$ and let $q \in \mathbb{R}^m$ be a vector such that $|q| = 1$ and $q^*\kappa = 0$. Put $p = lq$ and*

$$l_1 = l + \varepsilon A^* pq^*, \qquad \kappa_1 = \kappa + \varepsilon q p^* b. \qquad (4.6)$$

Then $\alpha_1(\lambda) = (\varepsilon\lambda + 1)\alpha(\lambda)$, where α_1 is the determinant of the matrix N_1 in (4.4), and for all $(x(\cdot), u(\cdot)) \in \mathfrak{M}[a]$, $a \in \mathbb{R}^n$, inequality (4.5) holds; moreover, if $l^*x + \kappa u \equiv 0$, then $p^*a = 0$ and $l_1^*x + \kappa_1 u \equiv 0$.

One can say that the transformation (4.3) shifts the roots of the determinant $\alpha(\lambda)$ of (2.3) that lie on the imaginary axis to the left half-plane. If $\det \kappa = 0$ (or, equivalently, if $\deg \alpha < n$), one can say that α has a root

$\lambda = \infty$ of multiplicity $n - \deg \alpha$. The transformation (4.6) shifts one of the roots $\lambda = \infty$ of α to the left half-plane.

Let (2.2) be the canonical representation of the functional Φ. If α has no roots at the imaginary axis (including roots $\lambda = \infty$), then it is easy to see that problem $LQ(\mathfrak{M}[a], \Phi)$ is nonsingular and the optimal regulator (0.6) is defined by the equality $h^* = \kappa^{-1} l^*$. In the singular case, once the pure imaginary roots λ_j of α are known, we can construct matrices $l_1 = l_1(\varepsilon)$, $\kappa_1 = \kappa_1(\varepsilon)$ by successive transformations (4.3) and (4.6) so that the determinant $\alpha_1(\lambda)$ of matrix (4.4) will have no roots on the imaginary axis (including $\lambda = \infty$). The regulators $u = -\kappa_1(\varepsilon)^{-1} l_1(\varepsilon)^* x$, $\varepsilon \to +0$, will then form a strongly optimizing sequence for the problem $LQ(\mathfrak{M}[a], \Phi)$; inequality (4.5) yields an effective upper bound for Φ over the processes produced by these regulators (for details, see [13], [24]).

§5. Optimal control under noise

Many situations requiring optimal filtration or optimal control with perturbations may be reduced to the deterministic stationary linear-quadratic optimal control problem, considered above, thanks to the natural duality between stochastic and deterministic problems. Some examples will now illustrate the application of the results of §§2–4 to optimal control and filtration.

Let $(\Omega, \mathfrak{P}, P)$ be a probability space, and let E_2^n be the standard Hilbert space of vector-valued random variables $\theta : \Omega \to \mathbb{R}^n$ with finite variance and zero expectation. Replacing $\mathbb{R}^n$ by E_2^n in the definition of the spaces $\mathfrak{D}_+^n$ and $(\mathfrak{D}_+^n)'$, we get spaces $P\mathfrak{D}_+^n$ and $(P\mathfrak{D}_+^n)'$ of classical and generalized stochastic processes respectively. Let PC^n be the space of bounded measurable functions $f : [0, \infty) \to E_2^n$, and let PC_0^n be the subspace of all functions $f \in PC^n$ such that $f(t) \to 0$ as $t \to \infty$. As before, we have a natural identification of classical and generalized processes, under which $P\mathfrak{D}_+^n \subset PC_0^n \subset PC^n \subset (P\mathfrak{D}_+^n)'$.

Let the controlled object be given by equations (0.13), (0.15), where $\xi \in (P\mathfrak{D}_+^n)'$ is the "state", $\eta \in (P\mathfrak{D}_+^m)'$ the "control", $\zeta_1 \in (P\mathfrak{D}_+^n)'$ the "noise in the controlled object", $\chi \in (P\mathfrak{D}_+^k)'$ the "observable output", and $\zeta_2 \in (P\mathfrak{D}_+^k)'$ the "noise in the observable output".

We first consider the simplest stochastic optimal control problem: the case of completely observable state, $\chi = \xi$. We can then assume that $\zeta_1 = a\zeta_0$, where $a \in \mathbb{R}^n$ is a given vector and ζ_0 is the normalized white noise. Define the space $P\mathfrak{M}[a]$ of *admissible stochastic processes* as the set of pairs (ξ, η) of functions $\xi \in PC^n$, $\eta \in PC^m$ that satisfy equation (0.13) and the condition

$$\sup_{t \geq 0} M(|\xi(t)|^2 + |\eta(t)|^2) < \infty$$

and can be represented in the form

$$\xi(t) = \xi_0(t) + \int_0^t x(t-s)\zeta_0(s)\,ds, \qquad (5.1)$$
$$\eta(t) = \eta_0(t) + \int_0^t u(t-s)\zeta_0(s)\,ds,$$

where $\xi_0 \in PC_0^n$, $\eta_0 \in PC_0^m$, and $x(\cdot)$ and $u(\cdot)$ are locally integrable deterministic functions. The problem of minimizing the functional (0.16) over the set of admissible stochastic processes $P\mathfrak{M}[a]$ will be denoted by $LQ(P\mathfrak{M}[a], P\Phi)$. Notice that the definition of admissible stochastic processes in the problem $LQ(P\mathfrak{M}[a], P\Phi)$ is a natural generalization of the notion of the processes generated by stabilizing feedback (0.14), where $\chi = \xi$. For example, we can easily see that equations (0.13) and $\eta = h^*\xi$, where $A + bh^*$ is a Hurwitz matrix, determine a stochastic process $(\xi, \eta) \in P\mathfrak{M}[a]$ such that the pair $(x(\cdot), u(\cdot))$ in (5.1) is the admissible process $(x, u) \in \mathfrak{M}[a]$ determined by regulator (0.6).

The following result enables us to apply the previous results to the problem $LQ(P\mathfrak{M}[a], P\Phi)$.

LEMMA 5.1. *Let $\tau \equiv 0$ in (1.3). Then $LQ(\mathfrak{M}[a], \Phi)$ and $LQ(P\mathfrak{M}[a], P\Phi)$ are equivalent. In other words,*

1°. *any admissible stochastic process $(\xi, \eta) \in P\mathfrak{M}[a]$ can be represented in the form (5.1), where $(x, u) \in \mathfrak{M}[a]$; conversely, for any admissible process $(x, u) \in \mathfrak{M}[a]$ equalities (5.1) with $\xi_0 \equiv 0$, $\eta_0 \equiv 0$ determine an admissible stochastic process $(\xi, \eta) \in P\mathfrak{M}[a]$.*
2°. *If the processes $(\xi, \eta) \in P\mathfrak{M}[a]$ and $(x, u) \in \mathfrak{M}[a]$ are related by (5.1), then*

$$P\Phi(\xi, \eta) = \Phi(x, u).$$

Lemma 5.1 implies, in particular, that the problem $LQ(P\mathfrak{M}[a], P\Phi)$ has an optimal process if and only if the problem $LQ(\mathfrak{M}[a], \Phi)$ does. Moreover, if $u = h_\varepsilon^* x$ is a strongly optimizing sequence of regulators in $LQ(\mathfrak{M}[a], \Phi)$, then $\{\eta = h_\varepsilon^*\xi\}$ is a strongly optimizing sequence of regulators in $LQ(P\mathfrak{M}[a], P\Phi)$.

We now consider the general case, when only the output (0.15) can be measured. Let us say that a stochastic process $e \in PC^q$ is of convolution type in the noise $\zeta = \mathrm{col}(\zeta_1, \zeta_2)$ if it can be represented as

$$e(t) = e_0(t) + \int_0^t E(t-s)\zeta(s)\,ds, \qquad \zeta = \mathrm{col}(\zeta_1, \zeta_2), \qquad (5.2)$$

where $e_0 \in PC_0^q$ and E is a deterministic matrix-valued function with components in $\mathfrak{L}_2^1$. Notice that (5.2) has the same meaning as (5.1) in the problem $LQ(P\mathfrak{M}[a], P\Phi)$: it defines the stationary stable dependence of the process e on the external perturbations ζ_1, ζ_2. We call a process e of convolutionary type an *admissible estimate in system* (0.13), (0.15) if the deterministic function $E(\cdot)$ in (5.2) is such that for $e_0 \equiv 0$ and for arbitrary deterministic functions $\zeta_1 \in \mathfrak{L}_2^n$, $\zeta_2 \in \mathfrak{L}_2^k$, $\xi \in \mathfrak{L}_2^n$ the equations (0.13), (0.15) with $\eta \equiv 0$ and $\chi \equiv 0$ (i.e., the equations $P\xi = A\xi + \zeta_1$, $c^*\xi + \zeta_2 = 0$) imply $e(t) \equiv 0$.

Define the space $F\mathfrak{M}$ of *admissible stochastic processes in the problem with incomplete observability* as the set of pairs (ξ, η) of processes $\xi \in PC^n$, $\eta \in PC^m$ such that they are of convolution type in the noise ζ, (0.13) holds, and $\eta(\cdot)$ is an admissible estimate in system (0.13), (0.15). We assume that the noise $\zeta = \mathrm{col}(\zeta_1, \zeta_2)$ in (0.13), (0.15) is a fixed white noise with correlation function

$$M\zeta(t)\zeta(s)^* = \begin{bmatrix} G_0 & g_0 \\ g_0^* & \Gamma_0 \end{bmatrix} \delta(t-s), \qquad (5.3)$$

where $G_0 = G_0^*$, g_0, and $\Gamma_0 = \Gamma_0^*$ are constant matrices of orders $n \times n$, $n \times k$, and $k \times k$ respectively. The problem of minimizing $P\Phi$ over the set $F\mathfrak{M}$ is called the *stochastic optimal control problem with incomplete observability* ($LQ(F\mathfrak{M}[a], P\Phi)$ for short).

Before stating the solution of the problem $LQ(F\mathfrak{M}[a], P\Phi)$, let us consider one more auxiliary problem, which is of independent interest. Let $(\eta, \xi) \in F\mathfrak{M}$ be a fixed admissible process in the problem $LQ(F\mathfrak{M}, P\Phi)$, $a_0 \in \mathbb{R}^n$. The problem of minimizing the functional

$$F\Phi = \overline{\lim_{t\to\infty}} M|a_0^*\xi(t) - \nu(t)|^2 \qquad (5.4)$$

over the set $F(\zeta) = \{\nu\}$ of admissible scalar estimates ν in system (0.13), (0.15) is called the *optimal filtration problem* ($LQ(F(\zeta), F\Phi)$ for short).

It turns out that this problem can be reduced to a deterministic problem $LQ(\mathfrak{M}[a], \Phi)$ by a suitable choice of matrices A, b and of the form $\mathfrak{G}$ in (1.2), (1.3), while the problem $LQ(F\mathfrak{M}, P\Phi)$ can be "split" into two independent problems: stochastic optimal control with complete observability and optimal filtration.

LEMMA 5.2. *Assume that (A^*, c) is a stabilizable pair and that*

$$\mathfrak{M}_0[a_0] = \{(x, u) \in \mathfrak{L}_2^n \times \mathfrak{L}_2^k : px = A^*x + cu + a_0\delta(t)\},$$

$$\Phi_0(x, u) = \int_0^\infty (x(t)^* G_0 x(t) + 2x(t)^* g_0 u(t) + u(t)^* \Gamma_0 u(t))\,dt,$$

where $a_0 \in \mathbb{R}^n$. Then for any process $(\xi, \eta) \in F\mathfrak{M}$ the problems $LQ(F(\zeta), F\Phi)$ and $LQ(\mathfrak{M}_0[a_0], \Phi_0)$ are equivalent. In other words,

$1°$. *if $\nu(\cdot) \in PC^1$ is an admissible estimate in system (0.13), (0.15), then*

$$\nu(t) - a_0^*\xi(t) = \nu_0(t) - \int_0^t \{x(t-s)^*\zeta_1(s) + u(t-s)^*\zeta_2(s)\}\,ds, \qquad (5.5)$$

where $\nu_0 \in PC_0^1$, $(x(\cdot), u(\cdot)) \in \mathfrak{M}_0[a_0]$. Conversely, any admissible process $(x, u) \in \mathfrak{M}_0[a_0]$ has an admissible estimate $\nu \in F(\zeta)$ of the form (5.4).

$2°$. *If $\nu \in F(\zeta)$ and $(x, u) \in \mathfrak{M}_0[a_0]$ are related by* (5.5), *then*

$$\Phi_0(x(\cdot), u(\cdot)) = F\Phi(\nu(\cdot)).$$

$3°$. *If $(x, u) \in \mathfrak{M}_0[a_0]$ is the process generated by a regulator* (0.6), *then the equations*

$$p\hat{\xi} = A\hat{\xi} + b\eta + h(c^*\hat{\xi} + d^*\eta - \chi) + \hat{\xi}(0)\delta(t), \tag{5.6}$$

$$\nu = a_0^*\hat{\xi} \tag{5.7}$$

define an admissible estimate $\nu \in F(\zeta)$ of the form (5.5) *for all $\hat{\xi}(0) \in E_2^n$.*

REMARK. Lemma 5.2 implies that the determination of an optimal regulator or optimizing sequence in $LQ(F(\zeta), F\Phi)$ is the same as in $LQ(\mathfrak{M}_0[a_0], \Phi_0)$. Relations (5.6) and (5.7) are written in the form of a Kalman-Bucy filter. When $LQ(\mathfrak{M}_0[a_0], \Phi_0)$ is a nonsingular problem and (0.6) is an optimal regulator, Equations (5.6) and (5.7) determine an optimal Wiener estimate of the process $a_0^*\xi$ based on observations (0.15).

THEOREM 5.3. *Let* (2.2) *be the canonical representation of the functional Φ with $\tau \equiv 0$ in* (1.3), $\Phi_*(a) = a^*Ha$, *and $H = H^*$. Let*

$$\Phi_0(x(\cdot), u(\cdot)) = a_0^*H_0a_0 + \|l_0^*x + \kappa_0 u\|^2$$

be the canonical representation of Φ_0. Then

$1°$. *the infimum of $P\Phi$ over $F\mathfrak{M}$ is equal to the trace of the matrix $l^*H_0l + HG_0$.*

$2°$. *Let $\varepsilon \to +0$ be a parameter and let $l_1 = l_1(\varepsilon)$, $l_{01} = l_{01}(\varepsilon)$, $\kappa_1 = \kappa_1(\varepsilon)$, $\kappa_{01} = \kappa_{01}(\varepsilon)$ be matrices such that $l_1(\varepsilon) \to l$ as $\varepsilon \to 0$. Let $A - b\kappa_1^{-1}l_1^*$ and $A^* - c\kappa_{01}^{-1}l_{01}^*$ be Hurwitz matrices. Assume that*

$$\|l^*x + \kappa u\|^2 \leq a^*Ka + \|l_1^*x + \kappa_1 u\|^2, \quad \forall (x, u) \in \mathfrak{M}[a], \quad \forall a \in \mathbb{R}^n,$$

$$\|l_0^*x + \kappa_0 u\|^2 \leq a^*K_0a + \|l_{01}^*x + \kappa_{01} u\|^2, \quad \forall (x, u) \in \mathfrak{M}_0[a_0], \quad \forall a_0 \in \mathbb{R}^n,$$

as $\varepsilon \to 0$, where $K = K(\varepsilon) \to 0$ and $K_0 = K_0(\varepsilon) \to 0$ as $\varepsilon \to 0$. Then the stochastic processes (ξ, η) defined by system (0.13), (0.15) *and the equations*

$$\eta = -\kappa_1^{-1}l_1^*\hat{\xi}, \tag{5.8}$$

$$p\hat{\xi} = A\hat{\xi} + b\eta - l_{01}(\kappa_{01}^*)^{-1}(c^*\hat{\xi} + d^*\eta - \chi) + \hat{\xi}(0)\delta(t), \tag{5.9}$$

form an optimizing sequence ($\varepsilon \to 0$) in the problem $LQ(F\mathfrak{M}, P\Phi)$ for an arbitrary $\hat{\xi}(0) \in E_2^n$; moreover,

$$P\Phi(\eta, \xi) - \inf P\Phi \leq \mathrm{Tr}\{KG_0 + l_1^*K_0l_1\}.$$

REMARK. A sequence of matrices $l_1, l_{01}, \kappa_1, \kappa_{01}$ with the properties required in $2°$ of Theorem 5.3 can be constructed by the method of §4. Theorem 5.3 is a generalization to the singular case of the "decomposition principle", which can be formulated as follows: if $\eta = h^*\xi$ is an optimal regulator in an optimal control problem with a completely observable state, and $\hat{\xi}$ is an optimal estimate of the state ξ based on the observed output, then $\eta = h^*\hat{\xi}$ is an optimal regulator in the problem with incomplete observation.

§6. Proofs of Theorem 2.1 and Lemma 3.1

Let $\mathfrak{H}_2^k = \{\tilde{z} : z \in \mathfrak{L}_2^k\}$ be the Hardy vector space in the right half-plane, and let $\mathfrak{H}_*^k$ be the set of transforms $\tilde{z}$ of quasistable distributions $z \in (\mathfrak{D}_+^k)'$ (the elements of $\mathfrak{H}_*^k$ are functions analytic in the half-plane $\operatorname{Re}\lambda > 0$ whose modulus is majorized by a polynomial in $|\lambda|$ in any half-plane $\operatorname{Re}\lambda \geq \varepsilon > 0$ [19]). We use the following standard result on approximation in Hardy spaces.

LEMMA 6.1. *Let $W(\lambda)$ be a rational $k \times q$ matrix. Then the set $\mathfrak{H}_2^k \cap \{Wv : v \in \mathfrak{H}_*^q\}$ is the closure in $\mathfrak{H}_2^k$ of the set $\mathfrak{H}_2^k \cap \{Wv : v \in \mathfrak{H}_2^q\}$.*

PROOF. There exists a rational matrix W^+ such that $WW^+W = W$ and W^+W is a polynomial matrix (this follows easily from a well-known result [26] about the diagonalization of polynomial matrices). Let $v_j \in \mathfrak{H}_2^q$, $Wv_j \in \mathfrak{H}_2^k$, and $Wv_j \to z$ in $\mathfrak{H}_2^k$. Then $W(\lambda)v_j(\lambda) \to z(\lambda)$ for $\operatorname{Re}\lambda > 0$. Therefore $W^+(\lambda)W(\lambda)v_j(\lambda) \to W^+(\lambda)z(\lambda)$ uniformly on any compact subset of the half-plane $\operatorname{Re}\lambda > 0$ that contains no poles of W and W^+. Since the set of the poles is finite, $v = W^+z \in \mathfrak{H}_*^q$. We have $W(\lambda)v(\lambda) = \lim W(\lambda)W^+(\lambda)W(\lambda)v_j(\lambda) = z(\lambda)$ for almost all λ with $\operatorname{Re}\lambda > 0$. Therefore $z \in \mathfrak{H}_2^k \cap \{Wv : v \in \mathfrak{H}_*^q\}$. Conversely, let $v \in \mathfrak{H}_*^q$ and $Wv = z \in \mathfrak{H}_2^k$. Then $W^+z = W^+Wv \in \mathfrak{H}_*^q$. Take a rational function ρ with no zeros in $\operatorname{Re}\lambda > 0$ and with no poles in $\operatorname{Re}\lambda \geq 0$, such that $|\rho(\lambda)W^*(\lambda)| \leq$ const for $\operatorname{Re}(\lambda) = 0$ and $|\rho(\lambda)W^+(\lambda)| \to 0$ as $|\lambda| \to \infty$. Then $\rho W^+z \in \mathfrak{H}_2^q$. Since ρ is an outer function, we can find a sequence of functions $f_j \in \mathfrak{H}_2^1$ that are bounded for $\operatorname{Re}\lambda > 0$ and satisfy $\|(f_j\rho - 1)z\|_i \to 0$ as $j \to \infty$; here $\|\cdot\|_i$ is the norm in $\mathfrak{H}_2^k$ (see, for example, [27, Chapter 7]. Therefore, $v_j = f_j\rho W^+z \in \mathfrak{H}_2^q$ and $Wv_j = f_j\rho z \to z$ in $\mathfrak{H}_2^k$. Lemma 6.1 is proved.

LEMMA 6.2. *Let l and κ be real matrices of orders $n \times k$ and $k \times m$ respectively, and let $\operatorname{rank}(\operatorname{col}(l, \kappa^*)) = k$. Then the following assertions are equivalent.*

(I) *The matrix (2.3) is right invertible for $\operatorname{Re}\lambda > 0$.*
(II) *For each $a \in \mathbb{R}^n$ and any $\varepsilon > 0$ there exists $(x, u) \in \mathfrak{M}[a]$ such that $\|l^*x + \kappa u\| < \varepsilon$.*
(III) *For each $a \in \mathbb{R}^n$ the set $\{l^*x + \kappa u : (x, u) \in \mathfrak{M}[a]\}$ is dense in $\mathfrak{L}_2^k$.*

PROOF. The implication (III) $\Longrightarrow$ (II) is obvious. Taking Laplace transforms, we get the implication (I) $\Longrightarrow$ (III) as a special case of Lemma 6.1 with $W(\lambda) = \kappa + (l^* + \kappa r^*)(\lambda I_n - B)^{-1}b$, where $B = A + br^*$ is a Hurwitz matrix. We now prove that (II) implies (I). Assume, on the contrary, that $\det N(\lambda_0) = 0$, $\operatorname{Re}\lambda_0 > 0$. Then $p^*A = \lambda_0 p^* + q^*l^*$ and $p^*b = q^*\kappa$, where $p \in \mathbb{C}^n$ and $q \in \mathbb{C}^k$ are vectors, $|p| + |q| \neq 0$. Hence, for $(x, u) \in \mathfrak{M}[a]$ we get $(d/dt)|p^*x|^2 = 2(\operatorname{Re}\lambda_0)|p^*x|^2 + 2\operatorname{Re} x^*pq^*(l^*x + \kappa u) \geq -\varepsilon |l^*x + \kappa u|^2$, where $\varepsilon > 0$ is a constant. Integrating, we obtain $\|l^*x + \kappa u\|^2 \geq \varepsilon^{-1}|p^*a|^2$. Hence, by (II) we get $p^*a = 0$, $\forall a \in \mathbb{R}^n$, i.e., $p = 0$. By the definition of p and q we have $q \neq 0$, $q^*l^* = 0$, and $q^*\kappa = 0$, contrary to the assumption. Lemma 6.2 is proved.

We now prove Lemma 3.1. For $V(x, t) = x^*Hx - 2\psi(t)^*x$ it follows from (3.1), (3.2) that for $(x, u) \in \mathfrak{M}[a]$, $a \in \mathbb{R}^n$, we have

$$dV(x, t)/dt = |l^*x + \kappa u - \varphi|^2 - |\varphi|^2 + \mathfrak{G}(x, u) - 2\tau_x^* x - 2\tau_u^* u.$$

Integrating, we obtain (2.2), where Φ_* is defined by (3.3). Conversely, let (2.2) hold for all $a \in \mathbb{R}^n$. Then $\Phi_* = \Phi_*(a) = \inf\{\Phi(v) : v \in \mathfrak{M}[a]\}$ is a quadratic function in the argument $a \in \mathbb{R}^n$ and $\Phi_*(0) = -|\varphi|^2$. Therefore (3.3) holds, where $H = H^*$ is an $n \times n$ matrix and $\psi_0 \in \mathbb{R}^n$. Rewriting (2.2) with $x(\cdot)$ replaced by $cx(\cdot)$, $u(\cdot)$ by $cu(\cdot)$, and a by ca ($c \in \mathbb{R}^1$ is a parameter), and equating the coefficients of c^2 and c on both sides of the resulting equality, we obtain

$$\int_0^\infty (\mathfrak{G}(x, u) - |l^*x + \kappa u|^2) \, dt = a^*Ha, \quad \forall (x, u) \in \mathfrak{M}[a], \forall a \in \mathbb{R}^n, \tag{6.1}$$

$$\int_0^\infty \{\tau_x^* x + \tau_u^* u + \varphi^*(l^*x + \kappa u)\} \, dt = -\psi_0^* a, \quad \forall (x, u) \in \mathfrak{M}[a], \forall a \in \mathbb{R}^n. \tag{6.2}$$

Formula (6.1) implies

$$\int_0^\infty (\mathfrak{G}(x, u) - |l^*x + \kappa u|^2) \, dt = x(t)^*Hx(t), \quad \forall (x, u) \in \mathfrak{M}[a], \forall a \in \mathbb{R}^n. \tag{6.3}$$

Differentiating (6.3) with respect to t, we can get

$$\mathfrak{G}(x, u) + 2x^*H(Ax + bu) = |l^*x + \kappa u|^2, \quad \forall x \in \mathbb{R}^n, \forall u \in \mathbb{R}^m, \tag{6.4}$$

which is equivalent to (3.1).

We now consider the identity (6.2). Let $B = A + br^*$ be a Hurwitz matrix, where r is a real $n \times m$ matrix, $\eta = \tau_x + l\varphi + r(\tau_u + \kappa^*\varphi)$. Then $\eta \in \mathcal{L}_2^n$ and there exist a unique vector $\psi(0) \in \mathbb{R}^n$ and a function $\psi \in \mathcal{L}_2^n$ such that $p\psi + B^*\psi = \eta + \psi(0)\delta(t)$ [21, Lemma 17]. Replacing the term $(\tau_x + l\varphi)^*x$ in the integrand of (6.2) by $(p\psi + B^*\psi - \psi_0\delta(t) - r^*(\tau_u + \kappa^*\varphi))x$, we obtain the identity

$$\int_0^\infty (B^*\psi - \kappa^*\varphi - \tau_u)(r^*x - u) \, dt = (\psi(0) - \psi_0)^* a,$$
$$\forall a \in R^n, \quad \forall (x, u) \in \mathfrak{M}[a]. \tag{6.5}$$

Since $B = A + br^*$ is a Hurwitz matrix, $\{r^*x - u : (x, u) \in \mathfrak{M}[a]\} = \mathfrak{L}_2^m$ for all $a \in \mathbb{R}^n$. Therefore, by (6.5), $\psi(0) = \psi_0$ and the second equation of (3.2) follows. Taking into account that $p\psi + B^*\psi = \eta + \psi(0)\delta(t)$ we obtain the first equation of (3.2). Lemma 3.1 is proved.

Now let us prove Theorem 2.1 for $\tau \equiv 0$. By [23, Theorem 4], $V(a) = \inf\{\Phi(v) : v \in \mathfrak{M}[a]\} \neq -\infty$, for all $a \in \mathbb{R}^n$. We know (see, for example, [25]) that $V(a) = a^*Ha$ is a quadratic form, where $H = H^*$ is a real matrix. By the Bellman principle, $\mathfrak{G}(x(t), u(t)) + (d/dt)V(x(t)) \geq 0$ for all $a \in \mathbb{R}^n$ and all $(x, u) \in \mathfrak{M}[a]$. This gives (6.4), where l and κ are real matrices of orders $n \times k$ and $k \times n$ respectively such that $\text{rank}(\text{col}(l, \kappa^*)) = k$. The identity (6.4) implies (6.1). It follows from (6.1) and the definition of H that the matrices l and κ satisfy the assumptions of Lemma 6.2 (II). Hence, the matrix (2.3) is right invertible for $\text{Re}\,\lambda > 0$. At the same time, it follows from (6.4) that

$$\mathfrak{G}(\tilde{x}, \tilde{u}) = |l^*\tilde{x} - \kappa\tilde{u}|^2, \qquad \forall (\tilde{x}, \tilde{u}) \in \mathfrak{M}[i\omega],\ \forall \omega \in \mathbb{R}^1 \cup \infty. \tag{6.6}$$

By assumption, the form $\mathfrak{G}$ is positive definite on $\mathfrak{N}[i\omega]$ almost for all $\omega \in \mathbb{R}^1$, so the matrix (2.3) is left invertible for almost all $\lambda = i\omega$, $\omega \in \mathbb{R}^1$. Hence, (6.1) is the canonical representation of Φ.

Now let $\tau \neq 0$. Notice that if Φ is semibounded on $\mathfrak{M}[a]$ for some $a \in \mathbb{R}^n$, then Φ is semibounded for all $a \in \mathbb{R}^n$ and, in particular, for $a = 0$ [23]. For $a = 0$, $(x, u) \in \mathfrak{M}[a]$, we have $\Phi(x, u) = \|l^*x + \kappa u\|^2 + 2\tau(x, u)$. Since Φ is semibounded, we obtain $\tau(x, u) = \hat{\varphi}(l^*x + \kappa u)$, where $\hat{\varphi}$ is a continuous functional on $\mathfrak{L}_2^m$ (see, for example, [25]). Since τ is linear in (x, u, a), this implies (6.2), where $\varphi(\cdot) \in \mathfrak{L}_2^m$ and $\psi_0 \in \mathbb{R}^n$. As already shown, this implies that there is a function $\psi \in \mathfrak{L}_2^n$ such that (3.2) holds. Thus, (2.2) is the canonical representation of Φ (Φ_* was defined in (3.3)).

It remains to note that if (2.2), (3.3) is a canonical representation, then H and ψ_0 are uniquely determined (see the proof of Lemma 3.1). Hence l^* and κ are also uniquely determined by (6.4) up to multiplication by a unitary matrix. For, given l and κ, the function φ is uniquely determined. Theorem 2.1 is proved.

§7. Proof of Theorem 2.2

Let (2.2) be the canonical representation of Φ on $\mathfrak{M}[a]$. Then a generalized process $(x, u) \in \mathfrak{M}_*[a]$ is optimal for the problem $LQ(\mathfrak{M}_*[a], \Phi)$ if and only if $l^*x + \kappa u = \varphi$. Taking Laplace transforms in this equation and in (1.2) we obtain

$$\lambda \tilde{x} = A\tilde{x} + b\tilde{u} + a, \qquad l^*\tilde{x} + \kappa\tilde{u} = \tilde{\varphi}. \tag{7.1}$$

Since the matrix (2.3) is invertible for $\text{Re}\,\lambda > 0$, it follows that for all $\tilde{\varphi} \in \mathfrak{H}_2^m$, $a \in \mathbb{R}^n$, and for any solution $(\tilde{x}, \tilde{u})$ of (7.1), we have $\tilde{x} \in \mathfrak{H}_*^n$ and $\tilde{u} \in \mathfrak{H}_*^m$. Hence, there exists an optimal generalized process in

$LQ(\mathfrak{M}_*[a], \Phi)$, and its transform $(\tilde{x}, \tilde{u})$ is determined by (7.1). A direct verification immediately shows that for any $m \times n$ matrix R, (7.1) implies that

$$\tilde{u}(\lambda) = R\tilde{x}(\lambda) + (U(\lambda) - RX(\lambda))a + (I_m - RA_\lambda^{-1}b)(l^*A_\lambda^{-1}b + \kappa)^{-1}\tilde{\varphi}(\lambda),$$

where $A_\lambda = \lambda I_n - A$, $\operatorname{Re}\lambda > 0$. Therefore it suffices to show that if R, α, β are defined as in Theorem 2.2, then α and β are polynomials.

For $k \in \mathbb{Z}$ define an operation $[\cdot]^k$ on the set of Laurent expansions of rational functions in the neighborhood of $\lambda = \infty$ as follows: $[\sum f_j \lambda^j]^k = \sum_{j \geq k} f_j \lambda^j$. Note that $[\lambda f]^k = \lambda[f]^k + \lambda^k f_{k-1}$, $[I_n]^k = \chi_k I_n$, where $\chi_k = 0$ for $k > 0$ and $\chi_k = 1$ for $k \leq 0$, $[cf]^k = c[f]^k$ for a constant c. By the definition of X and U in Theorem 2.2, we have

$$A_\lambda X(\lambda) - bU(\lambda) = I_n, \qquad l^*[X] + \kappa U = 0.$$

Hence,

$$A_\lambda [X]^k(\lambda) - b[U]^k(\lambda) = \chi_k I_n - \lambda^k X_{k-1}, \qquad l^*[X]^k + \kappa[U]^k = 0.$$

Since the matrix (2.3) is invertible, for almost all $\lambda \in \mathbb{C}^1$ we obtain

$$[X]^k(\lambda) = X(\lambda)(\chi_k I_n - \lambda^k X_{k-1}), \qquad [U]^k(\lambda) = U(\lambda)(\chi_k I_n - \lambda^k X_{k-1}), \quad (7.2)$$

whence

$$[U]^0 = U - U_{-1}X, \quad [X]^0 = (I_n - X_{-1})X. \tag{7.3}$$

To prove (7.3) it is sufficient to compare the coefficients of λ^{k-1} on both sides of (7.2) for all $k \in \mathbb{Z}$.

Formula (7.3) implies that $\alpha(\lambda) = [U]^0(\lambda) - R_0[X]^0(\lambda)$ is a polynomial matrix. Choose constant matrices Γ_1 and Γ_2 such that $l^*\Gamma_2 + \kappa\Gamma_1 = I_m$. A direct calculation shows that $\beta(\lambda) = \beta(\lambda)(\kappa\Gamma_1 + l^*\Gamma_2) = \Gamma_1 + \alpha(\lambda)b\Gamma_1 - \alpha(\lambda)A_\lambda\Gamma_2 - R\Gamma_2$ is a polynomial matrix. This proves Theorem 2.2.

§8. Proof of Lemma 2.3

Let L_0 and D_0 be real matrices of orders $q \times n$ and $q \times m$ respectively. Define the space $\mathfrak{M}_*[a, L_0, D_0]$ as the set of all pairs (x, u) of quasistable distributions $x \in (\mathfrak{D}_+^n)'$, $u \in (\mathfrak{D}_+^m)'$, satisfying (1.2) and the condition $L_0 x + D_0 u \in \mathfrak{L}_2^q$. The following result is easily deduced from Lemma 6.1.

LEMMA 8.1. *The set* $\mathfrak{Y}_* = \{L_0 x + D_0 u : (x, u) \in \mathfrak{M}_*[a, L_0, D_0]\}$ *is the closure of the set* $\mathfrak{Y} = \{L_0 x + D_0 u : (x, u) \in \mathfrak{M}[a]\}$ *in* $\mathfrak{L}_2^q$.

PROOF. Let $B = A + br^*$ be a Hurwitz matrix and let $v = v(x, u) = u - r^*x$. Taking Laplace transforms, we rewrite the statement of Lemma 8.1 in the form

$$\mathfrak{H}_2^q \cap \{W\tilde{v} : \tilde{v} \in \mathfrak{H}_{2*}^m\} = \operatorname{clos}(\mathfrak{H}_2^q \cap \{W\tilde{v} : \tilde{v} \in \mathfrak{H}_2^m\}),$$

where $W(\lambda) = (L_0 + D_0 r^*)(\lambda I_n - B)^{-1} b + D_0$. Lemma 8.1 is proved.

Before proving Lemma 2.3, we note the following:

(a) $\mathfrak{M}_*[a] \subset \mathfrak{H}^n_* \times \mathfrak{H}^m_*$, since the generalized processes $(x, u) \in \mathfrak{M}_*[a]$ satisfy (7.1), where $\tilde{\varphi} \in \mathfrak{H}^m_2$, and the matrix (2.3) is invertible for $\operatorname{Re} \lambda > 0$.

(b) Under the assumptions of Lemma 2.3, $\|Lx+Du\|^2 = \|l^*x+\kappa u\|^2$ for all $(x, u) \in \mathfrak{M}_0$, where (2.2) is the canonical representation of the functional (2.6). This follows from (6.1) for $\mathfrak{G}(x, u) = |Lx + Du|^2$.

(c) If $(x, u) \in \mathfrak{M}_*[a, L, D]$, $a = 0$, and $Lx+Du = 0$, then $x = 0$ and $u = 0$. This can be proved as follows. From (6.6) for $\mathfrak{G}(x, u) = |Lx + Du|^2$ we have $\ker(LA_{i\omega}^{-1}b + D) = \{0\}$ for almost all $\omega \in R^1$, where $A_\lambda = \lambda I_n - A$. Hence, $\ker(LA_\lambda^{-1}b+D) = \{0\}$ for almost all $\lambda \in \mathbb{C}^1$. It therefore follows from $Lx+Du = 0$ that $(LA_\lambda^{-1}b+D)\tilde{u}(\lambda) = 0$ for $\operatorname{Re}\lambda > 0$; hence, $\tilde{u}(\lambda) = 0$ for almost all λ with $\operatorname{Re}\lambda > 0$, i.e., $u = 0$.

Let $(x, u) \in \mathfrak{M}_*[a, L, D]$. By Lemma 8.1, there exists a sequence $(x_j, u_j) \in \mathfrak{M}[a]$ such that $Lx_j+Du_j \to Lx+Du$ in $\mathfrak{L}^k_2$. Then $L(x_j-x_k)+D(u_j-u_k) \to 0$ as $j, k \to \infty$. Since $(x_j - x_k, u_j - u_k) \in \mathfrak{M}_0$, this implies (see (b)) that $l^*(x_j - x_k) + \kappa(u_j - u_k) \to 0$ as $j, k \to \infty$. Hence $\{l^*x_j + \kappa u_j\}$ is a Cauchy sequence in $\mathfrak{L}^m_2$ and $l^*x_j+\kappa u_j \to y_* \in \mathfrak{L}^m_2$, where $\|y_* - \varphi\|^2 = \Phi(x, u) - \Phi_*$. Applying Lemma 8.1 to $L_0 = \operatorname{col}(L, l^*)$, $D_0 = \operatorname{col}(D, \kappa)$, we conclude that there exists $(\bar{x}, \bar{u}) \in \mathfrak{M}_*[a] \cap \mathfrak{M}_*[a, L, D]$ such that

$$L\bar{x} + D\bar{u} = Lx + Du, \qquad l^*\bar{x} = \kappa\bar{u} + y_*.$$

Now, by (c), $\bar{x} = x$ and $\bar{u} = u$, i.e., $(x, u) \in \mathfrak{M}_*[a]$, $\Phi_* + \|l^*x+\kappa u-\varphi\|^2 = \Phi(x, u)$. It remains to prove the converse: $\mathfrak{M}_*[a] \subset \mathfrak{M}_*[a, L, D]$. By (a), this can be done in the same way. Lemma 2.3 is proved.

§9. Proof of Theorem 2.4

Without loss of generality, we may assume in (1.2) that A is a Hurwitz matrix and $a = 0$. To reduce the problem to this case, we need only apply a feedback transformation $u \to u_1 + r^*x$, where $A + br^*$ is Hurwitz, and a translation $(x, u) \to (x - x_\alpha, u - u_\alpha)$, where $(x_\alpha, u_\alpha) \in \mathfrak{M}[a]$ is a fixed admissible process. In this case it will suffice to show that if $(x_j, u_j) \in \mathfrak{M}_0$ is an optimizing sequence in the problem $LQ(\mathfrak{M}_0, \Phi)$, then $u_j \to u_*$ in $W^m_2\langle s, k\rangle$, where $(x_*, u_*) \in \mathfrak{M}_*[a]$ is a generalized optimal process.

We first prove a weaker assertion: $u_j \to u_*$ in $(\mathfrak{D}^m_+)'$. Indeed, $l^*x_j + \kappa u_j \to l^*x_* + \kappa u_*$ in $\mathfrak{L}^m_2$; hence, $\beta(p)(l^*(x_j - x_*) + \kappa(u_j - u_*)) \to 0$ in $(\mathfrak{D}^m_+)'$ where $p = d/dt$ and β is the polynomial defined in Theorem 2.2. As follows from the proof of Theorem 2.2, the equality (2.5) holds for all $\varphi = l^*x + \kappa u$, $(x, u) \in \mathfrak{M}_*[a]$. Therefore, $u_j - u_* - R(x_j - x_*) \to 0$ in

$(\mathfrak{D}_+^m)'$. Taking into account that $p(x_j - x_*) = A(x_j - x_*) + b(u_j - u_*)$, we obtain $(p - A - bR)(x_j - x_*) \to 0$ in $(\mathfrak{D}_+^n)'$; thus, $x_j \to x_*$ in $(\mathfrak{D}_+^n)'$, so that $u_j \to u_*$ in $(\mathfrak{D}_+^m)'$.

To complete the proof of Theorem 2.4 it is now sufficient to show that

$$\int_0^\infty f(t)^* u(t)\, dt \leq c \|f\|_{s,k} \cdot \|l^* x + \kappa u\|$$

for arbitrary $f(\cdot) \in \mathfrak{D}_+^m$, $(x, u) \in \mathfrak{M}_0$, where $c > 0$ is a constant dependent only on A, b, l, and κ. By the definition of $\Pi(i\omega)$ in (0.9) and by (6.6),

$$\|l^* x + \kappa u\|^2 = \frac{1}{2\pi} \int_{-\infty}^{+\infty} \tilde{u}(i\omega)^* \Pi(i\omega) \tilde{u}(i\omega)\, d\omega.$$

Hence, using the definition of the numbers s and k, we obtain

$$\|l^* x + \kappa u\|^2 \geq \int_{-\infty}^{+\infty} |p(\omega) \tilde{u}(i\omega)|^2\, d\omega,$$

where $p(\omega) = c(1 + i\omega)^{-s}(\omega - \omega_1)^k \cdots (\omega - \omega_q)^k$, ω_j are the degeneracy points of the frequency inequality (0.7) and $c \neq 0$ is a constant. The basic idea of the proof is to find, for each $f \in \mathfrak{D}_+^m$, a regular rational function $\tilde{f}_0 = \tilde{f}_0(\lambda)$ with no poles in the right half-plane $\mathrm{Re}\, \lambda \leq 0$, such that

$$\int_{-\infty}^{+\infty} \left| \frac{\tilde{f}(i\omega) - \tilde{f}_0(i\omega)}{p(i\omega)} \right|^2 d\omega \leq c_1 \|f\|_{s,k}^2, \tag{9.1}$$

where c_1 is a constant dependent only on $\omega_1, \ldots, \omega_q$, s, k, c, and m. Then

$$\int_0^\infty f(t)^* u(t)\, d(t) = \frac{1}{2\pi} \int_{-\infty}^{+\infty} \tilde{f}^*(i\omega) u(i\omega)\, d\omega$$

$$= \frac{1}{2\pi} \int_{-\infty}^{+\infty} (\tilde{f}(i\omega) - \tilde{f}_0(i\omega)) u(i\omega)\, d\omega$$

$$\leq \frac{1}{2\pi} \left(\int_{-\infty}^{+\infty} \left| \frac{\tilde{f}(i\omega) - \tilde{f}_0(i\omega)}{p(\omega)} \right|^2 d\omega \right)^{1/2} \left(\int_{-\infty}^{+\infty} |p(\omega) \tilde{u}(i\omega)|^2\, d\omega \right)^{1/2}$$

$$\leq c_1 \|f\|_{s,k} \|l^* x + \kappa u\|,$$

as required.

LEMMA 9.1. *Let* $f \in \mathfrak{D}_+^m$, $\omega_0 \in \mathbb{R}^1$, $k > 0$. *Then*

$$\int_{-\infty}^{+\infty} \left| \frac{\tilde{f}(i\omega) - \tilde{f}_{01}(i\omega)}{(\omega - \omega_0)^k} \right|^2 d\omega \leq 2\pi \left(\frac{1}{(k+1)!} \int_0^\infty t^{k-\frac{1}{2}} |f(t)|\, dt \right)^2, \tag{9.2}$$

where

$$\tilde{f}_{01}(\lambda) = \sum_{j=0}^{k-1} \left(\frac{\lambda - i\omega_0}{i} \right)^j \frac{1}{j!} \left\{ \left(\frac{d}{d\omega} \right)^j \tilde{f}(i\omega) \right\} \bigg|_{\omega = \omega_0}.$$

The proof is by direct calculation. For $t_1 \geq 0$ and $t_2 \geq 0$, we have
$$\int_0^{\min(t_1, t_2)} (t_1 - t)^{k-1}(t_2 - t)^{k-1} dt \leq t_1^{k-1} t_2^{k-1} \min(t_1, t_2) \leq t_1^{k-\frac{1}{2}} t_2^{k-\frac{1}{2}}.$$

Using the Parseval formula, we obtain
$$\frac{1}{2\pi} \int_{-\infty}^{+\infty} e_k(\omega, t_1)^* e_k(\omega, t_2) d\omega \leq \{(k-1)!\}^{-2} t_1^{k-\frac{1}{2}} t_2^{k-\frac{1}{2}},$$

where $e_k(\omega, t) = \omega^{-k}\left(e^{i\omega t} - \sum_{j=0}^{k-1} (i\omega t)^j/j!\right)$. Hence,

$$\int_{-\infty}^{+\infty} \left|\frac{\tilde{f}(i\omega) - \tilde{f}_{01}(i\omega)}{(\omega - \omega_0)^k}\right|^2 d\omega$$
$$= \int_{-\infty}^{+\infty} d\omega \int_0^\infty \int_0^\infty e^{i\omega_0(t_1-t_2)} e_k(\omega - \omega_0, t_1)^*$$
$$\times e_k(\omega - \omega_0, t_2) f(t_1)^* f(t_2) dt_1 dt_2$$
$$= \int_0^\infty \int_0^\infty e^{i\omega_0(t_1-t_2)} f(t_1)^* f(t_2) dt_1 dt_2 \int_{-\infty}^{+\infty} e_k(\omega, t_1) e_k(\omega, t_2) d\omega$$
$$\leq 2\pi \int_0^\infty \int_0^\infty \frac{t_1^{k-\frac{1}{2}}}{(k-1)!} \frac{t_2^{k-\frac{1}{2}}}{(k-1)!} |f(t_1)| \cdot |f(t_2)| dt_1 dt_2.$$

Lemma (9.1) is proved.

Notice also that if $f \in \mathfrak{D}_+^m$, then
$$\int_{-\infty}^{+\infty} |(i\omega)^s \tilde{f}(i\omega) - \tilde{f}_{02}(i\omega)|^2 d\omega = 2\pi \int_0^\infty \left|\left(\frac{d}{dt}\right)^s f(t)\right|^2 dt, \quad (9.3)$$

where
$$\tilde{f}_{02}(\lambda) = \sum_{j=0}^{s-1} \lambda^j \left\{\left(\frac{d}{dt}\right)^j f\right\}(0), \quad s \geq 0.$$

For an arbitrary function $f \in \mathfrak{D}_+^m$, define vectors $f_{\alpha,\beta} = f_{\alpha,\beta}(f) \in \mathbb{C}^m$ such that
$$f_{0,\beta} = (d/dt)^\beta f(0) \quad \text{and} \quad f_{\alpha,\beta} = \left\{\left(\frac{1}{i}\frac{d}{d\omega}\right)^\beta \tilde{f}(i\omega)\right\}\bigg|_{\omega=\omega_\alpha}$$

for $\alpha \neq 0$, where $\beta \in \{0, 1, \ldots, s-1\}$ for $\alpha = 0$ and $\beta \in \{0, 1, \ldots, k-1\}$ for $\alpha \neq 0$. It is readily seen that $\sum |f_{\alpha,\beta}| \leq c_2 \|f\|_{s,k}$, where c_2 is a constant depending only on $\omega_1, \ldots, \omega_q$, s, k, and m. An arbitrary set of vectors $\{f_{\alpha,\beta}\}$ uniquely determines a vector-valued polynomial $g(\lambda)$ of degree less than $kq + s$ such that the function $\tilde{f}_0(\lambda) = (1-\lambda)^{-kq-s} g(\lambda)$ satisfies the estimates $\tilde{f}_0(i\omega) - \tilde{f}(i\omega) = O(|\omega|^{-s})$ as $\omega \to \infty$ and $\tilde{f}_0(i\omega) - \tilde{f}(i\omega) = O(|\omega - \omega_\alpha|^k)$ as $\omega \to \omega_\alpha$, $\alpha = 1, 2, \ldots, q$. The coefficients g_j of g are linearly dependent on the vectors $f_{\alpha,\beta}$ (so that $|g_j| \leq c_3 \|f\|_{s,k}$ for some constant c_3 depending only on $\omega_1, \ldots, \omega_q$, s, k, and m).

As pointed out above, once f has been determined, we can estimate the integral (9.1) in a neighborhood of $\omega_0 \in \{\omega_1, \ldots, \omega_q\}$ by formula (9.2) and in a neighborhood of $\omega = \infty$ by formula (9.3). This implies inequality (9.1), where c_1 depends only on $\omega_1, \ldots, \omega_q$, s, k, c, and m. Theorem 2.4 is proved.

§10. Proof of Lemma 3.2

In this section we consider matrix-valued measurable functions defined on subsets of the complex plane, symmetric with respect to the real axis and such that $f(\bar{\lambda}) = \overline{f(\lambda)}$, where the bar denotes componentwise conjugation. Let $\mathfrak{L}_2(i\mathbb{R})$ be the set of functions defined almost everywhere on the imaginary axis, such that

$$\|f\|_i^2 = \int_{-\infty}^{+\infty} \operatorname{Sp}\{f(i\omega)^* f(i\omega)\} \, d\omega < \infty,$$

where Q is the set of rational functions, $\mathfrak{H}_A$ the set of functions analytic in the half-plane $\operatorname{Re}\lambda > 0$, and $\mathfrak{H}_\infty$ the subset of functions in $\mathfrak{H}_A$ bounded in the domain $\operatorname{Re}\lambda > 0$.

LEMMA 10.1. *Let $W \in Q$ be a $p \times q$ matrix.*

1°. *If $v \in \mathfrak{L}_2(i\mathbb{R})$ and $u \in Q$ are functions such that $Wv - u \in \mathfrak{L}_2(i\mathbb{R})$, then $W\hat{v} - u \in \mathfrak{L}_2(i\mathbb{R})$ for some function $\hat{v} \in Q \cap \mathfrak{L}_2(i\mathbb{R})$.*
2°. *If $v \in \mathfrak{H}_2^q$ is a function such that $Wv \in \mathfrak{L}_2(i\mathbb{R}) \cap \mathfrak{H}_A$, then $Wv \in \mathfrak{H}_2^p$.*
3°. *If $v \in \mathfrak{H}_2^q$, $u \in Q$, and $Wv - u \in \mathfrak{L}_2(i\mathbb{R})$, then $W(v - v_0) \in \mathfrak{H}_2^p$ for some function $v_0 \in \mathfrak{H}_2^q \cap Q$.*

PROOF. 1°. It is sufficient to show that if

$$W(\lambda) = \sum_{j=-\infty}^{k} (\lambda - \lambda_0)^{-j} W_j \quad \text{and} \quad u(\lambda) = \sum_{j=-\infty}^{k} (\lambda - \lambda_0)^{-j} u_j$$

are the Laurent expansions of W and u in the neighborhood of $\lambda_0 \in i\mathbb{R}$ (or $W = \sum_{j \leq k} \lambda^{j-1} W_j$, $u = \sum_{j \leq k} \lambda^{j-1} u_j$ are the expansions in the neighborhood of $\lambda_0 = \infty$), then we can find matrices $z_0, \ldots, z_{k-1}$ such that

$$u_j = \sum_{s=0}^{k-j} W_{j+s} z_s, \qquad j = 1, 2, \ldots, k. \tag{10.1}$$

In that case we construct $\hat{v}$ such that

$$\hat{v}(\lambda) = z_0 + (\lambda - \lambda_0) z_1 + \cdots + (\lambda - \lambda_0)^{k-1} z_{k-1} + O((\lambda - \lambda_0)^k)$$

as $\lambda \to \lambda_0$, $\lambda_0 \in i\mathbb{R}$, and $\hat{v}(\lambda) = z_0 \lambda^{-1} + \cdots + z_{k-1} \lambda^{-k} + o(\lambda^{-k})$ as $\lambda \to \infty$. To fix ideas, consider the case $\lambda_0 = 0$. If $p_0, p_1, \ldots, p_{k-1}$ are vectors such that $p(\lambda) W(\lambda)$ is bounded for $\lambda \to 0$, where $p(\lambda) = p_0^* + \lambda p_1^* + \cdots + \lambda^{k-1} p_{k-1}^*$, then the product $p(\lambda) u(\lambda)$ is bounded as $\lambda \to 0$ (because by assumption $pWv - pu$

is square integrable in the neighborhood of $\lambda = 0$). This condition can be rewritten as $\hat{p}^* \hat{W} = 0 \implies \hat{p}^* \hat{u} = 0$, where

$$p = \begin{bmatrix} p_0 \\ p_1 \\ \cdots \\ p_{k-1} \end{bmatrix}, \quad \hat{W} = \begin{bmatrix} W_k & W_{k-1} & \cdots & W_1 \\ 0 & W_k & \cdots & W_2 \\ \cdots & \cdots & \cdots & \cdots \\ 0 & \cdots & \cdots & W_k \end{bmatrix}, \quad \hat{u} = \begin{bmatrix} u_1 \\ u_2 \\ \cdots \\ u_k \end{bmatrix}.$$

Therefore, we have found a matrix $\hat{z} = \text{col}(z_{k-1}, \ldots, z_0)$ such that $\hat{u} = \hat{W}\hat{z}$. The last equality is equivalent to (10.1).

$2°$. There exists a scalar function $\rho \in \mathfrak{H}_\infty \cap Q$ such that $\rho^{-1} \in \mathfrak{H}_A$ and $\rho W \in \mathfrak{L}_2(i\mathbb{R})$. For this function, $\rho W v \in \mathfrak{H}_2^p$. Since ρ is an outer function and $Wv \in \mathfrak{L}_2(i\mathbb{R})$, we can find a sequence $\{f_j\} \subset \mathfrak{H}_\infty$ of scalar functions such that $\|(\rho f_j - 1)Wv\|_i \to 0$. Since $\rho f_j Wv \in \mathfrak{H}_2^p$, this implies that $Wv = \lim(\rho f_j Wv) \in \mathfrak{H}_2^p$.

$3°$. By $1°$, there exists $\hat{v} \in Q \cap \mathfrak{L}_2(i\mathbb{R})$ such that $W\hat{v} - u \in \mathfrak{L}_2(i\mathbb{R})$. Let k be the maximal multiplicity of the poles of $W(\lambda)$ for $\lambda \in \mathbb{C}^1 \cup \{\infty\}$. Construct a rational function $v_0 \in \mathfrak{H}_2^q$ such that either $v(\lambda) - v_0(\lambda) = o((\lambda - \lambda_0)^k)$ as $\lambda \to \lambda_0$ for poles λ_0 of W in the half-plane $\text{Re}\,\lambda > 0$, or $\hat{v}(\lambda) - v_0(\lambda) = o((\lambda - \lambda_0)^k)$ as $\lambda \to \lambda_0$ for poles λ_0 of W on the imaginary axis, or $\hat{v}(\lambda) - v_0(\lambda) = o(\lambda^{-k})$ as $\lambda \to \infty$. Then $W(v-v_0) \in \mathfrak{L}_2(i\mathbb{R}) \cap \mathfrak{H}_A$. Since $v - v_0 \in \mathfrak{H}_2^q$, part $2°$ implies $W(v - v_0) \in \mathfrak{H}_2^p$. Lemma 10.1 is proved.

We now prove Lemma 3.2. By Theorem 2.1 and Lemma 3.1, if the cost functional is lower semibounded, then the function $\tilde{\theta}$ defined in (3.4) is in $\mathfrak{H}_2^{n+m}$ for a $\psi_0 \in \mathbb{R}^n$. Then we can put $\tilde{\theta}_0 = E^{-1} E_0 \psi_0$. Conversely, let $E^{-1}\tilde{\tau} - \tilde{\theta}_0 \in \mathfrak{L}_2(i\mathbb{R})$, $\tilde{\theta}_0 \in Q$. Then $E^{-1}(\tilde{\tau} - \tilde{\tau}_0) \in \mathfrak{H}_2^{n+m}$ for $\tau_0 \in \mathfrak{H}_2^{n+m} \cap Q$. Therefore, there exists $\psi_0 \in \mathbb{R}^n$ such that $E^{-1}(\tilde{\tau}_0 + E_0\psi_0) \in \mathfrak{H}_2$. Hence $E^{-1}(\tilde{\tau} + E_0\psi_0) = E^{-1}(\tilde{\tau}_0 + E_0\psi_0) + E^{-1}(\tilde{\tau} - \tilde{\tau}_0) \in \mathfrak{H}_2$. Lemma 3.2 is proved.

Proofs of Lemmas 4.1, 5.1, 5.2, and Theorem 5.3 can be found in [13], [24].

§11. Example 1

Consider the controlled object described by the equations

$$a(p)\varphi = b(p)\theta + c(p)f, \quad y = d(p)\varphi, \qquad (11.1)$$

where φ, θ, f, and y are scalars, φ is the control parameter, θ the control, f the noise, y the observed output, a, b, c, and d are given polynomials with real coefficients:

$$a(p) = p^2 + a_1 p + a_0, \quad b(p) = b_1 p + b_0,$$
$$c(p) = c_1 p + c_0, \quad d(p) = d_1 p + d_0,$$

where $p = d/dt$ is the differentiation operator. We must determine a control law for the object (11.1) such that the dependence of the output φ on the noise f is minimal; in other words, we are required to construct a regulator

$$\alpha(p)\theta = \beta(p)\varphi, \tag{11.2}$$

where α and β are polynomials, such that the reactivity of the output φ with respect to the input f, i.e.,

$$R = \overline{\lim}_{t\to\infty} \sup_{f\in\mathfrak{L}_2^1, f\neq 0} \frac{|\varphi(t)|^2}{\int_0^\infty |f(s)|^2 ds} \tag{11.3}$$

is minimal. The polynomials α and β must satisfy the conditions

$$\deg\alpha > \deg\beta \tag{11.4}$$

(i.e., (11.2) is an integrating regulator), and

$$\alpha(\lambda)a(\lambda) - \beta(\lambda)b(\lambda) \neq 0, \qquad \operatorname{Re}\lambda \geq 0 \tag{11.5}$$

(system (11.1)–(11.2) is stable). Regulators (11.2) with properties (11.4) and (11.5) are called *admissible*. If a regulator (11.2) is admissible, then one can easily see that the limit (11.3) is finite and does not depend on the initial data in (11.1) and (11.2), so that $R = R(\alpha, \beta)$. Below, the minimization of R on the set of admissible regulators will be referred to briefly as problem (11.1), (11.3).

The control problem for an object (11.1) in the case when

$$a_j > 0, \qquad b_1 = c_1 = d_1 = 0, \qquad b_0 c_0 d_0 \neq 0, \tag{11.6}$$

and the target of the control being absolute invariance of φ with respect to f, has been discussed in the context of the theory of invariance [28]–[30]. Later, we show that if (11.6) holds, then the infimum of the functional (11.3) equals zero but it cannot be attained in the class of admissible regulators. Therefore, in this case absolute invariance is not possible, but only "attainable in the limit" — a fact that aroused considerable debate at one time. The above results enable one to construct a sequence of regulators (11.2) with $\alpha = \alpha_k$, $\beta = \beta_k$, $k \to \infty$, such that $R(\alpha_k, \beta_k) \to \inf R$. In particular, if (11.6) holds, then $R(\alpha_k, \beta_k) \to 0$.

Let us formulate the final result under the assumption that the perturbation, the control, and the observed output of the object (11.1) are nontrivial, that is,

$$|d_0| + |d_1| \neq 0, \qquad |c_0| + |c_1| \neq 0, \qquad |b_0| + b_1| \neq 0.$$

The formulas are somewhat cumbersome but nevertheless have some practical value.

PROPOSITION 11.1. *Let* (∗) *denote the condition: the polynomials* a *and* bd *have no common roots in the half-plane* $\{\lambda \in C^1 : \operatorname{Re}\lambda \geq 0\}$. *Then the set of admissible regulators in problem* (11.1), (11.3) *is not empty if and only if* (∗) *holds. If* (∗) *holds then*

(I) (Formula for the infimum of the reactivity). $\operatorname{Inf} R = R_0 + R_1$, *where*

$$R_0 = \begin{cases} 0 & \text{if } b_0 b_1 \geq 0, \\ -2 b_0 b_1 \tau_1^2 \sigma_1^{-2} & \text{if } b_0 b_1 < 0. \end{cases}$$

$$R_1 = \begin{cases} R_{11} & \text{if } b_0 b_1 \geq 0, \\ R_{12} & \text{if } b_0 b_1 < 0. \end{cases}$$

$$R_{11} = \begin{cases} 0 & \text{if } d_0 d_1 \geq 0, \\ -2 d_0 d_1 \sigma_2^{-2} & \text{if } d_0 d_1 < 0. \end{cases}$$

$$R_{12} = \begin{cases} 0 & \text{if } d_0 d_1 \geq 0,\ c_0 c_1 < 0, \\ -8 c_0 c_1 b_0^2 b_1^2 \sigma_1^{-2} & \text{if } d_0 d_1 \geq 0,\ c_0 c_1 < 0, \\ -2 d_0 d_1 \tau_2^2 \sigma_2^{-2} (1 + 2 b_0 b_1 \sigma_3 \sigma_1^{-2} \tau_2^{-1})^2 & \text{if } d_0 d_1 < 0,\ c_0 c_1 \geq 0, \\ -8 c_0 c_1 b_0^2 b_1^2 \sigma_1^{-2} - 2 d_0 d_1 \tau_3^2 \sigma_2^{-2}\left(1 + \dfrac{2 b_0 b_1 \sigma_4}{\tau_3 \sigma_1}\right)^2 & \text{if } d_0 d_1 < 0,\ c_0 c_1 < 0. \end{cases}$$

$$\sigma_1 = a_0 b_1^2 - a_1 b_0 b_1 + b_0^2,$$
$$\sigma_2 = a_0 d_1^2 - a_1 d_0 d_1 + d_0^2,$$
$$\sigma_3 = a_0 d_1 c_1 - a_1 d_0 c_1 + d_0 c_0,$$
$$\sigma_4 = a_0 d_1 c_1 - a_1 d_0 c_1 - d_0 c_0,$$
$$\tau_1 = c_1 b_0 - b_1 c_0,$$
$$\tau_2 = c_1 d_0 - d_1 c_0,$$
$$\tau_3 = c_1 d_0 + d_1 c_0.$$

(II) (Formulas for an optimizing sequence of regulators). *An optimizing sequence can be represented in the form*

$$(p^2 + \alpha_1 p + \alpha_0)\theta = (\beta_1 p + \beta_0) y, \qquad (11.7)$$

where

$$\begin{aligned}
\alpha_1 &= -h_{11} - h_{22}, & \alpha_0 &= h_{11}h_{22} - h_{21}h_{12}; \\
\beta_1 &= -r_0 k_0 - r_1 k_1, & \beta_0 &= r_1 k_1 h_{11} - r_1 h_{21} k_0 - r_0 h_{22} k_0 - r_0 h_{12} k_1;
\end{aligned} \quad (11.8)$$

$$\begin{aligned}
h_{11} &= r_0 k_1 + k_0 d_0 + k_0 d_1 b_1 r_0; \\
h_{12} &= 1 + b_1 r_1 + d_1 k_0 + k_0 d_1^2 b_1 r_1; \\
h_{21} &= -a_0 + b_0 r_0 - a_1 b_1 r_0 + k_1 d_0 + k_1 d_1 b_1 r_1; \\
h_{22} &= -a_1 + b_0 r_1 - a_1 b_1 r_1 + k_1 d_1 + k_1 d_1 b_1 r_1;
\end{aligned} \quad (11.9)$$

$$(r_0, r_1) = \begin{cases} (-1/(\varepsilon b_1), -1/b_1) \\ \qquad \text{if } b_0 b_1 > 0, \\ (a_0/b_0 - 1/(\varepsilon^2 b_0), a_1/b_0 - 2/(\varepsilon b_0)) \\ \qquad \text{if } b_1 = 0, \\ (-1/(\varepsilon b_1) + a_1/(a_0 b_1) - \varepsilon/b_1, -1/b_1 + 1/(a_0 b_1)) \\ \qquad \text{if } b_0 = 0, \\ \left(\dfrac{b_0^2 + 2\varepsilon a_0 b_0 b_1 - a_0 b_1^2 - a_1 b_1 b_0}{\varepsilon b_1 \sigma_1}, \dfrac{\varepsilon b_0^2 + \varepsilon a_1 b_1 b_0 - 2 b_0 b_1 - \varepsilon a_0 b_1^2}{\varepsilon b_1 \sigma_1} \right) \\ \qquad \text{if } b_0 b_1 < 0. \end{cases}$$

$$(k_0, k_1) = \begin{cases} (-1/d_1, a_1/d_1 - c_0/(d_1 c_1)) & \text{if } d_0 d_1 > 0, c_0 c_1 > 0, \\ (-1/d_1, a_1/d_1 - 1/(\delta d_1)) & \text{if } d_0 d_1 > 0, c_1 = 0, \\ (-1/d_1, a_1/d_1 + c_0/(d_1 c_1)) & \text{if } d_0 d_1 > 0, c_0 c_1 < 0, \\ (-1/d_1, a_1/d_1 - \delta/d_1) & \text{if } d_0 d_1 > 0, c_0 = 0, \\ ([c_1 \delta^2 - a_0]/(a_0 d_1), [a_1 - \delta - \delta c_1]/d_1) & \text{if } d_0 = c_0 = 0, \\ ([1 - a_0]/(a_0 d_1), a_1/d_1 - \delta/d_1 - 1/(\delta d_1)) & \text{if } d_0 = c_1 = 0, \\ \left(\dfrac{a_1 \delta - 1 - \delta^2}{\delta d_0}, \dfrac{a_1 + \delta a_0 + \delta^2 a_1 - \delta - \delta a_1}{\delta d_0} \right) & \text{if } d_1 = c_0 = 0, \\ \left(\dfrac{a_1}{d_0} - \dfrac{2}{\delta d_0}, -\dfrac{1}{\delta^2 d_0} + \dfrac{2a_1}{\delta d_0} - \dfrac{a_1^2}{d_0} + \dfrac{a_0}{d_0} \right) & \text{if } d_1 = c_1 = 0. \end{cases}$$

$$(k_0, k_1) = \begin{cases} \left(-\dfrac{c_1\sigma_2 + 2d_0\tau_2}{c_1 d_1 \sigma_2}, \dfrac{c_1 a_1 \sigma_2 - c_0 \sigma_2 - 2d_0 \sigma_3}{c_1 d_1 \sigma_2} \right) \\ \qquad\qquad\qquad\qquad\qquad \text{if } d_0 d_1 < 0, \ c_0 c_1 > 0, \\[4pt] \left(\dfrac{2d_0 d_1 - \delta\sigma_2 - 2\delta d_0}{\delta \sigma_2 d_1}, \dfrac{\delta a_1 \sigma_2 + 2\delta d_0^2 a_1 - \sigma_2 - 2d_0^2 - 2d_0 d_1 \delta a_0}{\delta \sigma_2 d_1} \right) \\ \qquad\qquad\qquad\qquad\qquad \text{if } d_0 d_1 < 0, \ c_1 = 0, \\[4pt] \left(\dfrac{2\delta c_1 d_0 d_1 - c_1 \sigma_2 - 2d_0^2 c_1}{c_1 d_1 \sigma_2}, \dfrac{c_1(a_1 \sigma_2 + 2\delta a_0 d_1^2 - 3\delta \sigma_2 - 2a_0 d_0 d_1)}{c_1 d_1 \sigma_2} \right) \\ \qquad\qquad\qquad\qquad\qquad \text{if } d_0 d_1 < 0, \ c_0 = 0, \\[4pt] \left(-\dfrac{c_1 \sigma_2 + 2d_0 \tau_3}{c_1 d_1 \sigma_2}, \dfrac{c_1 a_1 \sigma_2 + c_0 \sigma_2 - 2d_0 \sigma_4}{c_1 d_1 \sigma_2} \right) \\ \qquad\qquad\qquad\qquad\qquad \text{if } d_0 d_1 < 0, \ c_0 c_1 < 0, \\[4pt] \left(\dfrac{\delta a_1 c_1 + \delta c_0 - c_1}{c_1 d_0}, \dfrac{\delta c_1 a_0 - c_0 \delta a_1 + a_1 c_1 - \delta c_1 a_1^2 + c_0}{c_1 d_0} \right) \\ \qquad\qquad\qquad\qquad\qquad \text{if } d_1 = 0, \ c_0 c_1 < 0, \\[4pt] \left(\dfrac{\delta a_1 c_1 - \delta c_0 - c_1}{c_1 d_0}, \dfrac{\delta c_1 a_0 + c_0 \delta a_1 + a_1 c_1 - \delta c_1 a_1^2 - c_0}{c_1 d_0} \right) \\ \qquad\qquad\qquad\qquad\qquad \text{if } d_1 = 0, \ c_0 c_1 > 0, \\[4pt] (\delta c_0 / (a_0 c_1 d_1) - 1/d_1, \ a_1/d_1 - c_0/(c_1 d_1 - \delta/d_1)) \\ \qquad\qquad\qquad\qquad\qquad \text{if } d_0 = 0, \ c_0 c_1 < 0, \\[4pt] (-\delta c_0 / (a_0 c_1 d_1) - 1/d_1, \ a_1/d_1 - c_0/(c_1 d_1) - \delta/d_1) \\ \qquad\qquad\qquad\qquad\qquad \text{if } d_0 = 0, \ c_0 c_1 < 0. \end{cases}$$

Here ε and δ are positive parameters tending to zero.

(III) *The cost functional (reactivity) defined on a regulator (11.7) with coefficients (11.8), (11.9) satisfies the following estimate*:

$$R - \inf R \leq \varepsilon \Delta_0 + \delta \Delta_1(\hat{\kappa}_0, \hat{\kappa}_1),$$

where

$$\Delta_0 = \begin{cases} c_1^2 & \text{if } b_0 b_1 > 0, \\ c_1^2 + (c_1 + \varepsilon[c_0 - a_1 c_1])^2 & \text{if } b_1 = 0, \\ c_0^2 a_0^{-2} + (c_1 - \varepsilon c_0 a_0^{-1})^2 & \text{if } b_0 = 0, \\ [2b_0 b_1(c_0 - a_1 c_1) - (b_0^2 - a_1 b_0 b_1 - a_0 b_1^2) c_1]^2 \sigma_1^{-2} & \text{if } b_0 b_1 < 0. \end{cases}$$

$$(\hat{\kappa}_0, \hat{\kappa}_1) = \begin{cases} (1, \varepsilon) & \text{if } b_0 b_1 < 0, \\ (1 - a_0 \varepsilon^2, \ 2\varepsilon - a_1 \varepsilon^2) & \text{if } b_1 = 0, \\ (\varepsilon a_1 / a_0 - 1 - \varepsilon^2, \ \varepsilon / a_0 - \varepsilon) & \text{if } b_0 = 0, \\ \left(\dfrac{b_0^2 + 2\varepsilon a_0 b_0 b_1 - a_0 b_1^2 - a_1 b_1 b_0}{\sigma_1}, \dfrac{\varepsilon b_0^2 - \varepsilon a_1 b_1 b_0 - 2 b_0 b_1 - \varepsilon a_0 b_1^2}{\sigma_1} \right) & \text{if } b_0 b_1 < 0, \end{cases}$$

$$\Delta_0(x, z) = \begin{cases} 0 & \text{if } d_0 d_1 c_0 c_1 \neq 0, \\ c_0^2 z^2 & \text{if } d_0 d_1 > 0, c_1 = 0, \\ c_1 z^2 & \text{if } d_0 d_1 > 0, c_0 = 0, \\ c_0^2 \sigma_2^{-2} [2 d_0 d_1 x - (\sigma_2 + 2 d_0^2) z]^2 & \text{if } d_0 d_1 < 0, c_1 = 0, \\ [2 d_0 d_1 c_1 x + (2 d_1^2 c_1 a_0 - 3 c_1 \sigma_2) z]^2 & \text{if } d_0 d_1 < 0, c_0 = 0, \\ c_1^2 z^2 + c_1^2 a_0^{-2} (\delta x - a_0 z)^2 & \text{if } d_0 = 0, c_0 = 0, \\ c_0^2 a_0^{-2} x^2 + c_0^2 a_0^{-2} (z - \delta x)^2 & \text{if } d_0 = 0, c_1 = 0, \\ c_1^2 z^2 + c_1^2 [x + (\delta - a_1) z]^2 & \text{if } d_1 = 0, c_0 = 0, \\ c_0^2 z^2 + c_0^2 [\delta x + (1 - \delta a_1) z]^2 & \text{if } d_1 = 0, c_1 = 0, \\ [c_1 x + (c_0 - a_1 c_1) z]^2 & \text{if } d_1 = 0, c_0 c_1 > 0, \\ [c_1 x - (c_0 + a_1 c_1) z]^2 & \text{if } d_1 = 0, c_0 c_1 < 0, \\ (c_1 z - c_0 a_0^{-1} x)^2 & \text{if } d_0 = 0, c_0 c_1 > 0, \\ (c_1 z + c_0 a_0^{-1} x)^2 & \text{if } d_0 = 0, c_0 c_1 < 0. \end{cases}$$

(IV) (Conditions of absolute invariance). *We have* $\inf R = 0$ *and* $\inf R$ *is attainable in the class of admissible regulators if and only if*

$$c_1 = 0, \qquad b_0 b_1 > 0, \qquad d_0 d_1 > 0. \tag{11.10}$$

and in that case the optimal regulator (11.2) *can be expressed as*

$$b(p) d(p) \theta = (\beta_1 p + \beta_0) y, \tag{11.11}$$

where β_1 *and* β_0 *are arbitrary real numbers such that* $\beta_1 < a_1$ *and* $\beta_0 < a_0$

REMARK. If the condition (∗) holds, formulas (11.7)–(11.9) for the parameters r_j, k_j, Δ_j, $\hat{\kappa}_j$ of the regulator make sense. For example (see formulas for $\hat{\kappa}_j$), $\sigma_1 \neq 0$ for $b_0 b_1 < 0$, since otherwise a and b would have a common root $\lambda_{ab} = -b_0 b_1^{-1}$ in the half-plane $\operatorname{Re} \lambda > 0$.

To prove Proposition 11.1 we can use Lemmas 5.1 and 5.2 and Theorem 5.3. We first write (11.1) in the form (0.13), (0.15), where $\eta = \theta$, $\chi = y$, $\xi = \operatorname{col}(\xi_0, \xi_1)$, $\xi_0 = \varphi$, $\xi_1 = p\varphi - b_1 \theta - c_1 f$, $\zeta_1 = \operatorname{col}(c_1 f, (c_0 - a_1 c_1) f)$, $\zeta_2 = d_1 c_1 f$, $d = d_1 b_1$, $b = \operatorname{col}(b_1, b_0 - a_1 b_1)$, $c = \operatorname{col}(d_0, d_1)$, $A = \begin{bmatrix} 0 & 1 \\ -a_0 & -a_1 \end{bmatrix}$. Moreover, we assume that f is not a deterministic function but a fixed generalized stochastic process, that is, white noise. Then the "noise" $\zeta = \operatorname{col}(\zeta_1, \zeta_2)$ satisfies (5.3), where

$$G_0 = \begin{bmatrix} c_1^2 & c_1 (c_0 - a_1 c_1) \\ c_1 (c_0 - a_1 c_1) & (c_0 - a_1 c_1)^2 \end{bmatrix};$$

$$g_0 = \begin{bmatrix} d_1 c_1^2 \\ d_1 c_1 (c_0 - a_1 c_1) \end{bmatrix}; \qquad \Gamma_0 = d_1^2 c_1^2.$$

We immediately see [16] that if (11.3) is an admissible regulator and φ, θ are defined by (11.1), (11.3), then $R(\alpha, \beta) = P\Phi(\eta, \xi)$, where $P\Phi$ is defined in (0.16) with $\mathfrak{G}(x, u) = |x_0|^2$ ($x = \text{col}(x_0, x_1) \in \mathbb{R}^2$, $u \in \mathbb{R}^1$). Moreover, in that case $(\eta, \xi) \in F\mathfrak{M}$, where $F\mathfrak{M}$ is the set of admissible processes in the appopriate stochastic optimal control problem with incomplete observation, as defined in §5. We will refer to this problem as LQ_* for short.

Therefore, we can say that problems (11.1), (11.3) and LQ_* require minimization of the same functional, but in LQ_* the functional is to be minimized over a larger set (since, of course, not all processes $(\eta, \xi) \in F\mathfrak{M}$ can be obtained as solutions of system (11.1), (11.2) with an admissible regulator in (11.2)). Actually, this difference between problems (11.1), (11.3) and LQ_* is inessential. Indeed, by Theorem 5.3, an optimal process (or optimizing sequence) in LQ_* can be obtained as a solution of system (0.13), (0.15), (5.8), (5.9). For a certain admissible regulator (11.2) this system is equivalent to system (11.2), (11.1). Indeed, if $(-\kappa_1)^{-1} l_1^* = (r_0, r_1)$, $(-\kappa_{01})^{-1} l_{01}^* = (\hat{\kappa}_0, \hat{\kappa}_1)$ in (5.8), (5.9), the required regulator (11.2) is of the form (11.7), where α_j, β_j are defined in (11.8), (11.9).

These remarks enable us to use Theorem 5.3 to prove parts (I)–(III) of Proposition 11.1. In accordance with Theorem 5.3, we must consider two auxiliary deterministic optimal control problems: the problem $LQ(\mathfrak{M}[a], \Phi)$, which in this case is

$$dx_0/dt = x_1 + b_1 u, \quad dx_1/dt = -a_0 x_0 - a_1 x_1 + (b_0 - a_1 b_1)u,$$
$$\Phi_1 = \int_0^\infty x_0^2 \, dt \to \inf \qquad (11.12)$$

and the problem

$$dx_0/dt = -a_0 x_1 + d_0 u, \quad dx_1/dt = x_0 - a_1 x_1 + d_1 u,$$
$$\Phi_2 = \int_0^\infty |c_1 x_0 + (c_0 - a_1 c_1)x_1 + d_1 c_1 u|^2 \, dt \to \inf. \qquad (11.13)$$

In both problems, $x_j(\cdot) \in \mathfrak{L}_2^1$ and initial values $x_0(0)$, $x_1(0)$ are fixed.

Let us analyze problem (11.12). At $\omega = \infty$ the degree of degeneracy of the frequency inequality equals two for $b_1 \neq 0$ and four for $b_1 = 0$. In addition, if $b_0 = 0$, there is a degeneracy of degree two at $\omega = 0$. Therefore, the problem is singular for all a_j, b_j. If $b_0 b_1 \geq 0$, the representation of Φ_1 in the form (11.12) is canonical (in particular, $\inf \Phi_1 = 0$ for all $x_0(0)$, $x_1(0)$). For $b_0 b_1 < 0$ the canonical representation of Φ_1 is

$$\Phi_1 = -\frac{2 b_0 b_1}{\sigma_1^2} |(b_0 - a_1 b_1)x_0(0) - b_1 x_1(0)|^2 + \int_0^\infty |\overline{\kappa}_0 x_0 + \overline{\kappa}_1 x_1|^2 dt,$$

where $\overline{\kappa}_j = \hat{\kappa}_j|_{\varepsilon=0}$.

The method of §4 yields an optimizing sequence of regulators in problem (11.12) of the form $u(t) = r_0 x_0(t) + r_1 x_1(t)$, and the estimates of Φ_1 of these corresponding controls give

$$\Phi_1 - \inf \Phi_1 \leq \begin{cases} \varepsilon x_0(0)^2 & \text{if } b_0 b_1 > 0, \\ \varepsilon(x_0(0)^2 + (x_0(0) + \varepsilon x_1(0))^2) & \text{if } b_1 = 0, \\ \varepsilon \left(\left[\dfrac{a_1 x_0(0) + x_1(0)}{a_0} \right]^2 + \left[x_0(0) - \varepsilon \dfrac{a_1 x_0(0) + x_1(0)}{a_0} \right]^2 \right) \\ \qquad\qquad\qquad\qquad\qquad \text{if } b_0 = 0, \\ \varepsilon \left[\dfrac{2 b_0 b_1 x_1(0) - (b_0^2 - a_1 b_0 b_1 - a_0 b_1^2) x_0(0)}{\sigma_1} \right]^2 \\ \qquad\qquad\qquad\qquad\qquad \text{if } b_0 b_1 < 0. \end{cases}$$

Let us remark that the infimum $\inf \Phi_1$ is simultaneously zero and is attainable if and only if either $x_0(0) = x_1(0) = 0$ or $x_0(0) = 0$ and $b_0 b_1 > 0$. Then the optimal regulator has the form $u = r_0 x_0 + r_1 x_1$.

Problem (11.6) is singular if and only if $c_0 c_1 d_0 d_1 = 0$ (if $c_1 d_1 = 0$ the frequency inequality has a degeneracy of degree two at $\omega = \infty$; if $c_0 d_0 = 0$ the same is true at $\omega = 0$). If $c_0 c_1 \geq 0$, $d_0 d_1 \geq 0$ the representation (11.3) of Φ_2 is canonical; otherwise, the canonical representation of Φ_2 is

$$\Phi_2 = \begin{cases} -2 c_0 c_1 x_1(0)^2 + \int_0^\infty [c_1 x_0 - (c_0 + a_1 c_1) x_1 + d_1 c_1 u]^2 dt, \\ \qquad\qquad\qquad\qquad \text{if } c_0 c_1 < 0, d_0 d_1 \geq 0, \\ \dfrac{2 d_0 d_1 \tau_2^2}{-\sigma_2^2} \left(x_0(0) + \dfrac{\sigma_3}{\tau_2} x_1(0) \right)^2 \\ \quad + \int_0^\infty \left[c_1 x_0 + (c_0 - a_1 c_1) x_1 + d_1 c_1 u + \dfrac{2 d_0 \tau_2}{\sigma_2} \left(x_0 + \dfrac{\sigma_3}{\tau_2} x_1 \right) \right]^2 dt, \\ \qquad\qquad\qquad\qquad \text{if } c_0 c_1 \geq 0, d_0 d_1 < 0, \\ -2 c_0 c_1 x_1(0)^2 - 2 d_0 d_1 \tau_3^2 \sigma_2^{-2} \left(x_0(0) + \dfrac{\sigma_4}{\tau_3} x_1(0) \right)^2 \\ \quad + \int_0^\infty [c_1 x_0 - (c_0 + a_1 c_1) x_1 + d_1 c_1 u + 2 d_0 \tau_3 \sigma_2^{-1} (x_0 + \sigma_4 \tau_3^{-1} x_1)]^2 dt, \\ \qquad\qquad\qquad\qquad \text{if } c_0 c_1 < 0, d_0 d_1 < 0. \end{cases}$$

An optimizing sequence may be constructed in the form $u(t) = k_0 x_0(t) + k_1 x_1(t)$. The estimates of Φ_2 for these controls give $\Phi_2 - \inf \Phi_2 \leq \Delta_0(x_0(0), x_1(0))$. If $x_0(0) = 1$, $x_1(0) = 0$, and $c_1 = 0$, $\inf \Phi_2$ is simultaneously zero and attainable if and only if $d_0 d_1 > 0$. The corresponding optimal regulator $u = k_0 x_0 + k_1 x_1$.

Parts (I), (II), and (III) of Proposition 11.1 now follow Theorem 5.3 by examination of problems (11.12) and (11.13) (in part (I) R_0 is $\inf \Phi_1$ for $x_0(0) = c_1$ and $x_1(0) = c_0 - a_1 c_1$; R_1 is $\inf \Phi_2$ for $x_0(0) = \bar{\kappa}_0$ and $x_1(0) = \bar{\kappa}_1$). To prove part (IV) we observe that if absolute invariance is possible, the infimum of the functional (0.16) in the stochastic optimal control problem with complete observation and that of the functional (5.4) in optimal filtration (where $a_0 = \mathrm{col}(1, 0)$) are both zero and attainable. By Lemmas 5.1 and 5.2, the same is true for the infima in problems (11.12) (with initial values $x_0(0) = c_1$, $x_1(0) = c_0 - a_1 c_1$) and (11.13) (with initial values $x_0(0) = 1$, $x_1(0) = 0$). Hence, if absolute invariance is possible, condition (11.10) holds. Conversely, if (11.10) holds, then the regulator $u = r_0 x_0 + r_1 x_1$ makes Φ_1 zero in (11.12) with $x_0(0) = 0$, $x_1(0) = c_0$, and the regulator $u = k_0 x_0 + k_1 x_1$ makes Φ_2 zero in (11.3) with $x_0(0) = 1$, $x_1(0) = 0$. It then follows from Theorem 5.3 that the regulator (11.7), which in this case has the form (11.11), guarantees absolute invariance. This completes the proof of Proposition 11.1.

§12. Example 2

Consider the controlled object described by the equations

$$a(p)\varphi = \theta + g, \qquad p = d/dt; \qquad (12.1)$$

here $a(p) = a_2 p^2 + a_1 p + a_0$; φ, θ, and g are scalar stochastic processes; θ is the control; the noise g is a stationary stochastic process with spectral density $\Phi_g(s) = \Phi_0/(c(s)c(-s))$, where $c(s) = s^2 + 2\mu s + \mu^2 + \nu^2$, with given positive constants a_j, μ, ν, and Φ_0. This equation describes the rolling of a ship with a damper [31]; in that case φ is the heel angle, θ is the controlling torque of the damper forces, g the perturbing torque, i.e., the angle of wave slope, the parameters a_j describe the inertial and hydrodynamical properties of the ship; Φ_0, μ, and ν depend on the swell intensity and the direction of the ship motion relative to the general run of the waves. Many problems of automatic pilot design for a turbulent atmosphere can also be reduced to Equation (12.1).

Let the admissible controls be generated by stationary linear regulators

$$\alpha(p)\theta = \beta(p)\varphi, \qquad (12.1)$$

which stabilize system (12.1), i.e., α and β are polynomials such that $\alpha \neq 0$ and $a(\lambda)\alpha(\lambda) - \beta(\lambda) \neq 0$ for $\mathrm{Re}\,\lambda \geq 0$. We must minimize the functional

$$J = J(\alpha, \beta) = \lim_{t \to \infty} M\{\varphi(t)^2 + \rho\theta(t)^2\}$$

for $\rho > 0$. Using the standard Wiener-Hopf method, we find an optimal regulator (12.2) [31], with $\beta = \beta_* = a\alpha - \psi$, where ψ is obtained by the factorization $a(p)a(-p) + 1/\rho = \psi(p)\psi(-p)$, $\psi(\lambda) \neq 0$ for $\mathrm{Re}\,\lambda \geq 0$; the function $\alpha(p) = \alpha_0 + \alpha_1 p$ is uniquely determined by the polynomial equation $a(-p) = \alpha(p)\psi(-p) + k(p)c(p)$, $\deg k < 2$. As shown in [31], this regulator is not suitable for practical use, since for $\alpha_1 \neq 0$ even a small change in the coefficients a_j may make the closed loop system (12.1) unstable. This effect, pointed out in [31]–[33] and elsewhere, is known as structural instability of the optimal regulator.

The natural solution in this situation is to limit the class of regulators. Let us call a regulator (12.2) admissible if and only if it is a stabilizing regulator

and
$$\alpha \not\equiv 0, \qquad \deg \alpha \geq \deg \beta. \tag{12.4}$$

Such regulators are called strictly realizable [16]. A strictly realizable regulator cannot be structural unstable [16]. We will consider minimization of the functional (12.3), referred to as problem (12.1)–(12.4), in this new class of admissible regulators.

PROPOSITION 12.1. *Problem* (12.1)–(12.4) *does not admit an optimal regulator. An optimizing sequence of regulators can be obtained in the form* (12.2) *with* $\alpha = \alpha_\varepsilon$, $\beta = \beta_\varepsilon$, $\varepsilon \to +0$, *where* α_ε *and* β_ε *are uniquely determined from the polynomial equation*

$$a(\lambda)\alpha_\varepsilon(\lambda) - \beta_\varepsilon(\lambda) = \psi(\lambda)(\varepsilon\lambda + 1)^3 \tag{12.5}$$

with $\deg(\beta_\varepsilon - \beta_*) < 2$, *and the following estimate holds*:

$$J(\alpha_\varepsilon, \beta_\varepsilon) - \inf J$$
$$\leq \varepsilon\rho\Phi_0[a_1^2 + ([1 - 2\mu\varepsilon]\alpha_1 + \varepsilon\alpha_0)^2 + (([1 - 2\mu\varepsilon]^2 - \varepsilon^2(\mu^2 + \nu^2))\alpha_1$$
$$+ 2\varepsilon(1 - \mu\varepsilon)\alpha_0 + \varepsilon^2(a_1 - \psi_1))^2],$$

where $\psi(\lambda) = \psi_2\lambda^2 + \psi_1\lambda + \psi_0$, β_*, *and* α_j *are defined as indicated previously.*

REMARK. The same result is obtained using the algorithms of [16].

To prove Proposition 12.1, one introduces a "state" $\xi = \text{col}(\varphi, \dot\varphi, g, \dot g)$, "control" $\eta = \theta$, and "noise in the equation of the controlled object" $\zeta = \text{col}(0, 0, 0, \sqrt{\Phi_0}f)$, where f is white noise, and "observable output" $y = (p + \sigma)\varphi$, where $\sigma > 0$ is an arbitrary constant. The initial problem is easily reduced to a stochastic optimal control problem under incomplete observation $LQ(F\mathfrak{M}, P\Phi)$ (see §5 and §11). Using Theorem 5 we construct an optimizing sequence of control laws (5.8), (5.9), which yields the optimizing sequence of regulators as in Proposition 12.1. The estimate in Proposition 12.1 follows from (5.10).

Note that the final result does not depend on $\sigma > 0$. If one takes $y = \varphi$ as the observable output, the optimizing sequence (5.8), (5.9) yields an optimizing sequence of integrating regulators (12.2), i.e., regulators with $\deg \alpha > \deg \beta$.

To construct an optimizing sequence we replace the functional (12.3) by

$$J_\varepsilon = J_\varepsilon(\alpha, \beta) = \lim_{t \to \infty} M\{\varphi(t)^2 + \rho\theta(t)^2 + \varepsilon\theta'''(t)\}^2 \qquad (\varepsilon > 0).$$

An optimal admissible regulator minimizing J_ε exists (this problem is nonsingular). The resulting sequence of admissible regulators $\alpha_\varepsilon(p)\theta = \beta_\varepsilon(p)\varphi$, $\varepsilon \to 0$, is optimizing [16]. However, the computational difficulties involved in practical implementation of this method are prohibitive.

References

1. R. E. Kalman, *Contributions to the theory of optimal control*, Bol. Soc. Math. Mexicana (2) **5** (1960), 102–119.
2. A. M. Letov, *Analytical design of regulators*, Avtomat. i Telemekh. (1960), no. 4, 436–446; no. 5, 561–571; no. 6, 661–669; English transl. in Automat. Remote Control **21** (1960).
3. A. I. Lur'e, *Minimal quadratic cost functional regulated system*, Izv. Akad. Nauk SSSR Tekhn. Kibernet. **4** (1963), 140–146. (Russian)
4. J. C. Willems, *Least squares stationary optimal control and the algebraic Riccati equation*, IEEE Trans. Automat. Control **16** (1971), 621–634.
5. A. Kh. Gelig, G. A. Leonov, and V. A. Yakubovich, *Stability of nonlinear systems with non-unique equilibrium state*, "Nauka", Moscow, 1978. (Russian)
6. V. A. Yakubovich, *Frequency theorem for periodic systems and the theory of analytical design of regulators*, Method of Lyapunov Functions, "Nauka", Novosibirsk, 1987, pp. 281–290. (Russian)
7. _____, *The frequency theorem for the case in which the state and control spaces are Hilbert spaces, and its application in certain problems in the synthesis of optimal control. II.*, Sibirsk. Mat. Zh. **16** (1975), no. 5, 1081–1102, 1132; English transl. in Siberian Math. J. **16** (1975).
8. J. C. Willems, A. Kitapci, and L. M. Silverman, *Singular optimal control: a geometric approach*, SIAM J. Control **24** (1986), 323–337.
9. M. H. J. Hautus and L. M. Silverman, *System structure and singular control*, Linear Algebra Appl. **50** (1983), 369–402.
10. H. I. Trentelman, *Families of linear-quadratic problems: continuity properties*, IEEE Trans. Automat. Control **32** (1987), 323–329.
11. A. G. Luk'yanov and V. I. Utkin, *Synthesis of optimal linear systems with degenerate cost functional*, Avtomat. i Telemekh. 7 (1982), 42–50; English transl. in Automat. Remote Control **43** (1982).
12. V. A. Yakubovich, *Singular optimal control for stationary system with quadratic functional*, Sibirsk. Mat. Zh. **26** (1985), no. 3, 189–200; English transl. in Siberian Math. J. **26** (1985).
13. A. V. Megretskiĭ and V. A. Yakubovich, *Linear-quadratic singular optimal control for a stationary system*, manuscript deposited at VINITI, Moscow, 1988. (Russian)
14. *Resolutions about "Pravda" articles "Enemies in Soviet disguise" and "Traditions of servility"*, Uspekhi Mat. Nauk (1937), no. 3, 275–278. (Russian)
15. *Wiping the scientific community of "Luzinism"*, Uspekhi Math. Nauk (1937), no. 3, 3–4. (Russian)
16. V. A. Yakubovich, *Optimality and invariance of linear stationary control systems*, Avtomat. i Telemekh. **6** (1984), 69–85; English transl. in Automat. Remote Control **45** (1984).
17. N. N. Luzin, *On the study of the matrix theory of differential equations*, Avtomat. i Telemekh. **5** (1940), 4–66.
18. A. I. Kukhtenko, *The main stages of development of invariance theory* I, Avtomat. i Telemekh. (1984), no. 2, 3–13; II, Avtomat. i Telemekh. (1985), no. 2, 3–14; III, Avtomat. i Telemekh. (1985), no. 6, 3–14; English transl. of I in Automat. Remote Control **45** (1984); of II in Automat. Remote Control **46** (1985); of III in Automat. Remote Control **46** (1985).
19. L. Schwartz, *Mathematics for the Physical Sciences*, Addison-Wesley, Reading, MA, Palo-Alto, CA, and London, 1966.
20. A. V. Megretskiĭ, *Generalized processes in linear-quadratic optimal control*, Vestnik Leningrad. Univ. Mat. Mekh. Astronom. (1988), no. 2, 114–116; English transl. in Vestnik Leningrad. Univ. Math. **21** (1988).
21. W. A. Andreew, Yu. F. Kasarinov, and V. A. Yakubovich, *The problem of optimal control synthesis for linear systems*, Electronische Informationsarbeitung und Kibernetik **8** (1972), 391–428.

22. V. A. Yakubovich, *A frequency theorem in control theory*, Sibirsk. Mat. Zh. **14** (1973), 384–420; English transl. in Siberian Math. J. **14** (1973).
23. D. Z. Arov and V. A. Yakubovich, *Conditions of semiboundness of quadratic functionals on Hardy spaces*, Vestnik Leningrad. Univ. Mat. Mekh. Astronom. (1982), no. 1, 11–18; English transl. in Vestnik Leningrad. Univ. Math. **15** (1982).
24. A. V. Megretskiĭ, *Singular linear-quadratic optimal control*, Dissertation, Leningrad, 1988. (Russian)
25. V. A. Yakubovich, *Conditional semiboundness of quadratic functional on a subspace of a Hilbert space*, Vestnik Leningrad. Univ. Mat. Mekh. Astronom. (1981), no. 19, 50–53; English transl. in Vestnik Leningrad. Univ. Math. **14** (1981).
26. F. R. Gantmakher, *Theory of matrices*, "Nauka", Moscow, 1966; English transl. of the 1st ed., Chelsea, New York, 1959.
27. K. Hoffman, *Banach Spaces of Analytic Functions*, Prentice Hall, Englewood Cliffs, N.J., 1962.
28. A. V. Mikhaĭlov, *On the design method for automatic regulators, proposed by Shchipanov*, Avtomat. i Telemekh. (1940), no. 5, 129–143.
29. L. N. Mikhaĭlov, *Some remarks on the theory of complete compensation of disturbances*, Avtomat. i Telemekh. (1940), no. 5, 145–154. (Russian)
30. S. A. Christianovich and F. R. Gantmakher, *Analysis of the basic theses of Prof. Shchipanov's study "Theory and design methods for automatic regulators"*, Avtomat. i Telemekh. (1940), no. 5, 123–128. (Russian)
31. I. P. Simakov, *On ill-definedness of the synthesis methods for optimal control systems under random loadings*, Avtomat. i Telemekh. (1974), no. 3, 186–189; English transl. in Automat. Remote Control **35** (1974).
32. P. V. Nadezhdin, *On practical (structural) instability of the systems, proposed in the paper I*, Avtomat. i Telemekh. (1973), no. 5, 196–198; English transl. in Automat. Remote Control **34** (1973).
33. Yu. P. Petrov, *Optimization of the control systems disturbed by wind and sea roughness*, Leningrad, 1973.

Translated by A. M. TSALIK

The Neumann Problem for Selfadjoint Elliptic Systems in a Domain with Piecewise-Smooth Boundary

S. A. NAZAROV AND B. A. PLAMENEVSKIĬ

§1. Introduction

During the last two decades the theory of elliptic boundary-value problems in domains with piecewise-smooth boundaries has seen considerable progress. Though fairly complete answers have been obtained for many questions, some basic results of the general theory are not very conclusive. The point is that the investigation of elliptic problems in domains with edges has essentially been reduced to a few model problems with "frozen" coefficients see [1–3], among others. These problems, which depend on real parameters, are posed in domains of the same type as the initial one, but of smaller dimensions. In the general theory, one adopts the hypothesis that the kernels and cokernels of the model problems are trivial. Only in a few special situations has it turned out to be possible to give complete answers without the above hypothesis [1, 4–9]. In addition, there are various examples where the hypothesis is not true under the usual conditions of the general theory. These examples include some problems of mathematical physics that are important in applications (in particular, the problem of boundary cracks in three-dimensional elastic bodies, see below, §5.3).

The aim of this paper is to study the Neumann problem for selfadjoint elliptic systems (of which the above example is a special case). It turns out that this case can be treated comprehensively, without any *a priori* assumptions. This is achieved by two devices: first, an additional condition (1.7) confines the investigation to a natural class of systems which includes all physically meaningful problems; second, a new scale of weighted functional spaces is introduced. To do this we must review the relation among the function spaces in which the operators of model problems are defined.

1. Domain. Let Ω be a domain in $\mathbb{R}^n$ with compact closure $\bar{\Omega}$ whose boundary $\partial\Omega$ contains edges of various dimensions which may meet at nonzero angles. In particular, in a three-dimensional domain we allow singularities of the type of vertices of cones with smooth directrices, smooth edges or vertices of polyhedra.

2. Boundary-value problem. We study a system $\mathscr{L}(x, \partial_x) = \|\mathscr{L}_{ij}(x, \partial_x)\|_{i,j=1}^k$ of differential operators in Ω with $C^\infty(\bar{\Omega})$ coefficients; here $\partial_x = (\partial/\partial x_1, \ldots, \partial/\partial x_n)$, $\text{ord}\,\mathscr{L}_{ij} = s_i + t_j$, $t_j = \alpha_j + \alpha_{\max}$, $s_i = \alpha_i - \alpha_{\max}$, α_j are positive numbers, $\alpha_{\max} = \max(\alpha_1, \ldots, \alpha_k)$, $s_1 + t_1 + \cdots + s_k + t_k = 2(\alpha_1 + \cdots + \alpha_k) = 2m$. We assume that

$$(\mathscr{L}u, v)_\Omega = a(u, v; \Omega) - (\mathscr{N}u, \mathscr{D}v)_{\partial\Omega}, \qquad (1.1)$$

where $u = (u_1, \ldots, u_k)$, $v = (v_1, \ldots, v_k)$ are vectors with $C^\infty(\bar{\Omega})$ components, $(\,,\,)_\Omega$ and $(\,,\,)_{\partial\Omega}$ are the scalar products in (vector-valued) spaces $L_2(\Omega)$ and $L_2(\partial\Omega)$, $\mathscr{D}v = \{\partial_\nu^h v_j \mid j = 1, \ldots, k,\ h = 0, \ldots, \alpha_j - 1\}$, ν is the vector of the outward normal, $\mathscr{N}$ is an $(m \times k)$-matrix of differential operators, and $a(u, v; \Omega)$ is a symmetric sesquilinear form (i.e., $a(u, v) = \overline{a(v, u)}$) such that

$$|a(u, v; \Omega)| \le c \sum_{j=1}^k \sum_{|\gamma| \le \alpha_j} \int_\Omega |\partial_x^\nu v_j(x)|^2\, dx.$$

Let Γ denote the union of all edges in $\partial\Omega$. We shall study the following boundary-value problem:

$$\mathscr{L}(x, \partial_x)u(x) = f(x),\quad x \in \Omega;\qquad \mathscr{N}(x, \partial_x)u(x) = g(x),\quad x \in \partial\Omega \setminus \Gamma \qquad (1.2)$$

which we call the Neumann problem. We assume that this problem is elliptic everywhere outside Γ and "nondegenerate" on Γ (for the rigorous formulation, see subsection 3). (Model problems for the Neumann problem in a domain with smooth nonintersecting edges were studied in [10–12].)

There are model problems of two types: (i) the problem in a cone (with a possibly nonsmooth director manifold) for a nonhomogeneous differential operator $A(\theta)$, where θ is a point on the sphere S^{d-1}; (ii) the problem in a bounded domain for an operator $\mathfrak{A}(\lambda)$ with a complex parameter λ.

In order to avoid unessential technicalities, we shall consider problem (1.2) in detail in a domain Ω under an additional assumption: by freezing the coefficients at any point $x^0 \in \Gamma$, we obtain a boundary-value problem in a domain $K^{n-d} \times \mathbb{R}^d$, where K^{n-d} is a cone that intersects the unit sphere S^{n-d-1} in a domain ω with the boundary that may contain smooth nonintersecting edges of different dimensions. This formulation yields model problems of both types: a problem for $A(\theta)$ in K^{n-d}, and a problem for $\mathfrak{A}(\lambda)$ in ω. These examples will show us how to overcome difficulties in the general case.

3. Boundary-value problem in a special domain. Let ω be a domain on the sphere S^{n-d-1} whose boundary $\partial\omega$ contains a union γ of smooth edges of various dimensions; for every point $x \in \gamma$ in a ν-dimensional edge there exists a neighborhood U and a diffeomorphism $\kappa : U \to \mathbb{R}^{n-d-1}$ that maps $U \cap \omega$ into the product $K^{n-d-\nu-1} \times \mathbb{R}^\nu$, where $0 \leq \nu < n-d-2$ and $K^{n-d-\nu-1}$ is a cone with a smooth director manifold.

We consider the following problem in $\Omega_0 \equiv K^{n-d} \times \mathbb{R}^d$:

$$L(\partial_x)u(x) = f(x), \quad x \in \Omega_0; \tag{1.3}$$

$$N(x, \partial_x)u(x) = g(x), \quad x \in \partial\Omega_0 \setminus \Gamma_0; \tag{1.4}$$

Γ_0 is a set of singular points on the boundary of Ω_0, that is, the product $k(\gamma) \times \mathbb{R}^d$, where $k(\gamma)$ is a closed conic "surface" with directrix γ; L is a formally selfadjoint matrix-valued elliptic operator whose entries are homogeneous differential expressions with constant coefficients; N is the operator of the Neumann boundary conditions in Green's formula

$$(Lu, v)_{\Omega_0} = a_0(u, v; \Omega_0) - (Nu, Dv)_{\partial\Omega_0} \tag{1.5}$$

where the operator D corresponds to Dirichlet conditions; compare subsection 1.

We shall assume that the problem (1.3), (1.4) is elliptic, that is, the boundary conditions cover L on $\partial\Omega_0 \setminus \Gamma_0$ (in the sense of [13]). In addition, we assume that the symmetric sesquilinear form

$$a_0(u, v; Q) = \int_Q \sum_{j,i=1}^k \sum_{|\sigma|=\alpha_i} \sum_{|\tau|=\alpha_j} a_{i,j}^{(\sigma,\tau)} \overline{\partial_x^\sigma v_i} \partial_x^\tau u_j \, dx \tag{1.6}$$

satisfies the following condition:

$$a_0(v, v; Q) = 0 \Longrightarrow v \in P|_Q, \tag{1.7}$$

where Q is an arbitrary domain in $\mathbb{R}^n$ and P is a certain subspace of polynomials with coefficients in $\mathbb{C}^k$. (Clearly, $pe^i \in P$, where e^i is a basis vector in $\mathbb{C}^k$ and p is a scalar polynomial of order $\operatorname{ord} p < \alpha_i$.)

Let D_{qi} and N_{qj} denote the entries of the $m \times n$ matrices D and N (see (1.5)), and define $\operatorname{ord} D_{qi} = r_q + \alpha_i$, $\operatorname{ord} N_{qi} = \sigma_q + \alpha_j$. It is obvious that $r_q + \alpha_i < \alpha_i$ and $r_q + \alpha_j + \sigma_q + \alpha_i = \alpha_j + \alpha_i - 1$; so $r_q < 0$ and $\sigma_q = -r_q - 1 \geq 0$. Thus $\operatorname{ord} N_{qj} \geq \alpha_j$ and pe^i is a solution of the homogeneous Neumann problem (1.3), (1.4).

4. The three-dimensional problem of elasticity theory.
To explain why condition (1.7) is quite natural, we consider the following example.

Let us treat u, f, and g in (1.3), (1.4) as the vectors of displacements, body forces, and external load, respectively. As usual,

$$\varepsilon_{jk}(u) = \frac{1}{2}\left(\frac{\partial u_j}{\partial x_k} + \frac{\partial u_k}{\partial x_j}\right), \quad \sigma_{jk}(u) = \sum_{p,q=1}^3 A_{j,k}^{p,q}\varepsilon_{p,q}(u) \tag{1.8}$$

are the components of the strain and stress tensors, A is the positive definite Hooke tensor of elastic moduli. System (1.3) consists of the three equations of equilibrium. The left-hand side of (1.4) has the type $Nu = \|\sum \sigma_{jk}(u)v_k\|$ and is the vector of stresses acting on an elementary surface with normal ν. The form

$$a_0(u, u; Q) = \int_Q \sum_{j,k=1}^{3} \sigma_{jk}(u, x)\varepsilon_{jk}(u; x)\, dx$$

is twice the elastic energy of the field u in Q.

Since the tensor A is positive definite, the equality $a_0(u, u; Q) = 0$ implies that $\varepsilon_{jk}(u) = 0$ in ω, $j, k = 1, 2, 3$. And this means that $u = b + c \times x$, where b and c are arbitrary ($\times$ denotes vector multiplication). Thus P in (1.7) is the space of rigid displacements.

5. Model problems. For $x \in \Omega_0$ let $x = (y, z)$, $y = (x_1, \ldots, x_{n-d})$, $z = (x_{n-d+1}, \ldots, x_n)$. Then the operators L and N may be written as $L(\partial_y, \partial_z)$ and $N(y, \partial_y, \partial_z)$ (we emphasize that coefficients of N are determined by the components of the normal $\nu(y)$ and depend only on $y|y|^{-1}$).

Taking Laplace transforms with respect to z in problem (1.3), (1.4) and changing variables $y \mapsto \eta = |\xi|^{-1}y$, where ξ is the variable dual to z, we obtain a problem in the cone $K \equiv K^{n-d}$, depending on the parameter $\theta = \xi|\xi|^{-1} \in S^{d-1}$,

$$\begin{aligned} L(\partial_\eta, \theta)U(\eta, \theta) &= F(\eta, \theta), & \eta \in K, \\ N(\eta, \partial_\eta, \theta)U(\eta, \theta) &= G(\eta, \theta), & \eta \in \partial K \setminus 0, \end{aligned} \quad (1.9)$$

where $|\xi|^{-\alpha_j}U_j(|\xi|y, \theta)$, $|\xi|^{\alpha_i}F_i(|\xi|y, \theta)$, $|\xi|^{\sigma_q}G_q(|\xi|y, \theta)$ are the Laplace transforms of u_j, f_i, g_q. The operator of the boundary-value problem (1.9) is denoted by $A(\theta)$.

We introduce spherical coordinates (r, φ) in the cone K, where $r = |y|$ and $\varphi = y|y|^{-1} \in \bar{\omega}$. The matrix elements $L_{ij}(\partial_y, 0)$ and $N_{qj}(y, \partial_y, 0)$ in spherical coordinates are $r^{-\alpha_i-\alpha_j}\mathbf{L}_{ij}(\varphi, \partial_\varphi, r\partial_r)$ and $r^{-\sigma_q-\alpha_j}\mathbf{N}_{qj}(\varphi, \partial_\varphi, r\partial_r)$. As usual [14, 15] the operator $\{L(\partial_y, 0), N(y, \partial_y, 0)\}$ is associated with a matrix pencil

$$\mathfrak{A}(\lambda) = \{\|\mathbf{L}_{ij}(\varphi, \partial_\varphi, \lambda + \alpha_j)\|_{i,j=1}^{k}, \quad \|\mathbf{N}_{qj}(\varphi, \partial_\varphi, \lambda + \alpha_j)\|_{q=1, j=1}^{m, k}\} \quad (1.10)$$

in ω on the sphere S^{n-d-1} with complex parameter λ.

We now define similar model problems of the "next generation" (they will be needed to study $A(\theta)$ and $\mathfrak{A}(\lambda)$). Choose a point $\varphi \in \gamma \subset \partial \omega$ and introduce new orthogonal coordinates (y', z') with the same origin; the z' axes include the old $z_1, \ldots, z_d$ axes, the z'_{d+1} axis points in the direction of φ, and the $z'_{d+2}, \ldots, z'_{d+\nu-1}$ axes point along the ν-dimensional edge containing φ. Then $y'_1, \ldots, y'_{n-d-\nu-1}$ form an orthogonal system in the space $\mathbb{R}^{n-d-\nu-1}$ containing the cone $K^{n-d-\nu-1}$ with a smooth director manifold. To simplify the notation we denote the operator L in the new coordinates by $L(\partial_{y'}, \partial_{z'})$ (without indicating the matrix of the coordinate transformation); the operator N with its coefficients frozen at a point $\varphi \in \gamma$ will be denoted by $N(\varphi, y'|y'|^{-1}, \partial_{y'}, \partial_{z'})$. Let $A_\varphi(\theta')$ denote the operator $\{L(\partial_{y'}, \theta'), N(\varphi, y'|y'|^{-1}, \partial_{y'}, \theta')\}$ of the boundary-value problem in $K^{n-d-\nu-1}$. The pencil $\mathfrak{A}_\varphi(\lambda)$ is obtained from $A_\varphi(0) = \{L(\partial_{y'}, 0), N(\varphi, y'|y'|^{-1}, \partial_{y'}, 0)\}$ in exactly the same way as the pencil $\mathfrak{A}(\lambda)$ is obtained from $A(0) = \{L(\partial_y, 0), N(y, \partial_y, 0)\}$.

Thus, associated with problem (1.3), (1.4), we have the operators $A(\theta)$, $\mathfrak{A}(\lambda)$, $A_\varphi(\theta')$, and $\mathfrak{A}_\varphi(\lambda)$ as model problems. (If $\varphi \in \partial\omega \setminus \gamma$, that is, if φ is a nonsingular point, we obtain a problem in a half-space instead of $A_\varphi(\theta'), \mathfrak{A}_\varphi(\lambda)$; see [16].)

§2. Investigation of the operators $\mathfrak{A}_\varphi(\lambda)$ and $A_\varphi(\theta')$

In order to construct the inverse operator for the boundary-value problem (1.2), we have to consider the model problems in the inverse order to that considered in §1. We therefore begin with a study of the pencil $\mathfrak{A}_\varphi(\lambda)$.

1. The model problem $\mathfrak{A}_\varphi(\lambda)$. To simplify notation, we replace $\mathfrak{A}_\varphi(\lambda)$ by the pencil (1.10), assuming in addition that $\partial\omega$ is smooth, $\gamma = \varnothing$, and write $n(\omega) = \dim \omega + 1$.

The pencil $\mathfrak{A}_\varphi(\lambda)$ is generated by the problem $A_\varphi(0)$. The fourth of the conditions (1.7) for the (new) problem is guaranteed by part (ii) of the following proposition and by the fact that linear changes of coordinates preserve the properties of the form a_0.

PROPOSITION 2.1. *Let $a_0^\theta(u, v; Q')$ be the form obtained from (1.6) by replacing the n-dimensional domain Q with an $(n-d)$-dimensional domain Q', $\partial_x = (\partial_y, \partial_z)$ with (∂_y, θ), and dx with dy. Then*

(i) $a_0^\theta(u', u'; Q') = 0$ for $|\theta| = 1 \Longrightarrow u' \equiv 0$ in Q';
(ii) $a_0^\theta(u', u'; Q') = 0$ for $\theta = 0 \Longrightarrow u' \in P'$, where P' is the set of polynomials in P that do not depend on z.

PROOF. Set $u(y, z) = u'(y)\exp(\theta z)$ and $Q = Q' \times T$, where T is a domain in $\mathbb{R}^{n-d}$. We have $a_0(u, u; Q) = \rho_T a_0^\theta(u', u'; Q')$, $\rho_T = \int_T \exp(\theta z)\, dz = 0$. Then $u \in P$ by (1.7). For $\theta \neq 0$, u is a polynomial only if $u' = 0$ and for $\theta = 0$, $u(y, z) = u'(y)$.

It is known [16] that the mappings

$$\mathfrak{A}(\lambda) : \prod_{j=1}^k W_2^{l+\alpha_j}(\omega) \to \prod_{j=1}^k W_2^{l-\alpha_j}(\omega) \times \prod_{q=1}^m W_2^{l-\sigma_q-1/2}(\partial\omega)$$

are isomorphisms for all $\lambda \in \mathbb{C}$, excluding the set of normal eigenvalues. In every strip $\|\operatorname{Re}\lambda\| < h$ there exist only finitely many points of the spectrum of the pencil $\lambda \mapsto \mathfrak{A}(\lambda)$.

Since the problem in the cone is formally selfadjoint, it follows that $\mathfrak{A}(\lambda)^* = \mathfrak{A}(-\lambda - n(\omega))$, where $\mathfrak{A}(\lambda)^*$ is the operator of the problem adjoint to $\mathfrak{A}(\lambda)$ in the sense of Green's formula. Hence the spectrum of the pencil $\mathfrak{A}$ is centrally symmetric about the point $-n/2$.

THEOREM 2.2. (i) *If $n(\omega)$ is odd or $\alpha_{\max} < n(\omega)/2$, then there are no points of the spectrum of $\mathfrak{A}$ on the straight line $l_0 = \{\lambda \in \mathbb{C} : \operatorname{Re}\lambda = -n(\omega)/2\}$.*

(ii) *If $n(\omega)$ is even and $\alpha_{\max} \geq n(\omega)/2$, then there is a unique eigenvalue $\lambda = -n(\omega)/2$ on l_0. The eigenspace consists of the restrictions to ω of polynomials X with coefficients in $\mathbb{C}^k$, where any homogeneous (scalar) polynomial of degree $\alpha_j - n(\omega)/2$ may be a component X_j of the vector $X = (X_1, \ldots, X_k)$ (if $\alpha_j < n(\omega)/2$, then $X_j = 0$). For each eigenvector there is an associated eigenvector, and there are no Jordan blocks of length greater than 2.*

PROOF. Let $\lambda \in l_0$ be an eigenvalue of $\mathfrak{A}$ and $\Phi = (\Phi_1, \ldots, \Phi_k)$ a corresponding eigenvector. Take $U_j(y) = r^{\lambda + \alpha_j} \Phi_j(\varphi)$. The vector-valued function $U = (U_1, \ldots, U_k)$ satisfies the homogeneous equations

$$L(\partial_y, 0)U = 0 \quad \text{in } K, \qquad N(y, \partial_y, 0) = 0 \quad \text{on } \partial K \setminus 0. \tag{2.1}$$

Applying Green's formula (1.5) for the domain $K_\varepsilon = \{y \in K : \varepsilon < |y| < \varepsilon^{-1}\}$, we get

$$0 = \int_{K_\varepsilon} LU \cdot \overline{U} dy$$

$$= a_0(U, U; K_\varepsilon) - \int_{\{y \in K : |y| = \varepsilon\}} NU \cdot \overline{DU} ds_y + \int_{\{y \in K : |y| = \varepsilon^{-1}\}} NU \cdot \overline{DU} ds_y. \tag{2.2}$$

As pointed out in §1.3, $\operatorname{ord} N_{qi} + \operatorname{ord} D_{qj} = \alpha_j + \alpha_i - 1$. Therefore, $NU \cdot \overline{DU}$ is a homogeneous function of degree $1 - n$. Hence the two integrals in the right-hand side of (2.2) cancel out, that is, $a_0(U, U; K_\varepsilon) = 0$. It follows from condition (1.7) that U is a vector-valued polynomial. Part (i) is now obvious and we proceed to part (ii). Since $\operatorname{ord} N_{qj} \geq \alpha_j$ and $\operatorname{ord} L_{ji} = \alpha_i + \alpha_j$ (see the end of §1.3), any polynomial X described in the statement of the Theorem satisfies equations (2.1). It remains only to verify the existence of associated eigenvectors.

Since the problem $\{L(\partial_y, 0), N(y, \partial_y, 0)\}$ is selfadjoint in the cone K, it follows from Theorem 8.1 of [15] that for any solution $U(y) = \{r^{\lambda + \alpha_j} \Phi_j(\varphi)\}_{j=1}^k$

of equations (2.1) there is a solution $W(y) = \{r^{-\bar{\lambda}-\eta(\omega)+\alpha_j}\Psi_j(\varphi, \log r)\}_{j=1}^k$ of the same equations (where $\Psi(\varphi, \log r)$ is a polynomial in $\log r$ whose coefficients are eigenvectors and associated eigenvectors of $\mathfrak{A}$) such that

$$\int_\omega (DU \cdot \overline{NW} - NU \cdot \overline{DW}) ds_y = 1. \tag{2.3}$$

Since $\lambda = -n(\omega)/2$, it follows that $-\bar{\lambda} - n(\omega) = -n(\omega)/2$. First assume that there are no associated eigenvectors. Then U and W in (2.3) are polynomials and therefore $NU = 0$ and $NW = 0$ which is impossible. Thus, there exist associated eigenvectors. Let

$$W(y) = \{r^{\alpha_j - n(\omega)/2}(\Phi_j(\varphi) \log r + \Psi_j(\varphi))\}_{j=1}^k,$$

where Φ is an eigenvector and Ψ an associated eigenvector. Applying formula (1.5), we obtain

$$0 = a_0(W, W; K_\varepsilon) - \int_{\{y \in K : |y|=\varepsilon\}} NW \cdot \overline{DW} ds_y + \int_{\{y \in K : |y|=\varepsilon^{-1}\}} NW \cdot \overline{DW} ds_y. \tag{2.4}$$

The components W_j satisfy the equalities $W_j(ty) = t^{\alpha_j - n/2}(W_j(y) + \log t V_j(y))$, and moreover $Y(y) = \{r^{\alpha_j - n/2}\Phi_j(\varphi)\}_{j=1}^k$ is a polynomial. Therefore,

$$\int_{\{y \in K : |y|=\varepsilon^{\pm 1}\}} NW \cdot \overline{DW} ds_y = \int_\omega NW \cdot \overline{DW} ds_y$$

$$\pm \log \varepsilon \int_\omega (NW \cdot \overline{DY} - NY \cdot \overline{DW}) ds_y + (\log \varepsilon)^2 \int_\omega NY \cdot \overline{DY} ds_y$$

$$= \int_\omega NW \cdot \overline{DW} ds_y \pm \log \varepsilon \int_\omega NW \cdot \overline{DY} ds_y. \tag{2.5}$$

Since W is not a polynomial, condition (1.7) and equalities (2.4), (2.5) imply that

$$\int_\omega (DY \cdot \overline{NW} - NY \cdot \overline{DW}) ds_y = \int_\omega DY \cdot \overline{NW} ds_y = (2 \log \varepsilon)^{-1} a_0(W, W; K_\varepsilon) \neq 0. \tag{2.6}$$

By [17] (see the biorthogonality relations (2.3)), the equality

$$\int_\omega (NY \cdot \overline{DW} - DY \cdot \overline{NW}) ds_y = 0$$

is a necessary condition for the existence of an associated eigenvector continuing the chain $\{\Phi, \Psi\}$. But by (2.6) this condition is not satisfied.

One can deduce from Theorem 2.2 that if $n(\omega)$ is even, the full algebraic multiplicity of the eigenvalue $\lambda = -n(\omega)/2$ is

$$\kappa = 2[(n(\omega)-1)!]^{-1} \sum_{j=1}^k \prod_{h=1}^{n(\omega)-1} (\alpha_j + h - n(\omega)/2). \tag{2.7}$$

2. The model problem $A_\varphi(\theta')$. The only essential difference between the operators $A_\varphi(\theta')$ and $A_\varphi(\theta)$ is that the cone for $A_\varphi(\theta)$ is considered as having a nonsmooth director manifold, while the cone for $A_\varphi(\theta')$ has a smooth director manifold. To simplify the notation, therefore, we shall study the operators $A_\varphi(\theta')$ and $A_\varphi(\theta)$ (the operator of problem (1.9)) simultaneously, on the additional assumption that the $(n(\omega) - 1)$-dimensional manifold $\partial \omega$ is smooth.

Let $\beta \in \mathbb{R}$, $l = 0, 1, \ldots$ and let $E_\beta^l(K)$ (as in [2, 3]) be the space of functions in K with the norm

$$\|u; E_\beta^l(K)\| = \left(\sum_{|\gamma| \leq l} \||\eta|^{\beta-l+|\gamma|}(1+|\eta|)^{l-|\gamma|} \partial_\eta^\gamma u; L_2(K)\|^2 \right)^{1/2}. \quad (2.8)$$

$E_\beta^{l+1/2}(\partial K)$ will denote the space of traces on ∂K of functions in $E_\beta^{l+1}(K)$.

REMARK 2.3. The origin of the norms in $E_\beta^l(K)$ may be explained as follows. Define a norm in the domain $K \times \mathbb{R}^\nu$ by

$$\|u; V_\beta^l(K \times \mathbb{R}^\nu)\| = \left(\sum_{|\sigma|+|\tau| \leq l} \int_K \int_{\mathbb{R}^\nu} |y|^{2(\beta-l+|\sigma|+|\tau|)} |\partial_y^\sigma \partial_z^\tau u(y,z)|^2 \, dy \, dz \right)^{1/2}. \quad (2.9)$$

This norm is equivalent to the norm

$$\left(\sum_{|\tau|+q \leq l} \int_0^\infty \int_{\mathbb{R}^\nu} r^{2(\beta-l)+n(\omega)-1} \|(r\partial_z)^\tau (r\partial_r)^q u(r,\cdot,z); W_2^{l-|\tau|-q}(\omega)\|^2 \, dr \, dz \right)^{1/2}, \quad (2.10)$$

where $r = |y|$. The passage from the operator of a boundary-value problem in $K \times \mathbb{R}^\nu$ to the operator $A(\theta)$ is made by taking Laplace transforms $\mathscr{L}_{z \to \xi}$ and applying the dilatation $y \mapsto \eta = |\xi|y$. These transformations, applied to (2.9), lead to the norm

$$\left(\sum_{|\sigma|+|\tau| \leq l} \int_K |\eta|^{2(\beta-l+|\sigma|+|\tau|)} |\partial_\eta^\sigma v(\eta)|^2 \, d\eta \right)^{1/2},$$

which is equivalent to (2.8). Yet another equivalent norm

$$\left(\sum_{|\tau|+q \leq l} \int_0^\infty r_\eta^{2(\beta-l)+n(\omega)-1+2|\tau|} \|(r_\eta \partial_{r_\eta})^q v(r_\eta,\cdot); W_2^{l-|\tau|-q}(\omega)\|^2 \, dr_\eta \right)^{1/2}$$

is obtained by applying the same transformations to (2.10).

It is convenient to provide the space $W_2^l(\omega)$ with an equivalent norm, which depends on a parameter $t > 0$:

$$\|v; W_2^l(\omega; t)\| = \left(\sum_{k=0}^l t^{2k} \|v; W_2^{l-k}(\omega)\|^2 \right)^{1/2}. \quad (2.11)$$

The norm (2.12) may now be rewritten as

$$\left(\sum_{q=0}^{l} \int_0^\infty r_\eta^{2(\beta-l)+n(\omega)-1} \|(r_\eta \partial_{r_\eta})^q v(r_\eta, \cdot); W_2^{l-q}(\omega; 1+r_\eta)\|^2 dr_\eta \right)^{1/2}. \quad (2.12)$$

In addition, the norm $\|\cdot; V_\beta^l(K)\|$, obtained from (2.9) by setting $\nu = 0$, is equivalent to the norm

$$\left(\sum_{q=0}^{l} \int_0^\infty r_\eta^{2(\beta-l)+n(\omega)-1} \|(r_\eta \partial_{r_\eta})^q v(r_\eta, \cdot); W_2^{l-q}(\omega; 1)\|^2 dr_\eta \right)^{1/2}. \quad (2.13)$$

If $l \geq \alpha_{\max}$, the mapping

$$A_\beta(\theta) \equiv A(\theta) : \prod_{j=1}^{k} E_\beta^{l+\alpha_j}(K) \to \prod_{j=1}^{k} E_\beta^{l-\alpha_j}(K) \times \prod_{q=1}^{m} E_\beta^{l-\sigma_q-1/2}(\partial K) \quad (2.14)$$

is continuous. By [14, 2, 3], (2.14) is a Fredholm operator only if there are no eigenvalues of $\mathfrak{A}(\lambda)$ on the line $\{\lambda \in \mathbb{C} : \operatorname{Re}\lambda = l - \beta - n(\omega)/2\}$.

THEOREM 2.4. *Choose $\delta > 0$ so small that the strip $0 < \operatorname{Re}\lambda + n(\omega)/2 \leq \delta$ does not intersect the spectrum of $\mathfrak{A}$. Then*

(i)
$$\operatorname{Ind} A_{l+\delta}(\theta) = -\operatorname{Ind} A_{l-\delta}(\theta),$$
$$\operatorname{Ind} A_{l+\delta}(\theta) = \operatorname{Ind} A_{l-\delta}(\theta) + \kappa,$$
$$(2.15)$$

where $\operatorname{Ind} A = \dim \ker A - \dim \operatorname{coker} A$, κ is the full algebraic multiplicity of the eigenvalue $\lambda = -n(\omega)/2$ (see Theorem 2.2 and formula (2.7)).

(ii) *The subspace $\ker A_{l-\delta}(\theta)$ is trivial.*

(iii) *If $n(\omega)$ is odd or $\alpha_{\max} < n(\omega)/2$, then both subspaces $\ker A_{l+\delta}(\theta)$ and $\operatorname{coker} A_{l+\delta}(\theta)$ are trivial.*

(iv) *If $n(\omega)$ is even and $\alpha_{\max} \geq n(\omega)/2$, then we have $\dim \ker A(\theta) = \kappa/2$ and $\dim \operatorname{coker} A_{l+\delta}(\theta) = 0$. One can choose a basis $\{\zeta^1, \ldots, \zeta^{\kappa/2}\}$ in $\ker A_{l+\delta}$ such that the components of the vector-valued functions $y \mapsto \zeta^h(y; \theta)$ satisfy the following relations:*

$$\zeta_j^h(\eta; \theta) - \chi(\eta) r^{\alpha_j - n(\omega)/2}$$
$$\times \left\{ \Phi_j^h(\varphi)\log r + \Psi_j^h(\varphi) + M_{h1}(\theta)\Phi_j^1(\varphi) + \cdots + M_{h\kappa/2}\Phi_j^{\kappa/2}(\varphi) \right\} \quad (2.16)$$
$$\in E_{l-\delta}^{l+\delta}(K),$$

where χ is a truncating function in $C_0^\infty(\mathbb{R}^{n(\omega)})$, $\chi(0) = 1$, $\{\Phi^h, \Psi^h\}$ are an eigenvector and an associated eigenvector corresponding to the eigenvalue $\lambda = -n(\omega)/2$ of $\mathfrak{A}$, (r, φ) are the spherical coordinates of η.

PROOF. (i) Obviously, $\operatorname{Ind} B_{l+\delta}(\theta) = -\operatorname{Ind} A_{l+\delta}(\theta)$, where $B_{l+\delta}(\theta)$ is the adjoint of $A_{l+\delta}(\theta)$ with respect to the duality in the product $L_2(K)^k \times$

$L_2(\partial K)^m$. The formal selfadjointness of $A(\theta)$ (in the sense of Green's formula) and theorems on local increase of smoothness (compare [13] and [15, 3]) imply that $\operatorname{Ind} A_{l-\delta}(\theta) = \operatorname{Ind} B_{l+\delta}(\theta)$. This proves the first formula in (2.10). The second formula follows from Lemmas 8.1, 8.2 of [15] (see also Theorem 2.6 in [18]).

(ii) The following Green's formula holds for the function $u = (u_1, \ldots, u_k) \in C_0^\infty(\bar{K} \setminus 0)$:

$$(L(\partial_\eta, \theta)u, u)_K = a_0^\theta(u, u; K) - (N(\eta, \partial_\eta, \theta)u, D(\eta, \partial_\eta, \theta)u)_{\partial K}. \quad (2.17)$$

By taking limits, one sees that this formula holds for vectors u with components u_j in $E_{l-\delta}^{l+\alpha_j}(K)$. If $u \in \ker A_{l-\delta}(\theta)$ then (2.17) implies that $a_0^\theta(u, u; K) = 0$ and therefore $u = 0$ (Proposition 2.1(i)).

(iii) By Theorem 2.2(i) l_0 contains no points of the spectrum of $\mathfrak{A}$. It therefore follows from (2.15) that $\operatorname{Ind} A_{l-\delta}(\theta) = \operatorname{Ind} A_{l+\delta}(\theta) = 0$. Moreover, since there are no points of the spectrum in the strip $|\operatorname{Re}\lambda + n(\omega)/2| \leq \delta$, it follows from Theorem 6.2 in [3] (behavior of the solutions in the cone) that $\ker A_{l+\delta}(\theta) = \ker A_{l-\delta}(\theta)$. It remains to use part (ii).

(iv) The assertions about the dimensions of the kernel and cokernel follow from (i) and (ii). Applying Theorem 2.2 of [14] (asymptotic behavior of the solution near the vertex of the cone), we obtain the following representations for any basis $\{\zeta^1, \ldots, \zeta^{\kappa/2}\}$ of $\ker A_{l+\delta}(\theta)$:

$$\zeta_j^h(\eta; \theta) - \chi(\eta) r^{\alpha_j - n(\omega)/2} \sum_{s=1}^{\kappa/2} \{T_{hs}(\theta)(\Phi_j^s(\varphi) \log r + \Psi_j^s(\varphi)) \\ + M_{hs}(\theta)\Phi_j^s(\varphi)\} \in E_{l-\delta}^{l+\alpha_j}(K), \quad (2.18)$$

where $T(\theta) = \|T_{hs}(\theta)\|$ and $M(\theta) = \|M_{hs}(\theta)\|$ are $(\kappa/2 \times \kappa/2)$-matrices. To verify the existence of the special basis (2.16), we must only show that the matrix T is nonsingular. If this is not the case, there exists a nontrivial linear combination ζ of functions $\zeta^1, \ldots, \zeta^{\kappa/2}$ from (2.18) such that

$$\zeta_j(\eta, \theta) - \chi(\eta) r^{\alpha_j - n(\omega)/2} \sum_{s=1}^{\kappa/2} \mu_s(\theta)\Phi_j^s(\varphi) \in E_{l-\delta}^{l+\alpha_j}(K). \quad (2.19)$$

Subtracting suitable polynomials from the functions ζ_j, we obtain elements of $E_{l-\delta}^{l+\alpha_j}(K)$ (see (2.19) and Theorem 2.2(i)). Therefore, Green's formula (2.17) is valid for ζ too. Consequently, $a_0^\theta(\zeta, \zeta; K) = 0$, that is, $\zeta = 0$. This contradiction implies that T is indeed a nonsingular matrix.

COROLLARY 2.5. *If $n(\omega)$ is odd or $\alpha_{\max} < n(\omega)/2$, then the mapping (2.14) is an isomorphism if $|\beta - l| < \delta$ and the following estimate holds:*

$$\|u; \mathscr{D}_\beta^l E(K)\| \leq \operatorname{const} \|A(\theta)u; \mathscr{R}_\beta^l E(K)\|, \quad (2.20)$$

where $\mathscr{D}_\beta^l E(K)$ and $\mathscr{R}_\beta^l E(K)$ are the products of the spaces in the left- and right-hand sides of (2.14), respectively.

COROLLARY 2.6. *Let* $n(\omega)$ *be even*, $\alpha_{\max} \geq n(\omega)/2$, *and* $\{F, G\} \in \mathscr{R}_{l-\delta}^l E(K)$. *Then problem* (1.9) *is solvable in the space* $\mathscr{D}_{l+\delta}^l E(K)$ *and any solution* $u \in \mathscr{D}_{l+\delta}^l E(K)$ *is of the form*

$$u(\eta) = u^0(\eta) + \sum_{h=1}^{\kappa/2} a_h \zeta^h(\eta); \qquad (2.21)$$

the particular solution u^0 *is such that*

$$u^0 - \chi \sum_{h=1}^{\kappa/2} c_h X^h \in \mathscr{D}_{l-\delta}^l E(K) \qquad (2.22)$$

and

$$\sum_{j=1}^{n(\omega)} \left\| u_j^0 - \chi \sum_{h=1}^{\kappa/2} c_h X_j^h ; E_{l-\delta}^{l+\alpha_j}(K) \right\| + \sum_{h=1}^{\kappa/2} |c_h| \leq \mathrm{const} \|\{F, G\}; \mathscr{R}_{l-\delta}^l E(K)\|; \qquad (2.23)$$

in these formulas the a_h *are arbitrary constants, the* $c_h = c_h(F, G)$ *are continuous functionals on* $\mathscr{R}_{l-\delta}^l E(K)$, *and* $\{X^1, \ldots, X^{\kappa/2}\}$ *is a basis in the space of polynomials defined in Theorem* 2.2(ii). *The inclusion* $u^0 \in \mathscr{D}_{l-\delta}^l E(K)$ *is equivalent to the equalities* $(F, \zeta^h)_K + (G, D(\eta, \partial_\eta, \theta)\zeta^h)_{\partial K} = 0$, $h = 1, \ldots, \kappa/2$ *(see* [15]).

Thus, under the conditions of part (ii) of Theorem 2.4, $\dim \ker A_\beta(\theta) = \kappa/2 > 0$ for $\beta \in (l, l+\delta)$, and $\dim \mathrm{coker}\, A_\beta(\theta) = \kappa/2 > 0$ for $\beta \in (l-\delta, l)$ (see Theorem 2.4), but the operator $A_l(\theta)$ is not Fredholm. Hence it is impossible to choose β such that the mapping (2.14) becomes an isomorphism. It is therefore necessary to change the scale of spaces in which the operator $A(\theta)$ is defined.

Similarly to [12], we define the space $E_{\beta,s}^l(K)$ of functions in K with the norm

$$\|u; E_{\beta,s}^l(K)\| = \left(\sum_{|\gamma| \leq l} \||\eta|^{\beta-l+|\gamma|+\Theta(s-|\gamma|)}(1+|\eta|)^{l-|\gamma|-\Theta(s-|\gamma|)} \partial_\eta^\gamma u; L_2(K)\|^2 \right)^{1/2} \qquad (2.24)$$

where $\Theta(t) = 1$ for $t \geq 0$ and $\Theta(t) = 0$ for $t < 0$. Thus, the weight factors for the derivatives $\partial_\eta^\gamma u$, $|\gamma| > s$, are the same in the norms (2.8) and (2.24), but for $|\gamma| \leq s$ the exponent of the weight near the point $\eta = 0$ in (2.24) is greater than that in (2.8) by one. It is clear that if $s \leq 0$ the spaces $E_{\beta,s}^l(K)$ and $E_\beta^l(K)$ coincide. The role of the spaces $E_{\beta,s}^l(K)$ will be clarified in Lemma 2.9.

REMARK 2.7. (i) As follows from [19, §2], when $-n(\omega)/2 > \beta - l - s > -1 - n(\omega)/2$, any function $u \in E^l_{\beta,s}(K)$ may be represented as $u(\eta) = v(\eta) + \chi(\eta)P^s(\eta)$, where $v \in E^l_{\beta,s}(K)$, P^s is a homogeneous polynomial of degree s, and χ is the same truncating function as in (2.16).

(ii) It follows immediately from (2.24) and (2.8) that the norm in $E^l_{\beta,s}(K)$ is equivalent to the following norm:

$$(\|(1+|\eta|)^{l-s-1}u; E^s_{\beta-l+s+1}(K)\|^2 + \|\nabla_{s+1}u; E^{l-s-1}_{\beta}(K)\|^2)^{1/2},$$

where $\nabla_{s+1}u$ is the row vector of all derivatives of order $s+1$.

LEMMA 2.8. *Let $p\,(\partial_\eta)$ be a (scalar) differential operator with constant coefficients*, ord $p = h$. *Then the mapping* $p: E^l_{\beta,s}(K) \to E^{l-h}_{\beta,s-h}(K)$ *is continuous*.

PROOF. If p is homogeneous, its continuity may be verified directly. To treat the nonhomogeneous case, one has only to use the inclusion $E^{l-h-1}_{\beta,s-h+1}(K) \subset E^{l-h}_{\beta,s-h}(K)$, which follows from the definition of the norms (2.24).

Let $n(\omega)$ be even and $\alpha_{\max} \geq n(\omega)/2$. In this case we modify the definition of the space $\mathscr{D}^l_\beta E(K)$: from now on, we assume that

$$\mathscr{D}^l_\beta E(K) = \prod_{j=1}^k E^{l+\alpha_j}_{\beta,\alpha_j - n(\omega)/2}(K), \qquad (2.25)$$

but $\mathscr{R}^l_\beta E(K)$ will again denote the right-hand side of (2.14), as before.

LEMMA 2.9. *Under the conditions of Corollary 2.6, the functions χX^h belong to the space (2.25) for $\beta = l - \delta$, and the solutions ζ^h do not belong to this space.*

PROOF. Recall that the component X^h_j of X^h is a homogeneous polynomial of degree $\alpha_j - n(\omega)/2$. Therefore $\chi X^h_j \in E^{l+\alpha_j}_\beta(K)$ for $\beta > l$. Since $\partial^\gamma_x X^h_j = 0$ for $|\gamma| > \alpha_j - n(\omega)/2$, it follows that $\chi X^h_j \in E^{l+\alpha_j}_{l-\delta,\alpha_j - n(\omega)/2}(K)$. The functions ζ^h_j admit representations (2.16). These functions do not belong to $E^{l+\alpha_j}_{l-\delta,\alpha_j - n(\omega)/2}(K)$ because the integrals

$$\int_0^1 |\eta|^{2(|\gamma|-\delta)}|\partial^\gamma_\eta \zeta^h(\eta)|^2 d\eta$$

are divergent for $|\gamma| > \alpha_j - n(\omega)/2$ and $\delta > 0$.

Consider the operator $A(\theta)$ in the space $\mathscr{D}^l_{l-\delta}E(K)$ of (2.25). Note that the entries of the matrix N of boundary conditions are

$$N_{qj}(\eta, \partial_\eta, \theta) = \sum_{h=1}^{n(\omega)} \nu_h(\eta|\eta|^{-1}) N^{(h)}_{qj}(\partial_\eta, \theta),$$

where $(\nu_1, \ldots, \nu_{n(\omega)})$ is a normal vector to $\partial K \setminus 0$, $N^{(h)}_{qj}$ are differential operators with constant coefficients, ord $N^{(h)}_{qj} \geq \alpha_j$. Therefore, Lemma 2.8 implies that the mapping

$$A(\theta) : \mathscr{D}^l_{l-\delta} \to \mathscr{R}^l_{l-\delta} E(K) \tag{2.26}$$

is continuous.

Let $\{F, G\} \in \mathscr{R}^l_{l-\delta}E(K)$. By Lemma 2.7 and (2.22) the particular solution u^0 of (1.9) is an element of the space $\mathscr{D}^l_{l-\delta}E(K)$ defined in (2.25). In addition, by (2.23), the estimate (2.20) holds for $\beta = l - \delta$. Again by Lemma 2.9, the solution u of (2.21) is not in $\mathscr{D}^l_{l-\delta}E(K)$ if at least one of the coefficients a_h is not zero. Thus the condition $u \in \mathscr{D}^l_{l-\delta}E(K)$ eliminates the arbitrary factor in the choice of a solution of (1.9). We have thus proved the following theorem.

THEOREM 2.10. *If $n(\omega)$ is even and $\alpha_{\max} \geq n(\omega)/2$, then the mapping (2.26) is an isomorphism for sufficiently small positive δ.*

This theorem and Corollary 2.5 provide us with function spaces in which the boundary-value problem (1.9) is uniquely solvable. We emphasize that the spaces $E^l_{\beta,s}(K)$ are used in Theorem 2.10 in the situation described in Remark 2.7(i).

REMARK 2.11. The norms of the functions $\eta \to (1 + |\eta|)^{l-s-1} u(\eta)$ and $\nabla_{s+1} u(\eta)$ in Remark 2.7(ii) may be rewritten in spherical coordinates as in (2.12). But $\nabla_{s+1} u$ cannot be replaced by the sequence of derivatives $\{r^{-s-1}(r\partial_r)^q \partial_\varphi^\sigma u, |q| + |\sigma| \leq s + 1\}$, because the operator ∇_{s+1} annihilates polynomials of degree s, while the sequence of derivatives does not necessarily vanish for such a polynomial.

§3. Investigation of the pencil $\mathfrak{A}$

In this section we study a pencil $\mathbb{C} \ni \lambda \mapsto \mathfrak{A}(\lambda)$ in a domain $\omega \in S^{n-\tilde{d}-1}$ with smooth edges in $\partial \omega$. We will introduce function spaces in which $\mathfrak{A}$ has a discrete spectrum of normal eigenvalues and will establish a result analogous to Theorem 2.2.

1. Formulation of the Main Theorem. First we introduce function spaces in the domain ω. Given a (scalar) function u, supported in $\partial \omega$, we define a norm that depends on a positive parameter t:

$$\|u; V^l_{\beta,s}(\omega; t)\| = \Biggl(\sum_{|\sigma| \leq l} \|\rho^{\beta-l+|\sigma|}(t\rho(1+t\rho)^{-1})^{\Theta(s-|\sigma|)} \\ \times (1+t\rho)^{l-|\sigma|} \partial^\sigma_\varphi u; L_2(\omega)\|^2 \Biggr)^{1/2}, \qquad (3.1)$$

where $\rho(\varphi)$ is the spherical distance between the point φ and the edge, and l, β, and s are the same as in (2.19). For functions supported away from the singularities of the boundary, the norm is equal to $\|u; W^l_2(\omega, t)\|$ (see (2.11)). The norm of a function with arbitrary support is "glued" together using a partition of unity from local norms, possibly with different exponents β and s on different edges; for our purposes it is sufficient to assume that β is the same for all edges. Let $V^l_\beta(\omega; t)$ be the space in which the local norm near an edge is (3.1) with $\Theta \equiv 0$ (in particular, $V^l_\beta(\omega; t) = V^l_{\beta,s}(\omega; t)$ if $s < 0$). As usual, $V^{l+1/2}_\beta(\partial\omega; t)$ is the space of traces on $\partial\omega \setminus \gamma$ of functions in $V^{l+1}_\beta(\omega; t)$.

REMARK 3.1. To clarify the origin of the norm (3.1), consider the following norm in $K^{n-d} \times \mathbb{R}^d$, where K^{n-d} is a cone with an edge

$$\Biggl(\sum_{|\sigma|+|\tau|+q \leq l} \int_{\mathbb{R}^d} dz \int_0^\infty r^{2(\beta^0-l)+n-d-1} dr \int_\omega \rho^{2(\beta-l)+n-d-1-\nu} \\ \times |(r\partial_z)^\sigma (r\partial_r)^q (\rho\partial_\varphi)^\tau u|^2 d\varphi \Biggr)^{1/2},$$

where r is the distance to the vertex of K^{n-d}, and ρ is the distance to a ν-dimensional edge on $\partial\omega$. The norm (3.1) with $\Theta = 0$ is obtained by taking Mellin transforms $\mathscr{M}_{r \mapsto \lambda}$ and Laplace transforms $\mathscr{L}_{z \mapsto \xi}$, noting that $t \sim |\lambda| + |\xi|$ for large $|\lambda| + |\xi|$.

PROPOSITION 3.2. *The norm* (3.1) *is equivalent to the following norm*:

$$\Biggl(\sum_{j=0}^{l-s-1} t^{2j} \Biggl(\sum_{|\alpha|=s+1} \|\partial^\alpha_\varphi u; V^{l-s-1-j}_\beta(\omega)\|^2 + \|u; V^s_{\beta-l+s+1+j}(\omega)\|^2 \Biggr) \\ + \sum_{q=l-s+1}^{l} t^{2q} \|u; V^{l-q}_\beta(\omega)\|^2 \Biggr)^{1/2}, \qquad (3.2)$$

where $V^l_{\beta,s}(\omega) \equiv V^l_{\beta,s}(\omega; 1)$. *In particular, if s is negative, then the equivalent norm is of the type*

$$(t^{2l} \|u; V^0_\beta(\omega)\|^2 + \|u; V^l_\beta(\omega)\|^2)^{1/2}.$$

PROOF. The square of (3.1) is equivalent to the sum of three expressions:

$$\sum_{|\sigma|\leq s} \sum_{j=0}^{l-s-1} t^{2+2j} \|\rho^{\beta-l+|\sigma|+1+j} \partial_\varphi^\sigma u; L_2(\omega)\|^2,$$

$$\sum_{|\sigma|\leq s} \sum_{j=l-s}^{l-|\sigma|-1} t^{2+2j} \|\rho^{\beta-l+|\sigma|+1+j} \partial_\varphi^\sigma u; L_2(\omega)\|^2,$$

$$\sum_{|\alpha|=s+1} \sum_{|\gamma|\leq l-s-1} \sum_{j=0}^{l-|\gamma|-s-1} t^{2j} \|\rho^{\beta-l+|\gamma|+s+1+j} \partial_\varphi^\gamma (\partial_\varphi^\alpha u); L_2(\omega)\|^2.$$

Changing the order of summation, we obtain (3.2).

We now introduce spaces of vector-valued functions in ω. For $l \geq \alpha_{\max}$, we define

$$\mathscr{R}_\beta^l V(\omega; t) = \prod_{j=1}^k V_\beta^{l-\alpha_j}(\omega; t) \times \prod_{q=1}^m V_\beta^{l-\sigma_q-1/2}(\partial\omega; t). \tag{3.3}$$

The space $\mathscr{D}_\beta^l V(\omega; t)$ is defined as the direct product of $V_{\beta,s+\alpha_j}^{l+\alpha_j}(\omega; t)$, $j = 1, \ldots, k$, where $\mathbf{s}$ is a vector with the same number of components as the number of edges in $\partial\omega$. Moreover, in the local norm (3.1) the exponent $s_h + \alpha_j$ is negative if $n - d - \nu_h - 1$ is odd and $s_h = -(n-d-\nu_h-1)/2$ if $n - d - \nu_h - 1$ is even (ν_h is the dimension of the edge).

THEOREM 3.3. (i) *The mapping*

$$\mathfrak{A}(\lambda): \mathscr{D}_\beta^l V(\omega; |\lambda|) \to \mathscr{R}_\beta^l V(\omega; |\lambda|) \tag{3.4}$$

is continuous, and $\|\mathfrak{A}(\lambda)\| \leq$ const *for all* $\lambda \in \mathbb{C}$.

(ii) *If* $\beta = l - \delta$ *and* δ *is a sufficiently small positive number, then the spectrum of the pencil* $\lambda \mapsto \mathfrak{A}(\lambda)$ *consists of normal eigenvalues. These eigenvalues (except possibly for a finite number) lie inside the double angle* $\{\lambda \in \mathbb{C} : |\operatorname{Re}\lambda| < k|\operatorname{Im}\lambda|\}$. *Outside this angle, for* $|\lambda| > R$, *we have*

$$\|\mathfrak{A}(\lambda)^{-1}; \mathscr{R}_{l-\delta}^l V(\omega; |\lambda|) \to \mathscr{D}_{l-\delta}^l V(\omega; |\lambda|)\| \leq \text{const}. \tag{3.5}$$

(iii) *If* $n - d$ *is odd or* $\alpha_{\max} < (n-d)/2$, *the line* $l_0 = \{\lambda \in \mathbb{C} : \operatorname{Re}\lambda = (d-n)/2\}$ *does not contain eigenvalues of* $\mathfrak{A}$ *for* $\beta = l - \delta$. *But if* $n - d$ *is even and* $\alpha_{\max} \geq (n-d)/2$, *then the only possible eigenvalue of* $\mathfrak{A}$ *on the line* l_0 *(for* $\beta = l - \delta$*) is* $\lambda = (d-n)/2$. *The eigenspace is the (possibly trivial) intersection of the domain of* $\mathscr{D}_{l-\delta}^l V(\omega)$ *and the space of traces on* ω *of the polynomials* X *specified in Theorem 2.2* (ii) *(with* $n(\omega)$ *replaced by* $n - d$*). For every eigenvector there exists an associated eigenvector, and there are no Jordan chains of length greater than* 2.

The first part of the Theorem is proved by a direct computation, using Lemma 2.8 together with the fact that $L_{ij}(\varphi, \partial_\varphi, \lambda+\alpha_j)$ is the sum of terms

$\lambda^h p_h(\varphi, \partial_\varphi)$, where $h + \mathrm{ord}\, p_h \leq \alpha_i + \alpha_j$ ($N_{qj}(\varphi, \partial_\varphi, \lambda + \alpha_j)$ has an analogous property). Part (iii) is proved in essentially the same way as Theorem 2.2; the necessary alterations will be pointed out in subsection 4. Part (ii) will be verified in the next two subsections.

2. Problem with a parameter near a smooth edge. Localization to the neighborhood of a point of an edge leads to a problem, depending on a parameter λ, in the domain $\Omega \equiv K^{n-d-\nu-1} \times \mathbb{R}^\nu$. Denote the operator of this problem by $\mathscr{A}(\partial_y, \partial_z, \lambda) = \{\mathscr{L}(\partial_y, \partial_z, \lambda), \mathscr{N}(\partial_y, \partial_z, \lambda)\}$. Taking Laplace transforms along the ν-dimensional edge we get an operator $\mathscr{A}(\partial_y, \eta', \lambda)$ in the cone $K^{n-d-\nu-1}$, where η' is the dual variable to z. Finally, considering λ as a real parameter and changing variables as in §1.5, we get an operator $\mathscr{A}(\partial_\eta, \theta)$ depending upon $\theta = \xi|\xi|^{-1} \in S^\nu$, where $\xi = (\xi', \lambda)$. We can now apply the results of subsection 2.3 to this operator. Returning to $\mathscr{A}(\partial_y, \partial_z, \lambda)$, we see that it is invertible.

We now define spaces in which $\mathscr{A}$ will be considered. We first define the space $V_{\beta,s}^l(\Omega; t)$ with a norm similar to (3.1):

$$\|u; V_{\beta,s}^l(\Omega; t)\| = \left(\left\|\sum_{|\sigma| \leq l} |y|^{\beta-l+|\sigma|}(t|y|)^{\Theta(s-|\sigma|)}\right.\right. \tag{3.6}$$
$$\left.\left. \times (1-t|y|)^{l-|\sigma|-\Theta(s-|\sigma|)} \partial_y^{\sigma'} \partial_z^{\sigma''} u; L_2(\Omega)\right\|^2\right)^{1/2},$$

where $\sigma = (\sigma', \sigma'')$ are multi-indices as before, $V_\beta^l(\Omega; t)$ is the space $V_{\beta,s}^l(\Omega; t)$ for negative s, and $V_\beta^{l+1/2}(\partial\Omega; t)$ is the space of traces on $\partial\Omega$ of the elements from $V_\beta^{l+1}(\Omega; t)$. The vector space $\mathscr{R}_\beta^l V(\Omega; t)$ is defined by (3.3) with ω replaced by Ω. The space $\mathscr{D}_\beta^l V(\Omega; t)$ is similar to $\mathscr{D}_\beta^l V(\omega; t)$ ($\partial\Omega$ has a single edge $\{0\} \times \mathbb{R}^\nu$). The mapping

$$\mathscr{A}(\partial_y, \partial_z, \lambda) : \mathscr{D}_\beta^l V(\Omega; |\lambda|) \to \mathscr{R}_\beta^l V(\Omega; |\lambda|) \tag{3.7}$$

is continuous and $\|\mathscr{A}(\partial_y, \partial_z, \lambda)\| \leq \mathrm{const}$ (compare Lemma 2.8 and Theorem 3.3(i)).

As explained at the beginning of this subsection, the analysis of the mapping (3.7) reduces to studying the operator $\mathscr{A}(\partial_\eta, \theta)$ in a cone. This operator acts in spaces with the norm (2.24) (see §2). In the next proposition we describe the passage from $E_{\beta,s}^l(K^{n-d-\nu-1})$ to $V_{\beta,s}^l(\Omega; t)$.

LEMMA 3.4. *Let $\xi' \mapsto \hat{u}(y, \xi')$ be the Laplace transform of the function $z \mapsto u(y, z)$ and $U((|\xi'|^2 + t^2)^{1/2} y, \xi') = (|\xi'|^2 + t^2)^{-h/2} \hat{u}(y, \xi')$, where t is a positive parameter, $h = \beta - l + N/2$, $N = n - d - \nu - 1$. Suppose that the function $\eta \mapsto U(\eta, \xi')$ belongs to $E_{\beta,s}^l(K^N)$ and the function $\xi' \mapsto \|U(\cdot, \xi'); E_{\beta,s}^l(K^N)\|$ belongs to $L_2(\mathbb{R}^\nu)$. Then $u \in V_{\beta,s}^l(\Omega; t)$ and*

$$\|u; V_{\beta,s}^l(\Omega; t)\| \leq \mathrm{const} \|U; L_2(\mathbb{R}^\nu, E_{\beta,s}^l(K^N))\|. \tag{3.8}$$

PROOF. Applying the change of variables $\eta \mapsto y = (|\xi'|^2 + t^2)^{-1/2}\eta$ to (2.24), we obtain

$$\|U; E_{\beta,s}^l(K^N)\|^2$$
$$= \int_{K^N} \sum_{|\gamma| \leq l} (|\xi'|^2 + t^2)^{\Theta(s-|\gamma|)} |y|^{2(\beta-l+|\gamma|+\Theta(s-|\gamma|))} \quad (3.9)$$
$$\times (1 + t^2|y|^2 + |\xi'|^2|y|^2)^{l-|\gamma|-\Theta(s-|\gamma|)} |\partial_y^\gamma \hat{u}(y, \xi')|^2 \, dy.$$

Introducing a multi-index $\sigma = (\sigma', \sigma'')$, where $\sigma' = \gamma$, let us evaluate the coefficient of $|\xi'|^{2|\sigma''|}|\partial_y^{\sigma'}\hat{u}|^2$ in the integrand in (3.9). If $|\sigma'| \leq s$, this coefficient is

$$|y|^{2(\beta-l+|\sigma|)}(t|y|)^2(1+t^2|y|^2)^{l-|\sigma|-1} \quad \text{for } \sigma'' = 0, \quad (3.10)$$
$$c|y|^{2(\beta-l+|\sigma|)}\{1+(t|y|)^2(1+t^2|y|^2)^{-1}\}(1+t^2|y|^2)^{l-|\sigma|} \quad \text{for } |\sigma''| > 0. \quad (3.11)$$

But if $|\sigma'| > s$ (i.e., $\Theta(s-|\sigma'|) = 0$), the coefficient is

$$c|y|^{2(\beta-l+|\sigma|)}(1+t^2|y|^2)^{l-|\sigma|}. \quad (3.12)$$

We now integrate (3.9) with respect to $\xi' \in \mathbb{R}^\nu$, pass from $\hat{u}(y, \xi')$ to $u(y, z)$ (the preimage of the Laplace transform), and estimate the resulting integral from below. If $|\sigma| > s$, we drop the second term from the expression in braces in (3.11); if $|\sigma| \leq s$ we drop the first term. The remaining expression in (3.11) and the expressions (3.10), (3.12) are precisely the weight factors for the derivatives $\partial_y^{\sigma'}\partial_z^{\sigma''} u$ in the norm (3.6).

REMARK 3.5. It follows from the proof of Lemma 3.3 that when $s < 0$ the norms on the left and on the right in (3.8) are equivalent (both sides of (3.8) depend on the parameter t, but the constants in the equivalence inequalities may be chosen to be independent of t).

LEMMA 3.6. *Let $\lambda \in \mathbb{R}$ and $\beta = l - \delta$. Then the operator (3.7) has a right inverse $\mathscr{R}_\beta^l V(\Omega; |\lambda|) \ni \{f, g\} \mapsto u \in \mathscr{D}_\beta^l V(\Omega; |\lambda|)$, and*

$$\|u; \mathscr{D}_\beta^l V(\Omega; |\lambda|)\| \leq \text{const}\|\{f, g\}; \mathscr{R}_\beta^l V(\Omega; |\lambda|)\|.$$

PROOF. Consider the following equation in Ω:

$$\mathscr{A}(\partial_y, \partial_z, \lambda)u = \{f, g\} \in \mathscr{R}_{l-\delta}^l V(\Omega; |\lambda|). \quad (3.13)$$

Taking Laplace transforms with respect to z, substituting $y \to \eta = |\xi|y$, and introducing new functions $U_j(\eta, \theta) = |\xi|^{\alpha_j}\hat{u}_j(|\xi|^{-1}\eta, \xi')$, $F_i(\eta, \theta) = |\xi|^{-\alpha_i}\hat{f}_i(|\xi|^{-1}\eta, \xi')$, $G_q(\eta, \theta) = |\xi|^{-\sigma_q}\hat{g}_q(|\xi|^{-1}\eta, \xi')$, where $i, j = 1, \ldots, k$, $q = 1, \ldots, m$, $\xi = (\xi', \lambda)$, we obtain the equation $\mathscr{A}(\partial_y, \theta)U = \{F, G\}$ in the cone $K^{n-d-\nu-1}$. By Remark 3.5,

$$\|\{F, G\}; \mathscr{R}_{l-\delta}^l E(K^{n-d-\nu-1})\| \leq c\|\{f, g\}; \mathscr{R}_{l-\delta}^l V(\Omega; |\lambda|)\|.$$

By Corollary 2.5 and Theorem 2.10, there exists a solution $U \in \mathscr{D}^l_{l-\delta}E(K^N)$ for all $\theta \in S^\nu$. Now return to the initial variables and use Lemma 3.4.

We have thus proved the existence of a solution of equation (3.13). To prove uniqueness, we have to pass from $V^l_{\beta,s}(\Omega;t)$ to $E^l_{\beta,s}(K^N)$ (in other words, we need a kind of converse of Lemma 3.4). If N is odd or $\alpha_{\max} < N/2$, then $\mathscr{D}^l_\beta V(\Omega;t)$ is a product of factors $V^{l+\alpha_j}_\beta(\Omega;t)$; hence the necessary passage is effected by Remark 3.5.

LEMMA 3.7. *Let N be even, $\alpha_j \geq N/2$, and $u_j \in V^{l+\alpha_j}_{l-\delta,\alpha_j-N/2}(\Omega;t)$. Then*

$$\|U_j; L_2(\mathbb{R}^\nu, E^{l+\alpha_j}_{l-\delta,\alpha_j-N/2}(K^N))\| \leq \operatorname{const}\|u; V^{l+\alpha_j}_{l-\delta,\alpha_j-N/2}(\Omega;t)\|$$

for all $t > 1$, where the constant is independent of t.

PROOF. Looking at the proof of Lemma 3.4, we notice that the squared norm $\|U_j; L_2(\mathbb{R}^\nu; E^{l+\alpha_j}_{l-\delta,\alpha_j-N/2}(K^N))\|^2$ is obtained by adding the expression

$$\sum_{\substack{\alpha_j-N/2\geq|\sigma'|+|\sigma''|,\\ |\sigma''|\geq 1}} \int_{\mathbb{R}^\nu}\int_{K^N} |y|^{2(|\sigma'|+|\sigma''|-\alpha_j-\delta)}$$

$$\times (1+t^2|y|^2)^{l+\alpha_j-|\sigma'|-|\sigma''|}|\xi'|^{2|\sigma''|}|\partial^{\sigma'}_y \hat{u}(y,\xi')|^2 dy\,d\xi' \quad (3.14)$$

to $\|u_j; V^{l+\alpha_j}_{l-\delta,\alpha_j-N/2}(\Omega;t)\|^2$ (see (3.11)). It is therefore sufficient to estimate the sum (3.14) in terms of $\|u_j; V^{l+\alpha_j}_{l-\delta,\alpha_j-N/2}(\Omega;t)\|^2$. This sum contains terms

$$\sum_{|\sigma'|+|\sigma''|=1}^{\alpha_j-N/2} \int_{\mathbb{R}^\nu}\int_{K^N} [t^2|y|^2(1+t^2|y|^2)^{-1}]|y|^{2(|\sigma'|+|\sigma''|-\alpha_j-\delta)}$$

$$\times (1+t^2|y|^2)^{l+\alpha_j-|\sigma'|-|\sigma''|}|\xi'|^{2|\sigma''|}|\partial^{\sigma'}_y \hat{u}(y,\xi')|^2 dy\,d\xi' \quad (3.15)$$

(see (3.6)). The sum (3.15) bounds the integrals (3.14) over the set $\mathbb{R}^\nu \times \{y \in K^N : t|y|\geq 1\}$. Now consider the zone

$$\Xi_t = \mathbb{R}^\nu \times \{y \in K^N : t|y| < 1\}.$$

The norm $\|u_j; V^{l+\alpha_j}_{l-\delta,\alpha_j-N/2}(\Omega;t)\|^2$ contains expressions

$$t^2 \sum_{|\sigma'|=0}^{\alpha_j-N/2} \int_{\mathbb{R}^\nu}\int_{K^N} |y|^{2(|\sigma'|+1-\alpha_j-\delta)}(1+t^2|y|^2)^{l+\alpha_j-|\sigma'|-1}|\partial^{\sigma'}_y \hat{u}(y,\xi')|^2 dy\,d\xi',$$

(3.16)

and
$$\sum_{|\sigma'|+|\sigma''|=\alpha_j-N/2} \int_{\mathbb{R}^\nu} \int_{K^N} |y|^{2(|\sigma'|+|\sigma''|+1-\alpha_j-\delta)}$$
$$\times (1+t^2|y|^2)^{l+\alpha_j-|\sigma'|-|\sigma''|-1} |\partial_y \partial_y^{\sigma'} \hat{u}(y,\xi')|^2 dy\, d\xi'. \tag{3.17}$$

Restricting the integrals to the above zone, we can omit the factors $(1+t^2|y|^2)$. As the integrals (3.16) converge, so do those integrals in (3.14) for which $|\sigma''| \geq 1$. Consequently, we can apply Hardy's inequality; hence the integral (3.17) taken over Ξ_t bounds those integrals in (3.14) for which $|\sigma'|+|\sigma''| = \alpha_j - N/2$ and the integration is performed over Ξ_t.

As a result, we see that the integrals in (3.14) with $|\sigma'|+|\sigma''| = \alpha_j - N/2$ are bounded above by $\mathrm{const}\|u_j V_{l-\delta,\alpha_j-N/2}^{l+\alpha_j}(\Omega;t)\|^2$. Using this fact, together with the convergence of the integrals (3.16) and Hardy's inequality, we estimate the integral in (3.14) with $|\sigma'|+|\sigma''| = \alpha_j - N/2 - 1$. Continuing in this way, we obtain the desired result.

The assertion we have just proved implies that the solution of equation (3.13) is unique. Indeed, taking Laplace transforms and changing variables, we pass from the solution $u \in \mathscr{D}_{l-\delta}^l V(\Omega;|\lambda|)$ of the homogeneous equation (3.13) to the solution U of the homogeneous problem (1.9); by Lemma 3.7, $u \in \mathscr{D}_{l-\delta}^l E(K^N)$. By Theorem 2.10, $\mathscr{A}(\partial_y, \theta)$ (see (2.26)) is an isomorphism; hence $U = 0$ and therefore $u = 0$. Together with Lemma 3.6, this means that the following statement is true.

THEOREM 3.8. *If $\lambda \in \mathbb{R} \setminus 0$, the operator (3.7) is an isomorphism. The inequality $\|\mathscr{A}(\partial_y, \partial_z, \lambda)\| + \|\mathscr{A}(\partial_y, \partial_z, \lambda)^{-1}\| \leq \mathrm{const}$ holds, and if $|\lambda| \geq 1$, the constant in this inequality does not depend on λ.*

3. Proof of part (ii) of Theorem 3.3. Now we are ready to use the Agranovich-Vishik scheme [16] to invert the pencil $\mathfrak{A}(\lambda)$ in the domain ω. For large real values of λ, the inverse $\mathfrak{A}(\lambda)^{-1}$ is obtained by "gluing" together the inverse operators of the model problems. The new model problems (that is, those not considered in [16]) near edges were studied in subsection 2 (Theorem 3.8). Strictly speaking, to implement the Agranovich-Vishik scheme we must verify one more property of the function spaces.

LEMMA 3.9. *Let $p(\varphi, \partial_\varphi)$ be a differential operator of order q in ω with coefficients in $C^\infty(\bar{\omega})$. Then the operator*
$$\lambda^k p : V_{\beta,\alpha_j-N/2}^{l+\alpha_j}(\omega;|\lambda|) \to V_\beta^{l-\alpha_j}(\omega;|\lambda|) \tag{3.18}$$
is compact if $q \leq \alpha_j + \alpha_i - k$, $k = 1, \ldots, \alpha_i + \alpha_j$. If $q < \alpha_j + \alpha_i - k$, the norm of the operator (3.18) is $o(1)$ as $|\lambda| \to \infty$.

By Lemma 3.9, the difference $\mathfrak{A}(\lambda) - \mathfrak{A}(\mu) : \mathscr{D}_{l-\delta}^l V(\omega;|\lambda|) \to \mathscr{R}_{l-\delta}^l V(\omega;|\lambda|)$ is a compact operator for any $\lambda, \mu \in \mathbb{C}$. If $\mu = \mathrm{Re}\,\lambda$, $|\mathrm{Im}\,\lambda| < \delta|\mu|$, and

$|\lambda| > R$, then the norm of this operator is $o(1)$ as $\delta \to 0$. The required properties of the spectrum of $\mathfrak{A}$ will now follow from a general result (Theorem 5.1 in [20]) on holomorphic operator-valued functions (see [16]).

PROOF OF LEMMA 3.9. We first verify the continuity of the operator

$$p: V_{\beta,\alpha_j-N/2}^{l+\alpha_j}(\omega;|\lambda|) \to V_\beta^{l-\alpha_j+1}(\omega;|\lambda|) \tag{3.19}$$

(for any fixed λ). Let u be a function with support near one of the edges. By the definition (3.1) of the norm (for $s < 0$)

$$\|pu; V_\beta^{l-\alpha_i+1}(\omega;|\lambda|)\| \le c(\lambda) \sum_{|\tau|\le l-\alpha_i+1} \|\rho^{\beta-l+\alpha_i-1+|\tau|}\partial_\varphi^\tau(pu); L_2(\omega)\|^2$$

$$\le c(\lambda) \sum_{|\tau|\le l-\alpha_j+1} \sum_{|\sigma|\le \alpha_i+\alpha_j-k} \|\rho^{\beta-l+\alpha_i-1+|\tau|}\partial_\varphi^{\tau+\sigma}u; L_2(\omega)\|^2. \tag{3.20}$$

The continuity of the mapping (3.19) will be proved if we can show that the exponents of the weight factors in (3.20) are not smaller than the exponents $\beta-l-\alpha_j+|\sigma|+|\tau|+\Theta(\alpha_j-|\sigma|-\tau-N/2)$ of the powers of ρ in the coefficients of the derivatives $\partial_\varphi^{\tau+\sigma}u$ in the norm of $V_{\beta,\alpha_j-N/2}^{l+\alpha_j}(\omega;|\lambda|)$ (compare with (3.1)); in other words, we have to show that

$$\alpha_j + \alpha_i - 1 \ge |\sigma| + \Theta(\alpha_j - |\sigma| - |\tau| - N/2). \tag{3.21}$$

Since $0 \le |\sigma| \le \alpha_j+\alpha_i-k$, $k \ge 1$, and $\alpha_i > 0$, this inequality follows from the chain of inequalities

$$\alpha_j + \alpha_i - 1 \ge |\sigma| + \Theta(\alpha_j - |\sigma| - N/2) \ge |\sigma| + \Theta(\alpha_j - |\sigma| - |\tau| - N/2).$$

Let ε be an arbitrary positive number, and let γ_ε denote the ε-neighborhood of the union γ of the edges on $\partial\omega$. Then $\|u; V_\beta^{l-\alpha_j}(\omega\cap\gamma_\varepsilon;|\lambda|)\| \le \varepsilon\|u; V_\beta^{l-\alpha_j+1}(\omega\cap\gamma_\varepsilon;|\lambda|)\|$. Since the imbedding $V_\beta^{l-\alpha_j+1}(\omega\setminus\gamma_\varepsilon;|\lambda|) \subset V_\beta^{l-\alpha_j}(\omega\setminus\gamma_\varepsilon;|\lambda|)$ is compact, the operator (3.18) is also compact.

We now proceed to the second assertion of Lemma 3.9. For functions supported outside a fixed neighborhood of γ, the estimate follows from interpolation inequalities due to Agranovich-Vishik [16]. Now let u be supported near an edge. We have

$$\|\lambda^k pu; V_\beta^{l-\alpha_i}(\omega;|\lambda|)\|^2 = \sum_{|\tau|\le l-\alpha_j} \|\lambda^k \rho^{\beta-l+\alpha_j+|\tau|}(1+|\lambda|\rho)^{l-\alpha_j-|\tau|}\partial_\varphi^\tau(pu); L^2(\omega)\|^2$$

$$\le c \sum_{|\tau|\le l-\alpha_j} \sum_{|\sigma|=0}^{\alpha_i+\alpha_j-k-1} \|\lambda^k \rho^{\beta-l+\alpha_j+|\tau|} \times (1+|\lambda|\rho)^{l-\alpha_j-|\tau|}\partial_\varphi^{\tau+\sigma}u; L^2(\omega)\|^2.$$

Comparing this expression with the norm (3.1) in $V_{\beta,\alpha_j-N/2}^{l+\alpha_j}(\omega;|\lambda|)$, we see that the following inequality has to be proved:

$$|\lambda|^{k+1}\rho^\beta(\rho^{-1}+|\lambda|)^{l-\alpha_j-|\tau|} \le c\rho^\beta(\rho^{-1}+|\lambda|)^{l+\alpha_j-|\tau|-|\sigma|}$$
$$\times (1+|\lambda|\rho^{-1})^{-\Theta(\alpha_j-|\tau|-|\sigma|-N/2)},$$

or, equivalently,

$$|\lambda|^{k+1} \le (\rho^{-1}+|\lambda|)^{\alpha_j+\alpha_i-|\sigma|}(1+(|\lambda|\rho)^{-1})^{-\Theta(\alpha_j-|\tau|-|\sigma|-N/2)}. \quad (3.22)$$

Since $\alpha_i+\alpha_j-|\sigma| \ge k+1$, formula (3.22) is obvious if $\Theta(\cdots) = 0$. But if $\Theta(\cdots) = 1$ the right-hand side of (3.22) is $|\lambda|(\rho^{-1}+|\lambda|)^{\alpha_i+\alpha_j-|\sigma|-1} \ge |\lambda|^{k+1}$. This means that the norm of the operator (3.18) is $O(|\lambda|^{-1})$.

4. Proof of part (iii) of Theorem 3.3. We indicate the changes that should be made in the proof of Theorem 2.2 to adjust it to the new situation. In the proof of Theorem 2.2 we used Green's formulas (2.2), (2.4) for the domain $K_\varepsilon^{n-d} = \{y \in K^{n-d} : \varepsilon < |y| < \varepsilon^{-1}\}$. Here, however, the cone K^{n-d} has edges, and the validity of Green's formulas is therefore not self-evident; difficulties may arise because of "bad" behavior of the solutions $U = \{r^{\lambda+\alpha_i}\Phi_i\}_{i=1}^k$ and $W = \{r^{\lambda+\alpha_i}(\Phi_i \log r + \Psi_i)\}_{i=1}^k$ of the homogeneous problem near the edges. In fact, formulas (2.2) and (2.4) remain valid for these solutions. If the functions Φ_j, Ψ_j were to belong to $V_{l-\delta}^{l+\alpha_j}(\omega;1)$, then all the integrals in (2.2), (2.4) would be convergent and the formulas would be true. However, in a neighborhood $\mathscr{O}_h$ of a ν-dimensional edge, we can only guarantee that $\Phi_j, \Psi_j \in V_{l-\delta,\alpha_j-N/2}^{l+\alpha_j}(\mathscr{O}_h, 1)$, where $N = n-d-\nu_h-1$. Hence, the derivatives of order up to $\alpha_j - N/2$ (inclusive) may increase near an edge faster than is allowed by the norm $\|\cdot\,;V_{l-\delta}^{l+\alpha_j}(\mathscr{O}_h;1)\|$; the highest-order derivatives are the same as in the norm of $V_{l-\delta}^{l+\alpha_j}(\mathscr{O}_h;1)$ (see (3.1) with $t = 1$). It follows that the forms $a_0(U,U;K_\varepsilon^{n-d})$ and $a_0(W,W;K_\varepsilon^{n-d})$ defined by (1.6), which contain only these derivatives, are convergent integrals. The surface integrals contain only linear combinations of the products $\partial_\varphi^\sigma \Phi_j \partial_\varphi^\tau \Phi_i$ for $|\sigma|+|\tau| < \alpha_i+\alpha_j-1$ and $|\sigma| < \alpha_i$; the coefficients in these combinations are smooth functions on $\bar{\omega}$. This fact, together with the inclusions $\Phi_j \in V_{l-\delta,\alpha_j-N/2}^{l+\alpha_j}(\mathscr{O}_h;1)$, implies that the surface integrals in (2.2) are finite. The integrals in (2.4) are analyzed using similar arguments involving both vectors Φ and Ψ.

In addition to formulas (2.2) and (2.4), we also need (2.3). The convergence of the integrals in (2.3) was in fact just established. However, in order to use Theorem 8.1 from [15] we must interpret the eigenvectors Φ of $\mathfrak{A}(\lambda)$ that belong to the eigenvalue $\lambda = -(n-d)/2$ as eigenvectors of the adjoint pencil. This is done with the help of the following two lemmas.

LEMMA 3.10. *Let* $\mathfrak{B}(\lambda) : \mathscr{R}_{l-\delta}^l V(\omega)^* \to \mathscr{D}_{l-\delta}^l V(\omega)^*$ *be the adjoint of the operator* $\mathfrak{A}(\lambda) : \mathscr{D}_{l-\delta}^l V(\omega) \to \mathscr{R}_{l-\delta}^l V(\omega)$ *with respect to the extension* of the duality in L_2. If $Y = (Y_1, \ldots, Y_k) \in V_{s+\delta}^{s-\alpha_1}(\omega) \times \cdots \times V_{s+\delta}^{s-\alpha_k}(\omega)$ (this product is contained in the space $\mathscr{D}_{l-\delta}^l V(\omega)^*$), $s \geq \alpha_{\max}$, and $X = (\psi_1, \ldots, \psi_k; \chi_1, \ldots, \chi_m) \in \mathscr{R}_{l-\delta}^l V(\omega)^*$ is a solution of the equation $\mathfrak{B}(\lambda)X = Y$, then $\psi_j \in V_{l+\delta}^{l+\alpha_j}(\omega)$, $j = 1, \ldots, k$.

This lemma is proved along the standard lines, using a special partition of unity, refined upon approaching an edge, and local estimates (compare [3, 15]).

LEMMA 3.11. *Any function* $\Phi_j \in V_{l-\delta, \alpha_j - N/2}^l(\mathscr{O}_h)$ *satisfies* $\Phi_j \in V_{\delta - N/2}^{\alpha_j - N/2}(\mathscr{O}_h)$ *(for even* $N = n - d - \nu - 1$ *and* $\alpha_j - N/2 \geq 0$*).*

PROOF. By [19, 21] we have the following representation for Φ_j:

$$\Phi_j(y, z) = \sum_{|\sigma| \leq \alpha_j - N/2} y^\sigma q_{j\sigma}(y, z) + \Phi_j^0(y, z), \quad (3.23)$$

where $\varphi = (y, z)$ are coordinates in the neighborhood $\mathscr{O}_h$ of the edge, $\Phi_j^0 \in V_{l-\delta}^{l+\alpha_j}(\mathscr{O}_h)$, $q_{j\sigma} \in V_{m-\delta-N/2}^m(\mathscr{O}_h)$ for $m = 1, 2, \ldots$. It remains to notice that all the terms on the right-hand side of (3.23) are elements of $V_{\delta - N/2}^{\alpha_j - N/2}(\mathscr{O}_h)$.

We can now verify that every Jordan chain of the original pencil $\mathfrak{A}(\lambda)$ that belongs to $\lambda = -(n-d)/2$ is a chain of the adjoint $\mathfrak{A}(\lambda)^*$ of $\mathfrak{A}(\lambda)$ with respect to Green's formula. Using Lemma 3.1, we can show, as usual [3, 15], that the algebraic multiplicities of the eigenvalue $\lambda = -(n-d)/2$ of the pencil $\mathfrak{A}(\lambda)^*$, which is defined on $\prod V_{l+\delta}^{l+\alpha_j}(\omega)$, are exactly these of the operator $\mathfrak{B}(\lambda)$. Hence they coincide with the algebraic multiplicities of the initial pencil $\mathfrak{A}(\lambda)$. By Lemma 3.11, the Jordan chains of $\mathfrak{A}(\lambda)$ and $\mathfrak{A}(\lambda)^*$ that belong to the value $-(n-d)/2$ are the same.

§4. Inversion of the operator $A(\theta)$

The operator $A(\theta)$ is defined only for the case $d > 0$. If $d = 0$ (the product $K^{n-d} \times \mathbb{R}^d$ is the cone K^n), we have enough information about the operator pencil $\mathfrak{A}$ (Theorem 3.3) to complete our investigation of problem (1.3), (1.4) (see the scheme of subsection 2.2).

1. Problem (1.3), (1.4) for $d = 0$. Let $\beta^0, \beta \in \mathbb{R}$ and let $\mathbf{s}$ be the vector used in §3.1 with the number of components equal to the number of edges in $\partial\omega$. Let $V_{\beta^0}^l(\beta, \mathbf{s}; K^n)$ be the space of functions in the cone $K^n = \mathbb{R}_+ \times \omega$ with the norm

$$\|u; V_{\beta^0}^l(\beta, \mathbf{s}; K^n)\| = \left(\sum_{q=0}^l \int_0^\infty r^{2(\beta^0 - l) + n - 1} \|(r\partial_r)^q u(r; \cdot); V_{\beta, \mathbf{s}-q}^{l-q}(\omega)\|^2 dr \right)^{1/2},$$

(4.1)

where $\mathbf{s} - q$ is the vector with components $s_1 - q, \ldots, s_M - q$ and M is the number of edges in $\partial\omega$.

The norm (4.1) has the same structure as (2.13); the role of the space $W_2^l(\omega)$ in the domain ω on the sphere is now assigned to the weight class $V_{\beta,s}^l(\omega)$.

We now define the spaces where the operator of (1.3), (1.4) acts. Set

$$\mathscr{D}_{\beta^0}^l V(\beta; K^n) = \prod_{j=1}^k V_{\beta^0}^{l+\alpha_j}(\beta; \mathbf{s} + \alpha_j; K^n),$$

where $\mathbf{s} + \alpha_j = (s_1 + \alpha_j, \ldots, s_M + \alpha_j)$. The range $\mathscr{R}_{\beta^0}^l V(\beta; K^n)$ of the operator (independent of $\mathbf{s}$) is the product

$$\prod_{j=1}^k V_{\beta^0}^{l-\alpha_j}(\beta; K^n) \times \prod_{q=1}^m V_{\beta^0}^{l-\sigma_q-1/2}(\beta; \partial K^n).$$

Here $V_{\beta^0}^l(\beta; K^n)$ is the space with the norm (4.1), but with $V_{\beta,s}^{l-q}(\omega)$ replaced by $V_\beta^{l-q}(\omega)$ (see subsection 3.1) and $V_{\beta^0}^{l-1/2}(\beta; \partial K^n)$ is the corresponding trace space.

The operator of problem (1.3), (1.4)

$$\{L, N\}: \mathscr{D}_{\beta^0}^l V(\beta; K^n) \to \mathscr{R}_{\beta^0}^l(\beta; K^n) \tag{4.2}$$

is continuous. To verify this, we need only write L and N in spherical coordinates and use Lemma 3.6, as done above, before formula (1.10).

LEMMA 4.1. *Let the vector* $\mathbf{s}$ *be chosen as in subsection* 3.1 *and let* $\beta = l - \delta$, $\delta \in (0, 1/2)$. *Then*

$$c\|v; V_{\beta,\mathbf{s}-\alpha_j}^{l+\alpha_j}(\omega, |\lambda|)\|^2 \leq \sum_{q=0}^{l+\alpha_j} |\lambda|^{2q} \|v; V_{\beta,\mathbf{s}+\alpha_j-q}^{l+\alpha_j-q}(\omega)\|^2 \tag{4.3}$$
$$\leq C\|v; V_{\beta,\mathbf{s}+\alpha_j}^{l+\alpha_j}(\omega, |\lambda|)\|^2,$$

where the constants c *and* C *do not depend on* λ *if* $|\lambda| \geq 1$.

PROOF. Let the function v be supported near a ν-dimensional edge. Recall that the component of $\mathbf{s}$ corresponding to this edge is $-N/2$ for even $N = n - \nu - 1$; but for odd N the number $s_h + \alpha_j$ is assumed to be negative. We will assume that N is even; otherwise, the result is contained in Proposition 3.1. By definition (3.1) the second expression in (4.3) is equivalent to

$$\sum_{|\sigma| \leq l+\alpha_j} \left\| \sum_{q=0}^{l+\alpha_j-|\sigma|} (|\lambda|\rho)^q \rho^{\Theta(\alpha_j-q-|\sigma|-N/2)} \rho^{-\delta-\alpha_j+|\sigma|} \partial_\varphi^\sigma v; L_2(\omega) \right\|^2 \tag{4.4}$$

(we have changed the order of summation). Formula (3.1), applied to the right-hand side of (4.3), implies that the second inequality in (4.3) will be proved if we show that

$$\sum_{q=0}^{l+\alpha_j-|\sigma|}(|\lambda|\rho)^q \rho^{\Theta(\alpha_j-q-|\sigma|-N/2)} \leq \text{const}(|\lambda|\rho)^{\Theta(\alpha_j-|\sigma|-N/2)} \qquad (4.5)$$
$$\times (1+|\lambda|\rho)^{l+\alpha_j-|\sigma|-\Theta(\alpha_j-|\sigma|-N/2)}.$$

If $|\sigma| > \alpha_j - N/2$, then Θ equals zero and (4.5) is obvious. If $|\sigma| \leq \alpha_j - N/2$, the right-hand side of (4.5) equals $c|\lambda|\rho(1+|\lambda|\rho)^{l+\alpha_j-|\sigma|-1}$, while the left-hand side may be rewritten as

$$\rho \sum_{q=0}^{\alpha_j-|\sigma|-N/2}(|\lambda|\rho)^q + (|\lambda|\rho)^{\alpha_j-|\sigma|-N/2+1}\sum_{q=0}^{l-1+N/2}(|\lambda|\rho)^q$$
$$\leq \text{const}\{\rho(1+|\lambda|\rho)^{\alpha_j-|\sigma|-N/2}$$
$$+ (|\lambda|\rho)^{\alpha_j-|\sigma|+1-N/2}(1+|\lambda|\rho)^{l-1+N/2}\}$$
$$\leq \text{const}\{\rho(1+|\lambda|\rho)^{l+\alpha_j-|\sigma|-1} + |\lambda|\rho(|\lambda|\rho)^{\alpha_j-|\sigma|-N/2}(1+|\lambda|\rho)^{l-1+N/2}\}.$$

Inequality (4.5) is now obvious for $|\sigma| \leq \alpha_j - N/2$ as well.

After proving the second inequality in (4.3), let us verify the first. Taking (4.4) into consideration and using (3.1) applied to the left-hand side of (4.3), we conclude that if $|\sigma| > \alpha_j - N/2$, the term in (4.4) that contains $\partial_\varphi^\sigma v$ is a bound for the corresponding term on the left-hand side of (4.3). If $|\sigma| = \alpha_j - N/2$, the necessary estimate follows from a comparison of the weight factors:

$$\rho^{-\delta-N/2}|\lambda|\rho(1+|\lambda|\rho)^{l-1+N/2} \leq \rho^{-\delta-N/2}\left\{\rho + |\lambda|\rho \sum_{q=0}^{l-1+N/2}(|\lambda|\rho)^q\right\}.$$

Now put $|\sigma| = \alpha_j - N/2 - k$, $k \geq 1$. Then the weight factors for $\partial_\varphi^\sigma u$ in the first and second expressions in (4.3) are, respectively,

$$\rho^{-\delta-k-N/2}|\lambda|\rho(1+|\lambda|\rho)^{l-1+k+N/2}, \qquad (4.6)$$
$$\rho^{-\delta-k-N/2}\{\rho(1+|\lambda|\rho)^k + (|\lambda|\rho)^{k+1}(1+|\lambda|\rho)^{l-1+N/2}\}. \qquad (4.7)$$

Note that: (i) if $\rho \geq 1$, the expression (4.7) majorizes the product (4.6); (ii) as $\rho \to 0$, the quantities (4.6) and (4.7) are infinitesimals to the same order.

When $k = 1$,

$$\int_\omega |\lambda|^4 \rho^{-N-2\delta+2}(1+|\lambda|\rho)^{2l-2+N}|\partial_\varphi^\sigma v|^2 d\varphi$$
$$+ \int_\omega |\lambda|^2 \rho^{-N-2\delta+2}(1+|\lambda|\rho)^{2l-2+N}|\partial_\varphi \partial_\varphi^\sigma v|^2 d\varphi \qquad (4.8)$$
$$\geq c \int_\omega |\lambda|^2 \rho^{-N-2\delta}(|\lambda|^2\rho^2 + 1)(1+|\lambda|\rho)^{2l-2+N}|\partial_\varphi^\sigma v|^2 d\varphi.$$

This needs some explanation. To estimate the second integral from below, we have used the inequality $(1+|\lambda|\rho)^{2l-2+N} \geq c(1+|\lambda|^{2l-2+N}\rho^{2l-2+N})$ and applied Hardy's inequality to each of two resulting integrals. This is possible by the above remark and the condition $2\delta \in (0, 1)$ (so that ρ appears in nonintegral powers).

When $k = 1$, the square of (4.6) is precisely the weight factor in the right-hand side of (4.8); the left-hand side of (4.8) appears in (4.4). We have thus estimated the derivatives of order $\alpha_j - N/2 - 1$ in the norm $\|v; V_{\beta,s+\alpha_j}^{l+\alpha_j}(\omega; |\lambda|)\|$. Since estimates of the derivatives of higher orders have already been established, we complete the proof by repeating the same argument for $k = 2, \ldots, \alpha_j - N/2$.

We now analyse the problem (1.3), (1.4) in the cone K^n. As usual [14], we take Mellin transforms

$$\tilde{v}(\lambda, \varphi) = (2\pi)^{-1/2} \int_0^\infty r^{-\lambda-1} v(r, \varphi) \, dr.$$

This gives the following family of problems, depending on a parameter $\lambda \in \Lambda(\beta^0) = \{\mu \in \mathbb{C} : \operatorname{Re}\mu = l - \beta^0 - n/2\}$,

$$\mathfrak{A}(\lambda)U(\lambda, \varphi) = \{F(\lambda, \varphi), G(\lambda, \varphi)\},$$

where $U(\lambda, \varphi) = \{\tilde{u}_j(\lambda-\alpha_j, \varphi)\}_{j=1}^k$, $F(\lambda, \varphi) = \{\tilde{f}_j(\lambda+\alpha_j, \varphi)\}_{j=1}^k$, $G(\lambda, \varphi) = \{\tilde{g}_q(\lambda+\sigma_q, \varphi)\}_{q=1}^m$. By Parseval's equality, Lemma 4.1, and Proposition 3.2, we have the equivalence of norms

$$\|u; \mathscr{D}_{\beta^0}^l V(l-\delta; K^n)\| \sim \left(\int_{\Lambda(\beta^0)} \|U(\lambda, \cdot); \mathscr{D}_{l-\delta}^l V(\omega; |\lambda|)\|^2 d\lambda\right)^{1/2},$$

$$\|\{f, g\}; \mathscr{R}_{\beta^0}^l V(l-\delta; K^n)\|$$
$$\sim \left(\int_{\Lambda(\beta^0)} \|\{F(\lambda, \cdot), G(\lambda, \cdot)\}; \mathscr{R}_{l-\delta}^l V(\omega; |\lambda|)\|^2 d\lambda\right)^{1/2}.$$

If the line $\Lambda(\beta^0)$ does not contain eigenvalues of the pencil $\lambda \mapsto \mathfrak{A}(\lambda)$, then by Theorem 3.3(i), there exists a uniformly bounded operator $\mathfrak{A}(\lambda)^{-1}$:

$\mathscr{R}^l_{l-\delta}V(\omega;|\lambda|) \to \mathscr{D}^l_{l-\delta}V(\omega;|\lambda|)$. Taking inverse Mellin transforms of the function $\Lambda(\beta^0) \ni \lambda \to \mathfrak{A}(\lambda)^{-1}\{F(\lambda,\cdot), G(\lambda,\cdot)\}$, we obtain a solution of the problem (1.3), (1.4).

THEOREM 4.2. *The operator of problem* (1.3), (1.4) *is an isomorphism* $\mathscr{D}^l_{\beta^0}V(l-\delta;K^n) \to \mathscr{R}^l_{\beta^0}V(l-\delta;K^n)$ *if and only if there are no points of the spectrum of the pencil* $\mathfrak{A}$ *on the line* $\Lambda(\beta^0)$.

Sufficiency was verified above; necessity can be verified as in Theorem 3.5 of [14].

2. Problem (1.9) in the cone K^{n-d}, $d > 0$. We introduce certain function spaces in K^{n-d}. Instead of the spaces $E^l_\beta(K)$ used in subsection 2.2, we must now use spaces $E^l_{\beta_0}(\beta, \mathbf{s}; K^{n-d})$ with the norm defined by

$$\|u, E^l_{\beta_0}(\beta, \mathbf{s}; K^{n-d})\|$$
$$= \left(\sum_{q=0}^{l} \int_0^\infty r^{2(\beta^0-l)+n-d-1} \|(r\partial_r)^q u(r,\cdot); V^{l-q}_{\beta, \mathbf{s}-q}(\omega; 1+r)\|^2 dr \right)^{1/2}.$$
(4.9)

(This norm stands in the same relationship to (2.12), as does the norm (4.1) to (2.13).) Definition (3.1) immediately implies that

$$\|v; V^l_{\beta, \mathbf{s}}(\omega)\|^2 \sim \sum_{|\sigma| \le l} \|\partial^\sigma_\varphi v; V^0_{\beta-l+|\sigma|, \mathbf{s}-|\sigma|}(\omega)\|^2.$$

Hence (4.9) is equivalent to the norm

$$\left(\sum_{|\sigma| \le l} \int_0^\infty r^{2(\beta^0-l+|\sigma|)+n-d-1}(1+r)^{2(l-|\sigma|)} \|\partial^\sigma_\eta u; V^0_{\beta-l+|\sigma|, \mathbf{s}-|\sigma|}(\omega)\|^2 dr \right)^{1/2}.$$
(4.10)

As usual, we use the notation $E^l_{\beta_0}(\beta; K^{n-d})$ for $\mathbf{s}$ with negative components and introduce trace spaces $E^{l-1/2}_{\beta^0}(\beta; \partial K^{n-d})$.

Applying Lemma 3.9 to the operators $L(\partial_\eta, \theta)$ and $N(\eta, \partial_\eta, \theta)$ written in spherical coordinates and using (4.9), we conclude that the operator

$$A_{\beta^0}(\theta) \equiv A(\theta): \prod_{j=1}^k E^{l+\alpha_j}_{\beta^0}(l-\delta, \mathbf{s}; K^{n-d})$$
$$\to \prod_{i=1}^k E^{l-\alpha_j}_{\beta^0}(l-\delta; K^{n-d}) \times \prod_{q=1}^m E^{l-\sigma_q-1/2}_{\beta^0}(l-\delta; \partial K^{n-d}),$$
(4.11)

of problem (1.9), where $l \ge \alpha_{\max}$, is continuous.

THEOREM 4.3. *Theorem 2.4 remains valid if we replace K by* K^{n-d}, $n(\omega)$ *by* $n-d$, *and the operator* (2.14) *by the operator* (4.11). (*The full algebraic* *multiplicity κ of the eigenvalue $\lambda = (d - n)/2$ is defined as in Theorem 3.3 (iii).)*

The proof proceeds along the same lines as in §2. The applicability of Green's formulas is guaranteed by the remarks in subsection 3.4; the relation (2.15) between the indices follows from the previous reasoning, together with Lemma 3.10. Finally, asymptotic formulas analogous to (2.16) are established in the same way as in [14] (by displacing the integration contour in the inverse Mellin transform).

It follows from this theorem (as in subsection 2.2) that if $n - d$ is even and $\leq 2\alpha_{\max}$, there exists no space in the scale $E^l_{\beta_0, s^0}(\beta, \mathbf{s}; K^{n-d})$ for which the operator (4.11) is an isomorphism. Indeed, if $\kappa > 0$ there are no such spaces; if $\kappa = 0$ the line $l_0 = \{\lambda \in \mathbb{C} : \operatorname{Re}\lambda = (d-n)/2\}$ has a neighborhood disjoint from the spectrum of $\mathfrak{A}$ (see Theorems 3.3(iii), 4.3), and the scale $E^l_{\beta_0}$ will be enough. In the case $\kappa > 0$, therefore, we have the spaces $E^l_{\beta^0, s^0}$ (see the final part of §2).

To define a norm in $E^l_{\beta_0, s^0}(\beta, \mathbf{s}; K^{n-d})$, we let $\mathscr{O}_h$ be a spherical neighborhood of the hth ν-dimensional edge in $\partial\omega$. For functions u with supports in the set $\{\eta \in \bar{K}^{n-d} : \eta = r\varphi, r > 0, \varphi \in \mathscr{O}_h\}$, we define

$$\|u; E^l_{\beta_0, s^0}(\beta, \mathbf{s}; K^{n-d})\|$$
$$= \left(\sum_{|\sigma| \leq l} \int_{K^{n-d}} r^{2(\beta^0 - l + |\sigma|)} \left(\frac{r}{1+r} \right)^{2\Theta(s^0 - |\sigma|)} \right.$$
$$\times \left\{ \rho^{2(\beta - l + |\sigma|)} \left(\frac{(1+r)\rho}{1+(1+r)\rho} \right)^{2\Theta(s_h - |\sigma|)} (1 + (1+r)\rho)^{2(l - |\sigma|)} \right\}$$
$$\left. \times |\partial^\sigma_\eta u(\eta)|^2 d\eta \right)^{1/2},$$
(4.12)

where $r = |\eta|$ is the distance to the vertex of the cone K^{n-d} and ρ is the (spherical) distance from the point $\varphi = \eta|\eta|^{-1} \in \omega$ to the edge $\partial\omega$. But if $\overline{\mathscr{O}} \subset \omega$ and $\operatorname{supp} u \subset \{\eta \in \bar{K}^{n-d} : \eta = r\varphi, r > 0, \varphi \in \mathscr{O}\}$, the norm of u is just $\|u; E^l_{\beta_0, s^0}(K^{n-d})\|$ as defined in (2.24).

For arbitrarily supported functions with arbitrary singularities, the norm is "glued" together with the help of a partition of unity.

Let us clarify the structure of the weight factors in (4.12). First recall that, when we defined the E spaces, the role of the "inner" norm on ω was played by the norm with parameter $1+r$ (see Remark 2.3 and, in particular, (2.12)). That is why the norm (4.12) includes the expression in braces, which is just the weight corresponding to $t = 1 + r$ in (3.1). The remainder of the factor outside the braces is the usual weight for $E^l_{\beta^0, s^0}$ spaces.

Now we define the spaces $\mathscr{D}_{\beta^0}^l E(K^{n-d})$ and $\mathscr{R}_{\beta^0}^l E(K^{n-d})$. The range $\mathscr{R}_{\beta^0}^l E(K^{n-d})$ always coincides with the right-hand side of (4.11). On the other hand, the definition of $\mathscr{D}_{\beta^0}^l E(K^{n-d})$ depends on the dimension $n-d$. If $n-d$ is odd (or $n-d > 2\alpha_{\max}$), the space is the same as on the left of (4.11). If $n-d$ is even, we set

$$\mathscr{D}_{\beta^0}^l E(K^{n-d}) = \prod_{j=1}^k E_{\beta^0, \alpha_j - (n-d)/2}^{l+\alpha_j}(l-\delta, \mathbf{s}; K^{n-d}), \qquad (4.13)$$

where $\mathbf{s}$ is the vector defined in subsection 3.1.

The following assertion can be proved using the same argument as at the end of §2.

THEOREM 4.4. *For a sufficiently small positive δ^0, the operator $A(\theta)$: $\mathscr{D}_{l-\delta^0}^l E(K^{n-d}) \to \mathscr{R}_{l-\delta^0}^l E(K^{n-d})$ of problem (1.8) is an isomorphism.*

REMARK 4.5. The numbers δ in (4.13) and δ^0 in Theorem 4.4 are not necessarily the same. However, they may be chosen the same, and we shall indeed do this from now on.

§5. Boundary-value problems in domains Ω_0 and Ω

At the end of §1 we have listed the model problems generated by the operator A_0 of the problem (1.3), (1.4). Our analysis of these problems ended with the inversion of $A(\theta)$. If $n-d$ is odd or $n-d > 2\alpha_{\max}$, the existence of the inverse A_0^{-1} follows from Theorem 4.4. However, if $n-d \leq 2\alpha_{\max}$ is even, Theorem 4.4 yields an estimate only for the seminorm of the solution of the equation $Au = \{f, g\}$. Under the additional assumption that u is compactly supported as a function in z, the seminorm turns out to be a norm.

This assumption does not interfere with the construction of regularizers for the operator of the initial problem (1.2) in a bounded domain (because regularizers are constructed by "gluing" together with the help of a partition of unity). The main theorem will be established in subsection 2; in subsection 3 we will consider some generalizations.

1. Function spaces. Let $V_{\beta^0, s^0}^l(\beta, \mathbf{s}; \Omega_0)$ be the space of functions in $\Omega_0 = K^{n-d} \times \mathbb{R}^d$ with the norm

$$\|u; V_{\beta^0, s^0}^l(\beta, \mathbf{s}; \Omega_0)\| = \left(\sum_{|\sigma| \leq l} \int_{K^{n-d}} \int_{\mathbb{R}^d} |y|^{2(\beta^0 - l + |\sigma|)} \left(\frac{|y|}{1+|y|} \right)^{2\Theta(s^0 - |\sigma|)} \right.$$
$$\times \{ \rho^{2(\beta - l + |\sigma|)} (\rho(1+\rho)^{-1})^{2\Theta(s_h - |\sigma|)}$$
$$\left. \times (1+\rho)^{2(l-|\sigma|)} \} |\partial_y^{\sigma'} \partial_z^{\sigma''} u(y, z)|^2 dy dz \right)^{1/2}$$

(5.1)

where u has support in $\mathbb{R}^d \times \mathbb{R}_+ \times \mathscr{O}_h$ (we are using the same notation as at the end of §4, and the additional notations $\sigma = (\sigma', \sigma'')$, $x = (y, z) \in \Omega_0$, $\varphi = y|y|^{-1}$). For functions with support in $\mathbb{R}^d \times \mathbb{R}_+ \times \mathscr{O}_h$, the norm coincides with the right-hand side of (3.6) for $t = 1$ and $\beta = \beta^0$, $s = s^0$, $\Omega = \Omega_0$. If the indices s^0 or s^h are negative, the term Θ in the corresponding exponent is certainly zero and these indices will be omitted from the notation of the spaces. The trace space (needed only for negative s) is denoted by $V^{l-1/2}_{\beta^0}(\beta; \Omega_0)$.

The structure of the norm (5.1) is similar to that of (4.12); since we are now concerned with the V (rather than E) spaces, the role of the parameter in the "inner" norm on ω is played by 1 and not by $1 + r$, as in (4.12). Of course, the factor $1 + \rho$ in (5.1) may be omitted.

Define

$$\mathscr{R}^l_{\beta^0} V(\beta; \Omega_0) = \prod_{i=1}^{k} V^{l-\alpha_j}_{\beta^0}(\beta; \Omega_0) \times \prod_{q=1}^{m} V^{l-\sigma_q-1/2}_{\beta^0}(\beta; \partial\Omega_0); \qquad (5.2)$$

and then

$$\mathscr{D}^l_{\beta^0} V(\beta; \Omega_0) = \prod_{j=1}^{k} V^{l+\alpha_j}_{\beta^0}(\beta; \mathbf{s}; \Omega_0), \qquad (5.3)$$

for odd $n - d$;

$$\mathscr{D}^l_{\beta^0} V(\beta; \Omega_0) = \prod_{j=1}^{k} V^{l+\alpha_j}_{\beta^0, \alpha_j - (n-d)/2}(\beta, \mathbf{s}; \Omega_0) \qquad (5.4)$$

for even $n - d$.

The operator $A_0 = \{L(\partial_x), N(x, \partial_x)\}$ of problem (1.3), (1.4) is a continuous mapping

$$A_0 : \mathscr{D}^l_{\beta^0} V(\beta; \Omega_0) \to \mathscr{R}^l_{\beta^0} V(\beta, \Omega_0). \qquad (5.5)$$

(This may be verified by using Lemmas 2.8 and 3.9, as before.)

LEMMA 5.1. (i) *The norm* $\|v; V^l_{\beta^0}(\beta; \Omega_0)\|$ *is equivalent to the norm*

$$\left(\int_{\mathbb{R}^d} |\xi|^{-2(\beta^0 - l) - (n-d)} \|\hat{v}(|\xi|^{-1}\eta, \xi); E^l_\beta(K^{n-d})\|^2 d\xi \right)^{1/2},$$

where $\hat{v}(y, \xi)$ *is the Laplace transform of the function* $z \mapsto v(y, z)$.

(ii) *Let the vector* $\mathbf{s}$ *and the number* s^0 *be consistent with the dimensions of the edges (see subsections 3.1, 4.2 and, in particular, formulas* (4.11), (4.13)).

Then the norm $\|v; V^{l+\alpha_j}_{l-\delta, s^0+\alpha_j}(l-\delta, \mathbf{s}+\alpha_j; \Omega_0)\|$ *is equivalent to the norm*

$$\left(\int_{\mathbb{R}^d} |\xi|^{2(\delta+\alpha_j)-(n-d)} \|\hat{v}(|\xi|^{-1}\eta, \xi); E^{l+\alpha_j}_{l-\delta, s^0+\alpha_j}(l-\delta, \mathbf{s}+\alpha_j; K^{n-d})\|^2 d\xi \right.$$

$$+ \sum_{|\sigma'| \leq s^0+\alpha_j} \int_{K^{n-d}} \int_{\mathbb{R}^d} |y|^{2(|\sigma|+1-\alpha_j-\delta)} (1+|y|)^{-2} \rho^{2(|\sigma|+1-\alpha_j-\delta)} \quad (5.6)$$

$$\left. \times |\partial_y^{\sigma''} v(y, z)|^2 dy\, dz \right)^{1/2}.$$

PROOF. Part (i) is in fact established in §5 of [2], because the space $V^l_{\beta^0}(\beta; \Omega_0)$ is defined without the exponent Θ (which leads to a "nonhomogeneous" norm).

We verify (ii). Let us assume that v has support in $\mathbb{R}^d \times \mathbb{R}_+ \times \mathscr{O}_h$. We have

$$-(n-d)/2 = s_0 < s_h = -(n-d-1-\nu_h)/2. \quad (5.7)$$

Using (4.12), we write the integrand in (5.6) as follows:

$$\int_{K^{n-d}} \sum_{|\sigma'| \leq l} |y|^{2(\beta^0-l-\alpha_j+|\sigma'|)} \left(\frac{|\xi||y|}{1+|\xi||y|} \right)^{2\Theta(s^0+\alpha_j-|\sigma'|)}$$

$$\times \left\{ \rho^{2(\beta-l-\alpha_j+|\sigma'|)} \left(\frac{(1+|\xi||y|)\rho}{1+\rho+|\xi||y|\rho} \right)^{2\Theta(s_h+\alpha_j-|\sigma'|)} \right.$$

$$\left. \times (1+\rho+|\xi||y|\rho^2)^{(l+\alpha_j-|\sigma'|)} \right\} |\partial_y^{\sigma'} \hat{v}(y, \xi)|^2 dy. \quad (5.8)$$

Let us remove the brackets in (5.8) above and calculate the coefficients of $|\xi^{\sigma''} \partial_y^{\sigma'} \hat{v}(y, \xi)|^2$. We have to consider three possible positions of the number $|\sigma'|$ relative to the exponents $s^0 + \alpha_j$ and $s_h + \alpha_j$; the values of Θ will be determined accordingly. Using (5.7), we transform (5.8) to the form

$$\int_{K^{n-d}} \left[\left\{ \sum_{|\sigma'|=0}^{s_h+\alpha_j} \sum_{|\sigma''|=1}^{l+|\alpha_j|-|\sigma'|} + \sum_{|\sigma'|=s_h+\alpha_j+1}^{l+\alpha_j} \sum_{|\sigma'|=0}^{l+\alpha_j-|\sigma'|} \right\} |y|^{2(\beta^0-l-\alpha_j+|\sigma|)} \right.$$

$$\times \rho^{2(\beta-l-\alpha_j+|\sigma|)} (1+\rho)^{2(l+\alpha_j-|\sigma|)} |\xi^{\sigma''} \partial_y^{\sigma'} \hat{v}(y, \xi)|^2 \quad (5.9)$$

$$+ \sum_{|\sigma'|=s^0+\alpha_j+1}^{s_h+\alpha_j} |y|^{2(\beta^0-l+\alpha_j+|\sigma'|)} \rho^{2(\beta-l-\alpha_j+|\sigma'|+1)}$$

$$\left. \times (1+\rho)^{2(l+\alpha_j-|\sigma'|-1)} |\partial_y^{\sigma'} \hat{v}(y, \xi)|^2 \right] dy,$$

where $|\sigma| = |\sigma'| + |\sigma''|$. The unessential factor $1 + \rho$ will now be omitted. By Parseval's equality for the transform $\mathscr{L}_{z \to \xi}$, we obtain

$$\|v; V^{l+\alpha_j}_{\beta^0, s^0+\alpha_j}(\beta; s_h + \alpha_j; \Omega_0)\|^2$$
$$= \int_{\mathbb{R}^d} \left\{ \sum_{|\sigma| < l} \int_{K^{n-d}} |y|^{2(\beta^0 - l - \alpha_j + |\sigma|)} \left(\frac{|y|}{1 + |y|}\right)^{2\Theta(s^0 + \alpha_j - |\sigma|)} \right. \quad (5.10)$$
$$\left. \times \rho^{2(\beta - l - \alpha_j + |\sigma| + \Theta(s_h + \alpha_j - |\sigma|))} |\xi^{\sigma''} \partial_y^{\sigma'} \hat{v}(y, \xi)|^2 \, dy \right\} d\xi$$

The sum (5.9) differs from the expression under the integral sign in (5.10) in two respects: first, it does not contain terms with indices $\sigma = (\sigma', \sigma'')$ such that $|\sigma'| \leq s^0 + \alpha_j$, $|\sigma''| = 0$; second, some of the terms in (5.10) involve additional factors ρ^2 and $|y|^2(1 + |y|)^{-2}$. Since these factors are bounded, the square of (5.6) is an upper bound for the integral (5.10).

To verify the equivalence of the norms (5.10) and (5.6), we still have to check the converse equality. We assume that $\beta^0 = \beta = l - \delta$, $\delta \in (0, 1/2)$ (this condition was not used until now). We first note that if $|\sigma| > s_h + \alpha_j$, the terms with the index σ in the integrands of (5.9) and (5.10) are the same. Let $s^0 + \alpha_j < |\sigma| \leq s_h + \alpha_j$. We have to estimate the integral of the term with the index σ in (5.9):

$$\int_{\mathbb{R}^d} \int_{K^{n-d}} |y|^{2(-\delta - \alpha_j + |\sigma|)} \rho^{2(-\delta - \alpha_j + |\sigma|)} |\xi|^{2|\sigma''|} |\partial_y^{\sigma'} \hat{v}(y, \xi)|^2 \, d\xi \, dy \quad (5.11)$$

only for $|\sigma''| > 0$ (because the weighting coefficients in (5.9) and (5.10) coincide when $\sigma'' = 0$).

Similarly to the proof of Lemma 4.1, we proceed by induction in $|\sigma| = s_h + \alpha_j, s_h + \alpha_j - 1, \ldots, s^0 + \alpha_j + 1$. For any $|\sigma|$, the expression (5.8) does not exceed

$$\int_{\mathbb{R}^d} \int_{K^{n-d}} |y|^{2(-\delta - \alpha_j + |\sigma|)} \rho^{2(-\delta - \alpha_j + |\sigma| + 1)} |\xi^{\sigma''} \partial_y^{\sigma'} \hat{v}(y, \xi)|^2 \, d\xi \, dy$$
$$+ \int_{\mathbb{R}^d} \int_{K^{n-d}} |y|^{2(-\delta - \alpha_j + |\sigma|)} \rho^{2(-\delta - \alpha_j + |\sigma| + 1)} |\xi^{\sigma''} \partial_\varphi \partial_y^{\sigma'} \hat{v}(y, \xi)|^2 \, d\xi \, dy.$$

(5.12)

Indeed, the first term in (5.12) majorizes (5.11) for $\rho > c > 0$, and by Hardy's inequality the second term majorizes (5.12) for $\rho < c$ (the variable in Hardy's inequality is the distance ρ to the edge on $\partial \omega$). Hardy's inequality is applicable due to the convergence of the integral

$$\int\limits_{\mathbb{R}^d}\int\limits_{K^{n-d}} \left(\frac{|y|}{1+|y|}\right)^{2\Theta(s^0+\alpha_j-|\sigma|-1)} |y|^{2(-\delta-\alpha_j+|\sigma|-1)} \qquad (5.13)$$
$$\times \rho^{2(-\delta-\alpha_j+|\sigma|)}|\xi|^{2(|\sigma''|-1)}|\partial_y^{\sigma''}\hat{v}(y,\xi)|^2 d\xi\, dy$$

which enters the norm (5.10) (recall that $|\sigma''| \geq 1$). The first integral in (5.12) is also contained in (5.10), but the second integral enters (5.10) only for $|\sigma| = s_h + \alpha_j$. We have thus estimated (5.11) for $|\sigma| = s_h + \alpha_j$. Now let $|\sigma| = s_h + \alpha_j - 1$. The second term of (5.12) is majorized by the integral (5.11), which we have just estimated. Thus the induction can be continued.

Finally, let $|\sigma| \leq s^0 + \alpha_j$. As before, we can assume that $|\sigma''| > 0$. We proceed by induction in $|\sigma| = s^0 + \alpha_j,\ s^0 + \alpha_j - 1, \ldots, 0$. We have to estimate the integral (5.11) for $|\sigma| = s^0 + \alpha_j$. In this case it differs from the corresponding term in (5.10)

$$\int\limits_{\mathbb{R}^d}\int\limits_{K^{n-d}} (|y|\rho)^{2(-\delta-\alpha_j+|\sigma|)}[|y|(1+|y|)^{-1}]^2 \rho^2 |\xi^{\sigma''}\partial_y^{\sigma'}\hat{v}(y,\xi)|^2 d\xi\, dy \qquad (5.14)$$

by two additional factors ρ^2 and $|y|^{-2}(1+|y|)^{-2}$. We first eliminate the last factor. By Hardy's inequality (with respect to the variable $|y|$) the integral

$$\int\limits_{\mathbb{R}^d}\int\limits_{K^{n-d}} (|y|\rho)^{2(-\delta-\alpha_j+|\sigma|+1)} \rho^2 |\xi^{\sigma''}\partial_y^{\sigma'}\hat{v}(y,\xi)|^2 dy\, d\xi \qquad (5.15)$$

is bounded by the sum of (5.14) and the expression

$$\int\limits_{\mathbb{R}^d}\int\limits_{K^{n-d}} (|y|\rho)^{2(-\delta-\alpha_j+|\sigma|+1)}|\xi^{\sigma''}\partial_y^{\sigma'}\partial_y\hat{v}(y,\xi)|^2 d\xi\, dy \qquad (5.16)$$

(the applicability of Hardy's inequality is guaranteed by the convergence of the integral (5.13)). The integrals (5.16) and (5.11) are the same for $|\sigma| = s^0 + \alpha_j + 1$; but the latter was considered in the previous inductive argument. It now remains to eliminate the factor ρ^2 from (5.15). As before, the integral (5.11) for $|\sigma| + s^0 + \alpha_j$ is bounded above by (5.12), which is obviously the sum of the integrals (5.15) and (5.16). This completes one step of the inductive process. By continuing this process, we complete the proof of the lemma.

Comparing formulas (5.9) and (5.1), we notice that contrary to (5.1), formula (5.9) does not contain terms with indices $\sigma = (\sigma', \sigma'')$ such that $|\sigma'| \leq s^0 + \alpha_j$, $|\sigma''| = 0$; that is why it was necessary to include the last sum in (5.6). Nevertheless, if we confine ourselves to functions $(y, z) \mapsto v(y, z)$ with certain restrictions on the locations of their supports, the Friedrichs inequality (with respect to the variables z) will enable us to prove that the norms

$$\|v;\ V^{l+\alpha_j}_{l-\delta,\, s^0+\alpha_j}(l-\delta,\, \mathbf{s}+\alpha_j;\, \Omega_0)\|$$

and

$$\left(\int_{\mathbb{R}^d} |\xi|^{2(\delta+\alpha_j)-(n-d)} \|\hat{v}(|\xi|^{-1}\eta, \xi); E^{l+\alpha_j}_{l-\delta, s^0+\alpha_j}(l-\delta, \mathbf{s}+\alpha_j; K^{n-d})\|^2 d\xi \right)^{1/2}$$
(5.17)

are equivalent. In other words, the following lemma holds.

LEMMA 5.2. *Let* $\operatorname{supp} v \subset \{(y, z) \in \bar{\Omega}_0 : |z| \leq c_0 + c_1|y|\}$. *Then*

$$\int_{\Omega_0} \sum_{|\sigma'| \leq s^0+\alpha_j} |y|^{2(\beta^0-l-\alpha_j+|\sigma|+1)} (1+|y|)^{-2} \rho^{2(\beta-l-\alpha_j+|\sigma'|+1)} |\partial_y^{\sigma'} v(y, z)|^2 dy\, dz$$

$$\leq \int_{\Omega_0} \sum_{|\sigma'| \leq s^0+\alpha_j} |y|^{2(\beta^0-l-\alpha_j+|\sigma|+1)} \rho^{2(\beta-l-\alpha_j+|\sigma'|+1)} |\partial_z \partial_y^{\sigma'} v(y, z)|^2 dy\, dz,$$
(5.18)

i.e., the norm in $V^{l+\alpha_j}_{l-\delta, s^0+\alpha_j}(l-\delta, \mathbf{s}+\alpha_j; \Omega_0)$ *is equivalent to* (5.17).

(Indeed, by (5.18) we can add the last sum in (5.6) to (5.9).)

We shall also need function spaces in the domain Ω for problem (1.2). We define them by formulas analogous to (5.2)–(5.4) and denote them by $\mathscr{D}^l_{\beta^0} V(\beta; \Omega)$ and $\mathscr{R}^l_{\beta^0} V(\beta; \Omega)$. As usual, the norms in the scalar spaces $V^l_{\beta^0}(\beta; \Omega)$, $V^{l-1/2}_{\beta^0}(\beta; \partial\Omega)$, $V^l_{\beta^0}(\beta, \mathbf{s}; \Omega)$, and $V^l_{\beta^0, s^0}(\beta, \mathbf{s}; \Omega)$ are "glued" together from local norms with the help of a partition of unity (the local norms have already been defined: they are norms of type (5.1), (3.6), and so on). If the boundary $\partial\Omega$ contains only one edge, s^0 is a number; otherwise it may be a vector.

2. Solvability of problems in Ω_0. We first consider problem (1.3), (1.4) in Ω_0.

PROPOSITION 5.3. (i) *Let* $n-d$ *be odd or* $n-d > 2\alpha_{\max}$. *Then the mapping* (5.5) *is an isomorphism for* $\beta^0 = \beta = l - \delta$.

(ii) *Let* $n-d$ *be even,* $n-d \leq 2\alpha_{\max}$ *and* $(f, g) \in \mathscr{R}^l_{l-\delta} V(l-\delta; \Omega_0)$. *Then there exists a solution* u *of the problem* (1.3), (1.4) *such that*

$$\sum_{j=1}^k \int_{\mathbb{R}^d} |\xi|^{2(\delta+\alpha_j)-(n-d)} \|\tilde{u}_j(|\xi|^{-1}\eta, \xi); E^{l+\alpha_j}_{l-\delta, s^0+\alpha_j}(l-\delta; \mathbf{s}+\alpha_j; K^{n-d})\|^2 d\xi$$

$$\leq \operatorname{const} \|\{f, g\}; \mathscr{R} V^l_{l-\delta}(l-\delta; \Omega_0)\|^2.$$
(5.19)

If the solution u *has compact support, then*

$$\|u; \mathscr{D}^l_{l-\delta} V(l-\delta; \Omega_0)\| \leq \operatorname{const} \|\{f, g\}; \mathscr{R}^l_{l-\delta} V(l-\delta; \Omega_0)\|. \quad (5.20)$$

PROOF. Taking Laplace transforms $\mathscr{L}_{z \to \xi}$ and changing variables as in subsection 1.5, we obtain problem (1.9) with the operator $A(\theta)$. By

Lemma 5.1(i), the right-hand side $\{F, G\}$ of the problem belongs to $\mathscr{R}^l_{l-\delta} E(l - \delta; K^{n-d})$ (for almost all $\xi \in \mathbb{R}^d$). By Theorem 4.4, we can invert the operator $A(\theta)$ to obtain a solution $U \in \mathscr{D}^l_{l-\delta} E(l - \delta; K^{n-d})$. Returning to the old variables and taking inverse Laplace transforms, we obtain a solution u of the problem (1.3), (1.4). If the conditions of part (i) are fulfilled, Lemma 5.1(i) gives (5.20). In case (ii), Lemma 5.1(ii) implies only the estimate (5.19), but the additional assumption that $\operatorname{supp} u$ is compact, together with Lemma 5.3, yields (5.20).

Proposition 5.3 shows that standard schemes may be used to construct regularizers for the operator

$$A : \mathscr{D}^l_{l-\delta} V(l - \delta; \Omega) \to \mathscr{R}^l_{l-\delta} V(l - \delta; \Omega) \qquad (5.21)$$

of problem (1.2). This gives the following result.

THEOREM 5.4. *The operator* (5.21) *is a Fredholm operator.*

3. Generalizations and corollaries. Here we discuss three examples. The first one, the problem of a boundary crack in a three-dimensional elastic body, is a direct illustration of Theorem 5.4. Next we consider deformation of an elastic body under the action of a rigid stamp. Since the boundary conditions are not Neumann conditions, we must change the function spaces: for certain components of the solution, homogeneous norms can be preserved. In the last problem, concerning a biharmonic operator, the role of the edge is played by an arc of a curve in a three-dimensional domain (in other words, this is the Sobolev problem). The model operator $A(\theta)$ will be associated with a problem in the complete angle $\mathbb{R}^2 \setminus 0$, preserving all the features of the Neumann problem.

(i) *Problem of a boundary crack.* Let G be a domain in the half-space $\mathbb{R}^3_- = \{x \in \mathbb{R}^3 : x_3 < 0\}$ bounded by a smooth surface ∂G, where ∂G has a plane part Σ on the boundary of $\mathbb{R}^3_-$. Assume that Σ contains a segment $I = \{x : x_2 = x_3 = 0, |x_1| < a\}$, and let h denote a function in $C^\infty[-a, a]$ such that $h(\pm a) = 0$, $h'(\pm a) \neq 0$, and $h(t) > 0$ for $|t| < a$. Let $M = \{x : x_2 = 0, |x_1| < a, 0 > x_3 > -h(x_1)\}$ be a boundary crack with two "banks" $M_\pm$. The boundary $\partial \Omega$ of the domain $\Omega = G \setminus \overline{M}$ contains three edges, $I_\pm$ and $l = \{x : x_2 = 0, |x_1| < a, x_3 = -h(x_1)\}$, and two vertices $P_\pm = (\pm a, 0, 0)$; both banks $M_\pm$ of the crack are in $\partial \Omega$. Consider the following problem of elasticity theory:

$$-\sum_{k=1}^{3} \frac{\partial}{\partial x_k} \sigma_{jk}(u; x) = f_j(x), \quad x \in \Omega, \ j = 1, 2, 3, \qquad (5.22)$$

$$N(x, \partial_x) u(x) \equiv \sigma(u; x) \nu(x) = g(x), \quad x \in \partial \Omega \setminus \Gamma, \qquad (5.23)$$

where $\sigma_{jk}(u)$ is the stress tensor with components (1.8), ν is the unit vector of the outward normal, and Γ is, as usual, the union of the edges and vertices.

It was shown in subsection 1.4 that condition (1.7) holds for (5.22), (5.23). We may therefore apply Theorem 5.4. Since $n = 3$, $d = 1$, and $\alpha_j = 1$, the norms in the function spaces are

$$\|v; V_\beta^l(\beta; \Omega)\| = \left(\sum_{|\gamma|\leq l} \|r^{\beta-l-|\gamma|}\partial_x^\gamma v; L_2(\Omega)\|^2\right)^{1/2}, \quad (5.24)$$

$$\|v; V_\beta^l(\beta, 0; \Omega)\| = \left(\sum_{|\gamma|\leq l} \left\|r^{\beta-l+|\gamma|}\left(\frac{r}{r_p}\right)^{\delta_{|\gamma|,0}}\partial_x^\gamma v; L_2(\Omega)\right\|^2\right)^{1/2}. \quad (5.25)$$

In (5.24) and (5.25) r denotes the smallest distance from a point x to the edges $I_\pm$ and l, $r_p = \min\{|x - P_+|, |x - P_-|\}$.

PROPOSITION 5.5. *Let* $f_j \in V_{l-\delta}^{l-1}(l-\delta; \Omega)$, $g_j \in V_{l-\delta}^{l-1/2}(l-\delta; \partial\Omega)$, $j = 1, 2, 3$, *and let* δ *be a small positive number. Then a solution* $u = (u_1, u_2, u_3)$ *of problem* (5.22), (5.23) *with components in* $V_{l-\delta}^{l+1}(l-\delta, 0; \Omega)$ *exists only if*

$$\int_\Omega fv\,dx + \int_{\partial\Omega} gv\,ds_x = 0, \quad \forall v \in \mathbf{L},$$

where $\mathbf{L} = \{v : v(x) = b + x \times c\}$ *is the linear manifold of rigid deformations and* b *and* c *are constant vectors. The solution* u *is defined up to the addition of an element from* $\mathbf{L}$.

This proposition is essentially a straightforward corollary of Theorem 5.4. The only thing to be clarified is the information about the kernel and the cokernel of the operator of the problem. Since $\mathbf{L}$ is contained in $V_{l-\delta}^{l+1}(l-\delta; 0; \Omega)$, it follows that $\mathbf{L}$ is a subset of the kernel. On the other hand, $V_{l-\delta}^{l+1}(l-\delta, 0; \Omega) \subset W_1^2(\Omega)$ and therefore, by (1.7), the elements of $\mathbf{L}$ exhaust the kernel. Instead of studying the cokernel, one considers the kernel of the adjoint problem in the sense of Green's formula, using the procedure of local increasing of smoothness (see subsection 3.4).

(ii) *Problem of a stamp.* Let G be the same domain as in (i), and Σ_0 the zone of contact with an absolutely rigid smooth stamp—a (plane) subdomain of Σ, $\overline{\Sigma}_0 \subset \Sigma$, $\partial\Sigma_0$ is a simple smooth closed contour. Consider the system of equations (5.22) with boundary conditions

$$N(x, \partial_x)u(x) = g(x), \quad x \in \partial\Omega \setminus \overline{\Sigma}_0, \quad (5.26)$$

$$\sigma_{j3}(u; x) = \psi_j(x), \quad j = 1, 2; \quad u_3(x) = \psi_3(x), \quad x \in \Sigma_0. \quad (5.27)$$

Therefore, we have a boundary-value problem in a domain with an edge $\partial\Sigma_0$. Since the component u_3 satisfies a Dirichlet condition on Σ_0, the function spaces retain a homogeneous norm for u_3 (polynomials with nonzero third components in the linear manifold P' defined in Proposition 2.1 are excluded from the kernel of the model operator because of the Dirichlet boundary condition). The norm in $V_\beta^l(\Omega)$ is defined by (5.24), where r is the distance to $\partial\Sigma_0$, and the norm in $V_{\beta,0}^l(\Omega)$ is defined by

$$\|v; V_{\beta,0}^l(\Omega)\| = \left(\sum_{|\gamma|\leq l} \|r^{\beta-l+|\gamma|+\delta_{|\gamma|,0}} \partial_x^\gamma v; L_2(\Omega)\|^2 \right)^{1/2}.$$

PROPOSITION 5.6. *Let* $f_j \in V_{l-\delta}^{l-1}(\Omega)$, $g_j \in V_{l-\delta}^{l-1/2}(\partial\Omega\setminus\Sigma_0)$, $j = 1, 2, 3$; $\psi_j \in V_{l-\delta}^{l-1/2}(\Sigma_0)$, $j = 1, 2$; $\psi_3 \in V_{l-\delta}^{l+1/2}(\Sigma_0)$, *and let* $\delta > 0$ *be a small number. Then a solution* u *of problem* (5.22), (5.27) *with components* $u_i \in V_{l-\delta,0}^{l+1}(\Omega)$, $i = 1, 2$, $u_3 \in V_{l-\delta}^{l+1}(\Omega)$, *exists only if conditions* (5.26) *hold, where the role of* **L** *is played by* $\{v : v(x) = b_1\mathbf{e}^1 + b_2\mathbf{e}^2 + c_3(x_2, -x_1, 0)\}$. *The solution* u *is defined up to an arbitrary additive constant in this linear manifold.*

(iii) *Sobolev problem.* Let Ω be a three-dimensional domain with a smooth boundary $\partial\Omega$ containing a segment $I = \{x : x_2 = x_3 = 0, |x_1| < a\}$. Consider the Sobolev problem

$$\begin{aligned} \Delta^2 u(x) &= f(x), \quad x \in \Omega\setminus I, \quad u(x) = 0, \quad x \in I; \\ u(x) &= (\partial u/\partial v)(x) = 0, \quad x \in \partial\Omega. \end{aligned} \quad (5.28)$$

Now $n = 3$, $d = 1$, $\alpha = 2$; the role of the cone K^{n-d} is played by the complete angle $\mathbb{R}^2\setminus 0$, and the operator $A(\theta) : E_{l-\delta,1}^{l+2}(\mathbb{R}^2) \to E_{l-\delta}^{l-2}(\mathbb{R}^2)$ is an isomorphism (the properties of this operator are completely similar to those of the operator in the Neumann problem). The spaces for problem (5.28) are defined by formulas (5.24) and (5.25), where r is the distance to I and r_p the distance to the nearest endpoint of I. Theorem 5.4 and the same arguments as in (i) (relating to the description of the kernel and cokernel) lead to the following result.

PROPOSITION 5.7. *For* $f \in V_{l-\delta}^{l-2}(l-\delta; \Omega)$ *with* $\delta \in (0, 1/2)$, *there exists a unique solution* $u \in V_{l-\delta}^{l+2}(l-\delta, 1; \Omega)$ *of the problem* (5.28).

We must explain the choice of the number δ. It can be shown that the eigenvalues of the pencil $\mathfrak{A}_\pm(\lambda)$ that corresponds to the ends $P^\pm$ of I and is defined on the sphere punctured at a point, are integers. This fact determines the interval to which δ is confined.

In conclusion, let us consider the behavior of the solution near the points $P^\pm$. Let $x^\pm$ be Cartesian coordinates with origins at these points. The functions $1 \pm x_1^\pm$, $|x^\pm|^{-1}$, $x_2^\pm|x^\pm|^{-1}$, $x_3^\pm|x^\pm|^{-1}$ are eigenvalues corresponding to the eigenvalue $\lambda = -1$ of the pencil $\mathfrak{A}^\pm$. Therefore, $U_\pm^1(x^\pm) = |x^\pm| \pm x_1^\pm$, $U_\pm^2(x^\pm) = x_2^\pm$, $U_\pm^3(x^\pm) = x_3^\pm$, are solutions of the model boundary-value problems in the cones (with vertices $V^\pm$). Hence the solution of problem (5.28) with a smooth function in the right-hand side is an element of $W_\infty^1(\Omega)$, but not necessarily of $C^1(\Omega)$.

References

1. V. A. Kondrat′ev, *On smoothness of the solution of a Dirichlet problem for an elliptic problem of second order in a piecewise smooth domain*, Differentsial′nye Uravneniya **6** (1970), no. 10, 1831–1843; English transl. in Differential Equations **6** (1970).
2. V. G. Maz′ya and B. A. Plamenevskiĭ, *Elliptic boundary-value problems on manifolds with singularities*, Problems in Mathematical Analysis, vol. 6, Spectral Theory, Boundary Value Problems, Izdat. Leningrad. Univ., Leningrad, 1977, pp. 85–142. (Russian)
3. _____, L_p-*estimates for solutions of elliptic boundary-value problems in domains with edges*, Trudy Mosk. Mat. Obshch. **37** (1978), 49–93; English transl. in Trans. Moscow Math. Soc. **1980**, no. 2.
4. _____, *The first boundary-value problem for the classical equation of mathematical physics in domains with piecewise-smooth boundaries*. I, Z. Anal. Anwendungen **2** (1983), no. 4, 335–359; II, Z. Anal. Anwendungen **4**, no. 6, 523–551.
5. _____, *On an oblique derivative problem in a domain with piecewise-smooth boundary*, Funkts. Anal. i Prilozhen. **5** (1971), no. 3, 102–103; English transl. in Functional Anal. Appl. **5** (1971).
6. V. G. Maz′ya, *On an oblique derivative problem in a domain with edges of various dimensions*, Vestnik Leningrad. Univ. Mat. Mekh. Astronom. (1973), no. 7, 34–39; English transl. in Vestnik Leningrad Univ. Math. **6** (1973).
7. V. G. Maz′ya and B. A. Plamenevskiĭ, *On boundary-value problems for an elliptic equation of second order in a domain with edges*, Vestnik Leningrad. Univ. Mat. Mekh. Astronom. (1975), no. 1, 102–108; English transl. in Vestnik Leningrad Univ. Math. **8** (1975).
8. G. Eskin, *Boundary-value problems for second-order elliptic equations in domains with corners*, Pseudodifferential Operators and Applications, Proc. Sympos. Pure Math., vol. 43, Amer. Math. Soc., Providence, R.I., 1985, pp. 105–131.
9. V. A. Solonnikov, *Solvability of a three-dimensional free boundary problem for the steady-state system of Navier-Stokes equations*, Zap. Nauchn. Sem. Leningrad. Otdel. Mat. Inst. Steklov. (LOMI) **84** (1979), 252–285; English transl. in J. Soviet Math. **21** (1983), no. 3.
10. N. Kh. Arutyunyan, S. A. Nazarov, and B. A. Shoikhet, *Estimates and asymptotic behavior of the stress-strain state of a three-dimensional body with a crack in elasticity theory and creep theory*, Dokl. Akad. Nauk SSSR **266** (1982), no. 6, 1365–1369; English transl. in Soviet Math. Doklady **26** (1982).
11. W. Zajączkowski and V. A. Solonnikov, *On the Neumann problem for second-order elliptic problems in a domain with edges on the boundary*, Zap. Nauchn. Sem. Leningrad. Otdel. Mat. Inst. Steklov. (LOMI) **127** (1983), 7–48; English transl. in J. Soviet Math. **27** (1984), no. 2.
12. S. A. Nazarov, *Estimates near an edge for the solution of the Neumann problem for an elliptic system*, Vestnik Leningrad. Univ. Mat. Mekh. Astronom. (1988), no. 1, 37–42; English transl. in Vestnik Leningrad Univ. Math. **21** (1988).
13. J.-L. Lions and E. Magenes, *Problèmes aux limites non homogènes et applications*. I–III, Dunod, Paris, 1968.
14. V. A. Kondrat′ev, *Boundary-value problems for elliptic equations in domains with conic or corner points*, Trudy Mosk. Mat. Obshch. **16** (1967), 219–292; English transl. in Trans. Moscow Math. Soc. **1967** (1968).
15. V. G. Maz′ya and B. A. Plamenevskiĭ, *On coefficients in asymptotic formulas for solutions of elliptic boundary-value problems in a domain with conic points*, Math. Nachr. **76** (1977), 29–60; English transl. in Amer. Math. Soc. Transl. (2), vol. 123, 1984, pp. 57–88.
16. M. S. Agranovich and M. I. Vishik, *Elliptic problems with a parameter and parabolic problems of general type*, Uspekhi Mat. Nauk **19** (1964), no. 3, 53–161; English transl. in Russian Math. Surveys **19** (1964).
17. V. G. Maz′ya and B. A. Plamenevskiĭ, *On coefficients in asymptotic formulas for solutions of elliptic boundary-value problems in cones*, Zap. Nauchn. Sem. Leningrad. Otdel. Mat. Inst. Steklov. (LOMI) **52** (1975), 110–127; English transl. in J. Soviet Math. **9** (1978), no. 5.

18. V. G. Maz'ya, S. A. Nazarov, and B. A. Plamenevskiĭ, *Asymptotic formulas for solutions of elliptic boundary-value problems under singular perturbations of the domain*, Tbilis. Gos. Univ., Tbilisi, 1981. (Russian)
19. V. G. Maz'ya and B. A. Plamenevskiĭ, *Weighted spaces with nonhomogeneous norms and boundary-value problems in domains with conic points*, Elliptische Differentialgleichungen, (Proc. Conf. Rostock, 1977), Wilhelm-Pieck Univ., Rostock, pp. 161–189; English transl. in Amer. Math. Soc. Transl. (2), vol. 123, 1984, pp. 89–107.
20. I. Z. Gokhberg and M. G. Kreĭn, *Introduction to the Theory of Linear Nonselfadjoint Operators*, GITTL, Moscow, 1965; English transl. in Transl. Math. Mono., vol. 18, Amer. Math. Soc., Providence, R.I., 1969.
21. V. G. Maz'ya and J. Rossman, *Über die Lösbarkeit und die Asymptotik der Lösungen elliptischer Randwertaufgaben in Gebieten mit Kanten*, Preprint 07/84, Akad. Wiss. DDR., Inst. für Math., Berlin, 1984.

Translated by N. ZOBIN

Limit Sets of Domains in Flows

S. YU. PILYUGIN

§1. Statement of main results

In 1987, at the Voronezh Winter School, V. I. Arnol'd described the following phenomenon. Consider the flow φ on the straight line in the figure, where A and B are stationary points. The ω-limit set $\omega(G)$ for the domain G (the thickened segment) is just the stationary point A. On the other hand, there are points in an arbitrary small neighborhood U of $\omega(G)$, whose trajectories tend to the stationary point B as $t \to \infty$. Thus $\omega(U)$ differs considerably from $\omega(G)$. Arnol'd conjectured that this phenomenon is "physically impossible".

In this paper we study the stability of limit sets of domains in flows (it is clear that the phenomenon described above is due to the instability in Lyapunov's sense of the stationary point A, which is the ω-limit set of G).

One of the main results (Theorem 1) will show that Arnol'd's conjecture is true in the following sense: for a C^0-generic flow and any point x, the ω-limit set of a sphere of radius r centered at x may be unstable for at most a countable set of values of r.

We now present the rigorous definitions and formulations. Let M be a smooth closed n-dimensional manifold with Riemann metric d, and let $F(M)$ be the space of continuous flows $\varphi : \mathbb{R} \times M \to M$ with the C^0-metric ρ_0:

$$\rho_0(\varphi, \psi) = \max_{t \in [-1, 1], x \in M} d(\varphi(t, x), \psi(t, x)),$$

where $\varphi, \psi \in F(M)$.

Given a set $G \subset M$, we define the ω-limit set $\omega(G)$ under the action of φ by

$$\omega(G) = \left\{ \lim_{k \to \infty} \varphi(t_k, x_k) : t_k \xrightarrow[k \to \infty]{} +\infty, x_k \in G \right\}$$

(the notation does not reflect the dependence of $\omega(G)$ on φ, as this does not lead to confusion). Clearly, $\omega(G)$ is compact and is invariant under φ.

Let $a > 0$, $A \subset M$, and $\varphi \in F(M)$. We denote the a-neighborhoods of A and φ by $N_a(A)$ and $N_a(\varphi)$, respectively. We denote by M^* the set of all compact subsets of M. Let $A, B \in M^*$. Define the deviation of A from B by
$$r(A, B) = \max_{x \in A} d(X, B),$$
and the Hausdorff distance by
$$R(A, B) = \max(r(A, B), r(B, A)).$$

As usual, a set in a topological space X is said to be of the second category if it contains the intersection of a countable family of open and dense subsets of X. A property will be called generic if it holds for all elements of some second category set in X.

Our main results pertain to the two most important classes of flows. The first class is that of the generic flows in $F(M)$. The second consists of the flows generated by systems of differential equations of class C^1 on M that satisfy Axiom A and the strict tranversality condition (these flows will be called A-flows for brevity; for the definitions see, for example, [11]). It is known [2] that the set of all A-flows is dense in $F(M)$.

THEOREM 1. *A generic flow $\varphi \in F(M)$ possesses the following property: for any point $x \in M$ there exists a countable set $B(x)$ in $(0, +\infty)$ such that, if $r \in (0, \infty) \backslash B(x)$, the set $\omega(N_r(x))$ is Lyapunov stable.*

A compact invariant set that is asymptotically stable with respect to a flow φ will be called an *attractor*.

THEOREM 2. *If φ is an A-flow then for any point $x \in M$ there is a finite set of numbers $C(x)$ such that, if $r \in (0, +\infty) \backslash C(x)$, the set $\omega(N_r(x))$ is an attractor.*

One question that arises in connection with the above problem is how can one characterize the sets for which the operation of taking ω-limit sets of neighborhoods, iterated arbitrarily many times, produces only a finite number of "abrupt" extensions? To answer this question, we have to study the structure of Lyapunov stable sets in generic flows and A-flows.

Following [3], we call a nonempty set I a *quasiattractor* for a flow φ if
$$I = \bigcap_{m > 0} I_m,$$
where each I_m is an attractor for φ.

THEOREM 3. *In a generic flow, every Lyapunov stable set is a quasiattractor.*

THEOREM 4. *In an A-flow, every Lyapunov stable set is an attractor.*

We now introduce the following notation. For $\delta > 0$ and $G \subset M$, we put $\omega_\delta^0(G) = \omega(G)$ and
$$\omega_\delta^{k+1}(G) = \omega(N_\delta(\omega_\delta^k(G))) \quad \text{for } k = 0, 1, \ldots.$$

THEOREM 5. *In a generic flow, for any set G and any $\varepsilon > 0$ there exists $\Delta > 0$ such that, for $\delta \in (0, \Delta)$ and $k, \ell \geq 1$,*
$$R(\omega_\delta^k(G), \omega_\delta^\ell(G)) < \varepsilon.$$

Theorem 5 shows that, in a generic flow φ, iteration of ω-limit sets of small neighborhoods for any set G "stabilizes" no later than at the second step.

THEOREM 6. *In an A-flow, for any set, there exists $\Delta > 0$ such that for $\delta \in (0, \Delta)$ and $k > 1$*
$$\omega_\delta^k(G) = \omega_\delta^1(G).$$

REMARK 1. Propositions similar to Theorems 1–6 (modulo natural modifications) are true for discrete dynamical systems (cascades).

REMARK 2. The proofs of weakened versions of Theorems 1 and 2 were sketched in [4].

This paper is structured as follows. In §2 we present some auxiliary constructions: we study prolongations of a point in a flow with respect to the dynamical system and the initial data, investigate the set of points of continuity of a monotone many-valued mapping, and describe the C^0-closure technique. Theorems 1 and 2 will be proved in §3, Theorems 3 and 4 in §4, and Theorems 5 and 6 in §5.

§2. Auxiliary constructions

We first define prolongations. Fix a point $x \in M$, a flow φ, and a number $\eta > 0$, and define the sets
$$P_{x,\eta} = \overline{\bigcup_{\psi \in N_\eta(\varphi)} \psi([0, +\infty), x)},$$
$$Q_{x,\eta} = \overline{\bigcup_{y \in N_\eta(x)} \varphi([0, +\infty), y)},$$
$$P_x = \bigcap_{\eta > 0} P_{x,\eta}, \quad Q_x = \bigcap_{\eta > 0} Q_{x,\eta}.$$

The sets P_x, Q_x are called the *prolongations* of x with respect to the dynamical system and with respect to the initial data, respectively.

The following lemma is proved in [5].

LEMMA 2.1. *There exists a second category set F_1 in $F(M)$ such that for a flow $\varphi \in F_1$ and any point $x \in M$*
$$P_x = Q_x.$$

We consider the following sets. Fix a point x and a flow φ and define
$$P_x^\omega = \left\{ \lim_{k \to \infty} \varphi_k(t_k, x) : t_k \xrightarrow[k \to \infty]{} +\infty, \, \rho_0(\varphi_k, \varphi) \xrightarrow[k \to \infty]{} 0 \right\},$$
$$Q_x^\omega = \left\{ \lim_{k \to \infty} \varphi(t_k, x_k) : t_k \xrightarrow[k \to \infty]{} +\infty, \, x_k \xrightarrow[k \to \infty]{} x \right\}.$$

LEMMA 2.2. *For a flow $\varphi \in F_1$ and any $x \in M$,*
$$P_x^\omega = Q_x^\omega.$$

PROOF. First, we observe that
$$Q_x = \varphi([0, +\infty), x) \cup Q_x^\omega, \tag{2.1}$$
$$P_x = \varphi([0, +\infty), x) \cup P_x^\omega. \tag{2.2}$$

We prove equality (2.1). Let $y \in Q_x$. Then $y \in Q_{x,\eta}$ for any $\eta > 0$. Choose a sequence of numbers $\eta_m > 0$, $\eta_m \xrightarrow[m \to \infty]{} 0$. For every fixed m, there exist sequences $\xi_k^m \in M$, $t_k^m > 0$, such that
$$d(\xi_k^m, x) < \eta_m, \quad y = \lim_{k \to \infty}(t_k^m, \xi_k^m).$$

Choose numbers $k(m)$ such that
$$d(\varphi(t_{k(m)}^m, \xi_{k(m)}^m), y) < \eta_m.$$

Then if $\tau_m = t_{k(m)}^m$, $\zeta_m = \xi_{k(m)}^m$, we have $y = \lim_{m \to \infty} \varphi(\tau_m, \zeta_m)$, $x = \lim_{m \to \infty} \zeta_m$. If the sequence τ_m is not bounded from above, then $y \in Q_x^\omega$, otherwise $y \in \varphi([0, +\infty), x)$. This proves $Q_x \subset \varphi([0, +\infty), x) \cup Q_x^\omega$; the converse inclusion is obvious.

It is easily seen that $Q_x^\omega \subset P_x^\omega$ always. We prove that
$$P_x^\omega \subset Q_x^\omega.$$

Take $y \in P_x^\omega$. If there is no $\tau \geq 0$ such that $y = \varphi(\tau, x)$, then it follows from $y \in Q_x$ and (2.1) that $y \in Q_x^\omega$. If $y = \varphi(\tau, x)$, $\tau \geq 0$, we must consider two cases.

CASE 1. x is a stationary point or a point in a closed trajectory of φ. Then, clearly, $\varphi \in Q_x^\omega$.

CASE 2. $\varphi(t, x) \neq x$, $t \neq 0$. Since
$$\varphi(\tau, x) = \lim_{k \to \infty} \varphi_k(t_k, x),$$
where $\rho_0(\varphi_k, \varphi) \xrightarrow[k \to \infty]{} 0$, $t_k \xrightarrow[k \to \infty]{} +\infty$, it follows that
$$\varphi(-1, x) = \lim_{k \to \infty} \varphi_k(t_k - \tau - 1, x) \in P_x^\omega. \tag{2.3}$$

If $y \notin Q_x^\omega$, then $x \notin Q_x^\omega$, and similarly $\varphi(-1, x) \notin Q_x^\omega$ (the set Q_x^ω is clearly invariant). If follows from our assumption about $\varphi(t, x)$ that
$$\varphi(-1, x) \notin \varphi([0, +\infty), x);$$
therefore, we infer from equality (2.1) that $\varphi(-1, x) \notin Q_x$. But this contradicts (2.2), (2.3), and the equality $P_x = Q_x$.

Now consider a mapping $\Phi : (a, b) \to M^*$, where M^* is the set of compact subsets of M. We call Φ an increasing mapping if $\Phi(t_1) \subset \Phi(t_2)$ for any $t_1, t_2 \in (a, b)$ such that $t_1 < t_2$.

LEMMA 2.3 ([6]). *If $\Phi : (a, b) \to M^*$ is an increasing mapping, then its set of points of discontinuity (in the Hausdorff metric) is at most countable.*

We shall often have occasion to use the C^0-closure technique. The proof of the following lemma is similar to the argument of [8, p. 1715].

LEMMA 2.4. *Consider a flow $\varphi \in F(M)$, sequences of numbers t_k, τ_k, sequences of points x_k, y_k, a point z, and a number $a > 0$ such that*
(1) $t_k \leq -1/2$, $\tau_k \geq 1/2$;
(2) $x_k \xrightarrow[k \to \infty]{} z$, $y_k \xrightarrow[k \to \infty]{} z$;
(3) $d(\varphi(t_k, x_k), z) \geq a$; $d(\varphi(\tau_k, y_k), z) \geq a$.
Then for any $\eta > 0$ and any neighborhood U of z one can find k_0 such that, for $k \geq k_0$, there exists a flow ψ with the properties
(1) *if $\varphi((\theta_1, \theta_2), x) \cap U = \varnothing$, $\theta_1, \theta_2 < 0$, then for $t \in (\theta_1, \theta_2)$,*
$$\varphi(t, x) = \psi(t, x);$$
(2) $\psi(t_k, z) = \varphi(t_k, x_k)$, $\psi(\tau_k, z) = \varphi(\tau_k, y_k)$;
(3) $\rho_0(\varphi, \psi) < \eta$.

An application of this lemma will be described in detail in the proof of Lemma 3.1. Subsequent reference to it will be shorter.

§3. Proofs of Theorems 1, 2

Before proving Theorem 1, we establish a few lemmas. Let G be a domain in M. We define its trajectory boundary with respect to a flow φ by
$$\partial^\tau G = \{x \in \partial G : \varphi(\mathbb{R}, x) \cap G = \varnothing\}.$$
In Lemmas 3.1, 3.2, we consider a flow $\varphi \in F_1$.

LEMMA 3.1. *If $\omega(G)$ is not Lyapunov stable, there exists a point $x \in \partial^\tau G$ such that*
$$P_x^\omega \not\subset \omega(G).$$

PROOF. Since $\omega(G)$ is unstable, there exist a number $\varepsilon > 0$ and two sequences $\xi_k \in M$, $\tau_k \xrightarrow[k \to \infty]{} +\infty$ such that
$$\xi_k \xrightarrow[k \to \infty]{} \omega(G), \quad d(\varphi(\tau_k, \xi_k), \omega(G)) = \varepsilon.$$
Choose a subsequence of ξ_k (which we again denote by ξ_k) such that, for some points $\xi \in \omega(G)$, $\zeta \in \partial N_\varepsilon(\omega(G))$,
$$\xi_k \xrightarrow[k \to \infty]{} \xi, \quad \varphi(\tau_k, \xi_k) \xrightarrow[k \to \infty]{} \zeta.$$
There exist sequences $x_k \in G$ and $t_k \to +\infty$ such that $\varphi(t_k, x_k) \xrightarrow[k \to \infty]{} \xi$. Choose a subsequence of x_k (which we denote by x_k) such that $x_k \xrightarrow[k \to \infty]{}$ $x \in \overline{G}$. We consider two cases.

CASE 1. $(x = \xi)$. Choose $\tau'_k \in (0, \tau_k)$ such that
$$d(\varphi(\tau'_k, \xi_k), \omega(G)) = \varepsilon/2.$$
It is clear that $\tau'_k \xrightarrow[k \to \infty]{} +\infty$. Fix an arbitrary $\eta > 0$. By Lemma 2.4, we can find k_0 such that for $k \geq k_0$ there exists a flow $\psi \in N_\eta(\varphi)$ with the properties
$$\psi(\tau'_k, \xi) = \varphi(\tau'_k, \xi_k), \qquad \psi(t_k, \xi) = \zeta.$$
This means that $\xi \in P_{x,\eta}$ for any $\eta > 0$. Hence $\xi \in P_x$.

CASE 2. $(x \neq \xi)$. Let $d(x, \xi) = 2a$. Choose numbers t'_k such that
$$d(\varphi(t'_k, x_k), \xi) = a.$$
For arbitrary $\eta > 0$, we can find k_0 such that for $k \geq k_0$ there exists a flow $\psi \in N_\eta(\varphi)$ with the properties
$$\psi(t'_k, x) = \varphi(t'_k, x_k),$$
$$\psi(t_k, x) = \xi,$$
$$\psi(\tau_k, \xi) = \zeta.$$
Hence it follows that in this case also $\xi \in P_x$.

It now follows that
$$\zeta \in P_x^\omega, \quad d(\zeta, \omega(G)) = \varepsilon. \tag{3.1}$$

We observe that if $\tilde{x} \in \overline{G} \setminus \partial^T G$, then $Q_{\tilde{x}}^\omega \subset \omega(G)$. Indeed, since $Q_{\tilde{x}}^\omega = Q_{\varphi(\tau, \tilde{x})}^\omega$ for every τ, we may assume that $\tilde{x} \in G$. But if $\tilde{x} \in G$, the inclusion $Q_{\tilde{x}}^\omega \subset \omega(G)$ is obvious. Since $P_x^\omega = Q_x^\omega$ by Lemmas 2.1 and 2.2, relations (3.1) can hold only if $x \in \partial^T G$. The lemma is proved.

Fix $x \in M$ and consider the mapping $\Phi : (0, +\infty) \to M^*$ that carries $r > 0$ to the set $\omega(N_r(x))$. (Recall that M^* is the set of all compact subsets of M with the Hausdorff metric R.)

LEMMA 3.2. *If r is a point of continuity of Φ, then $\omega(N_r(x))$ is Lyapunov stable.*

PROOF. Suppose that $\omega(N_r(x))$ is unstable. Then by Lemma 3.1 there are points $\xi \in \partial^T(N_r(x))$, $y \in P_\xi^\omega$, such that $y \notin \omega(N_r(x))$.

Since by the choice of $\varphi \in F$ we have $P_\xi^\omega = Q_\xi^\omega$, there exist sequences $x_k \xrightarrow[k \to \infty]{} \xi$, $t_k \xrightarrow[k \to \infty]{} \infty$, such that
$$y = \lim_{k \to \infty} \varphi(t_k, x_k).$$
Consider a sequence of numbers $r_m \xrightarrow[m \to \infty]{} r$, $r_m > r$. For any m there exists $k(m)$ such that
$$x_k \in N_{r_m}(x), \quad k \geq k(m);$$

but then $y \in \Phi(r_m)$ for any m. The obvious inequality

$$R(\Phi(r), \Phi(r_m)) \geq d(y, \Phi(r)) > 0$$

implies that r is not a point of continuity of Φ, proving the lemma.

We can now prove Theorem 1. The mapping Φ is increasing, as follows from its definition (just before Lemma 3.2). By Lemma 2.3, there exists a countable set $B(x)$ such that every $r \in (0, +\infty) \setminus B(x)$ is a point of continuity of Φ. By Lemma 3.2, the sets $\omega(N_r(x))$, $r \in (0, +\infty) \setminus B(x)$, are Lyapunov stable.

REMARK. There are certain conditions that guarantee Lyapunov stability of a particular set $\omega(N_r(x))$. It was proved in [9] that $F(M)$ contains a set F_2 of the second category that possesses the property that for $\varphi \in F_2$ there exists a second category set L in M such that, for $\xi \in L$,

$$P_\xi = \overline{\varphi([0, +\infty), \xi)}.$$

It follows from (2.2) that if $\varphi \in F_2$ and $\xi \in L$, then P_ξ^ω coincides with ω_ξ, which is the ω-limit set of ξ under φ. It is evident that $\omega_\xi \subset \omega(N_r(x))$ for $\xi \in \overline{N}_r(x)$. Therefore, if $\varphi \in F_1 \cap F_2$ and

$$\partial^T(N_r(x)) \cap L = \varnothing$$

for some $x \in M$ and $r > 0$, then $\omega(N_r(x))$ is Lyapunov stable (this follows from Lemma 3.1).

We now proceed to the proof of Theorem 2. The following fact is well known (see, for example, [1]). Let φ be a flow generated by a system of differential equations of class C^1 that satisfy axiom A and the strict transversality condition. Then the nonwandering set Ω of φ admits a unique representation as a union of finite sets $\Omega_1, \ldots, \Omega_m$, each of which is closed and invariant and contains a dense half-trajectory; these Ω_i, $i = 1, \ldots, m$, are called basis sets.

Let $W^s(\Omega_i)$ and $W^u(\Omega_i)$ be the stable and unstable "manifolds" of the basis set Ω_i. Let Ω_i and Ω_j be different basis sets. We write $\Omega_i \to \Omega_j$ if $W^u(\Omega_i) \cap W^s(\Omega_j) \neq \varnothing$.

It is known [10], that

$$M = \bigcup_{i=1}^m W^s(\Omega_i) = \bigcup_{i=1}^m W^u(\Omega_i) \tag{3.2}$$

and the following statements are equivalent:

(1) $\Omega_i \to \Omega_j$;
(2) $\overline{W^u(\Omega_i)} \supset W^u(\Omega_j)$;
(3) $\overline{W^u(\Omega_i)} \cap W^u(\Omega_j) \neq \varnothing$.

Let m^* be a collection of subsets of $\{1, \ldots, m\}$. For $a \in m^*$ we put

$$b(a) = a \cup \{j \in \{1, \ldots, m\} : \exists i \in a : \Omega_i \to \Omega_j\}.$$

The domain of attraction $\Omega(I)$ of an attractor I is defined by
$$D(I) = \left\{x \in M : \varphi(t, x) \xrightarrow[t\to\infty]{} I\right\}.$$

LEMMA 3.3. *Let $a \in m^*$. Then*
$$I = \bigcup_{i \in a} \overline{W^u(\Omega_i)}$$
is an attractor and its domain of attraction $D(I)$ is
$$\bigcup_{j \in b(a)} W^s(\Omega_j).$$

PROOF. It is clear that I is an invariant compact set that is the union of the sets $W^u(\Omega_j)$ over all $j \in b(a)$. We will first show that I is Lyapunov stable. Suppose the contrary. Let $\Delta > 0$ be the least distance from Ω_j, $j \notin b(a)$, to I. There exist $\varepsilon \in (0, \Delta/2)$, a sequence of points $\xi_k \xrightarrow[k\to\infty]{} I$, and a sequence of numbers $t_k > 0$ such that $d(\varphi(t_k, \xi_k), I) \geq \varepsilon$. Choose t_k such that
$$\begin{aligned} d(\varphi(t, \xi_k), I) &< \varepsilon, \quad t \in [0, t_k), \\ d(\varphi(t_k, \xi_k), I) &= \varepsilon. \end{aligned} \quad (3.3)$$
Clearly, $t_k \xrightarrow[k\to\infty]{} +\infty$. Let $y \in \partial N_\varepsilon(I)$ be a limit point of the sequence $\varphi(t_k, \xi_k)$. It follows from (3.2) that $y \in W^u(\Omega_j)$ for some basis set Ω_j. If $j \in b(a)$, then $y \in I$ by the definition of I. This contradicts $d(y, I) = \varepsilon$. Thus, $j \notin b(a)$. There exists $\tau < 0$ such that
$$d(\varphi(\tau, y), \Omega_j) < \varepsilon/2,$$
but then, for sufficiently large k,
$$d(\varphi(t_k + \tau, \xi_k), \Omega_j) < \varepsilon/2, \quad (3.4)$$
and at the same time $t_k + \tau \in [0, t_k)$, contrary to (3.3). Thus I is stable.

By the main theorem in [11], there exists $\delta > 0$ such that, if there exist points $x \in W^s(\Omega_j)$, $y \in W^u(\Omega_i)$ with $d(x, y) < \delta$, then $\Omega_i \to \Omega_j$. Therefore, if $x \in W^u(\Omega_j)$, $j \in b(a)$, then the ω-neighborhood of x contains no points of $W^s(\Omega_i)$, $i \in b(a)$. Thus, all the trajectories in $N_\delta(I)$ tend to I as $t \to \infty$, that is, I is an attractor. The equality for $D(I)$ is obvious.

LEMMA 3.4. *Let $G \subset M$ be an open set and let $a \in m^*$ be a set such that*
$$\overline{G} \subset \bigcup_{i \in b(a)} W^s(\Omega_i), \quad G \cap W^s(\Omega_i) \neq \varnothing, \quad i \in a.$$
Then
$$\omega(G) = \bigcup_{i \in a} \overline{W^u(\Omega_i)}.$$

PROOF. The set $I = \bigcup_{i \in a} \overline{W^u(\Omega_i)}$ is an attractor by Lemma 3.3, and $\overline{G}$ is a compact subset of $D(I)$; hence $\omega(G) \subset I$. We prove the inverse inclusion.

Consider one of the basis sets Q_i, $i \in a$. Fix a point $x \in G \cap W^s(\Omega_i)$. It follows from [12] that there is a trajectory $\gamma \in \Omega_i$ such that $x \in W^s(\gamma)$. Since G is open and the closed trajectories are dense in Ω_i, we may assume that γ is closed (we are considering the generic case, when Ω_i is not a stationary point). Then by the λ-lemma [1]

$$W^u(\gamma) \subset \omega(G).$$

But it is well known that $W^u(\Omega_i) \subset \overline{W^u(\gamma)}$ for all trajectories $\gamma \in \Omega_i$. This implies the lemma.

To complete the proof of Theorem 2, fix a point $x \in M$. Let Ω_i be one of the basis sets. Consider $r_i > 0$ such that

$$\begin{aligned} \overline{N_{r_i}(x)} \cap W^s(\Omega_i) &\neq \varnothing, \\ N_{r_i}(x) \cap W^s(\Omega_i) &= \varnothing. \end{aligned} \qquad (3.5)$$

It is evident that

$$\overline{N_r(x)} \cap W^s(\Omega_i) = \varnothing$$

for $r < r_i$ and

$$N_r(x) \cap W^s(\Omega_i) \neq \varnothing$$

for $r > r_i$. Therefore, (3.5) may be true for at most one number r_i. Let $C(x)$ be the corresponding set of numbers $r_1, \ldots, r_m$, for all basis sets $\Omega_1, \ldots, \Omega_m$, and let $r \in (0, +\infty) \backslash C(x)$. Let $a \in m^*$ be the set of i such that

$$\overline{N_r(x)} \cap W^s(\Omega_i) \neq \varnothing.$$

Since $r \notin C(x)$, it follows that for all $i \in a$

$$N_r(x) \cap W^s(\Omega_i) \neq \varnothing;$$

by the definition of a,

$$N_r(x) \subset \bigcup_{i \in a} W^s(\Omega_i);$$

therefore, by Lemma 3.4, the set $\omega(N_r(x))$ coincides with $\bigcup_{i \in a} \overline{W^u(\Omega_i)}$ and by Lemma 3.3 it is an attractor.

§4. Proofs of Theorems 3, 4

We introduce a certain characteristic of stable sets I in a flow φ. Define the number

$$\theta(I, \varphi) = \sup \left\{ \varlimsup_{k \to \infty} d(\varphi_k(t_k, x_k), I) \right\},$$

where the supremum is taken over all sequences of flow φ_k, points x_k, and numbers t_k such that

$$\rho_0(\varphi_k, \varphi) \xrightarrow[k \to \infty]{} 0, \quad x_k \xrightarrow[k \to \infty]{} I, \quad t_k \xrightarrow[k \to \infty]{} +\infty.$$

LEMMA 4.1. *If φ is a generic flow, then $\theta(I, \varphi) = 0$ for any stable set I.*

PROOF. Consider a flow φ in the set F_1 of Lemma 2.1. Suppose the statement of the lemma is false: there exists a Lyapunov stable set I such that $\theta(I, \varphi) > 0$. This means that there exists a number $a > 0$ and sequences $\varphi_k \to \varphi$, $t_k \to +\infty$, $x_k \to x_0 \in I$, such that

$$d(\varphi_k(t_k, x_k), I) \geq a$$

(this can always be achieved by considering a subsequence, if necessary). We will assume that $\varphi_k(t_k, x_k)$ converges to a point ξ. Then $d(\xi, I) \geq a$.

We show that $\xi \in P_{x_0}^\omega$. Using Lemma 2.4, construct a sequence of flows $\widetilde{\varphi}_k$ with the properties

$$\rho_0(\varphi_k, \widetilde{\varphi}_k) \xrightarrow[k \to \infty]{} 0,$$

$$\widetilde{\varphi}_k(t, x_0) = \varphi_k(t, x_k) \quad \text{for } t \geq 1.$$

Then $\rho_0(\widetilde{\varphi}_k, \varphi) \xrightarrow[k \to \infty]{} 0$, $\xi = \lim_{k \to \infty} \widetilde{\varphi}_k(t_k, x_0)$.

Now it follows from the stability of I that there exists $\delta > 0$ such that, for any $x \in N_\delta(I)$,

$$\varphi(t, x) \in N_{a/2}(I), \quad t \geq 0.$$

Then

$$Q_{x_0, \eta} \subset \overline{N_{a/2}(I)} \quad \text{for } \eta < \delta. \tag{4.1}$$

It follows from (4.1) that ξ does not belong to $Q_{x_0}^\omega$, contradicting the equality $Q_{x_0}^\omega = P_{x_0}^\omega$.

The following lemma follows from Lemma 3.2 in [13].

LEMMA 4.2. *Let I be an attractor and let K be a compact subset of the domain of attraction $D(I)$. Then for any $\varepsilon > 0$ there exists a neighborhood $H(k, \varepsilon)$ of φ such that, for any flow $\psi \in H(K, \varepsilon)$, $x_k \in K$, and $t_k \to +\infty$,*

$$\overline{\lim} \, d(\psi(t_k, x_k), I) < \varepsilon.$$

To prove the next lemma we need the concept of $(\varepsilon, 1)$-chain. Following [14], we will say that a flow φ has an $(\varepsilon, 1)$-chain beginning at x_0 and passing through x, if there exist points $x_1, \ldots, x_m$ and numbers $t_0, \ldots, t_{m-1}$ such that

$$t_i \geq 1, \quad i = 0, \ldots, m-1;$$
$$d(x_{i+1}, \varphi(t_i, x_i)) < \varepsilon, \quad i = 0, \ldots, m-1;$$
$$x \in \varphi([0, +\infty), x_m).$$

LEMMA 4.3. *Let I be a compact invariant set of an arbitrary flow φ. Then I is a quasiattractor if and only if $\theta(I, \varphi) = 0$.*

PROOF. 1. *Necessity.* Let I be a quasiattractor of φ, $I = \bigcap_k I_k$, where each I_k is an attractor for φ. Fix an arbitrary $\delta > 0$, and find an attractor I_m in $\{I_k\}$ such that $I_m \subset N_\delta(I)$. Choose a compact set K such that

$$I_m \subset \text{Int} \, K \subset K \subset N_\delta(I), \quad K \subset D(I_m).$$

Then there exists $\Delta > 0$ such that $N_\Delta(I) \subset \operatorname{Int} K$ and $N_\Delta(\varphi) \subset H(K, \delta)$ (the set $H(K, \delta)$ is defined in Lemma 4.2).

Consider arbitrary sequences φ_k, x_k, t_k satisfying the conditions $\rho_0(\varphi_k, \varphi) \to 0$, $x_k \in I$, $t_k \to +\infty$ as $k \to \infty$. There exists k_0 such that, if $k \geq k_0$, then $\varphi_k \in N_\Delta(I)$, $x_k \in N_\Delta(I)$. Therefore
$$\varlimsup d(\varphi_k(t_k, x_k), I) < \delta.$$
Consequently, $\theta(I, \varphi) < \delta$; but δ is arbitrary and so $\theta(I, \varphi) = 0$.

2. *Sufficiency.* We will apply a construction that was used for other purposes in [3]. Define Y_k, $k \geq 1$, to be the set of all points x for which there exist $(1/k, 1)$-chains beginning in I and passing through x.

It follows immediately from the definition that Y_k is open and positively invariant, i.e., $\varphi(t, x) \in Y_k$ for $x \in Y_k$ and $t \geq 0$.

We will show that for $\tau \geq 1$
$$N_{1/k}(\varphi(\tau, Y_k)) \subset Y_k. \tag{4.2}$$
Indeed, for $\xi \in \varphi(\tau, Y_k)$ there exist points $x_0, \ldots, x_m, y$ and numbers $t_0, \ldots, t_{m-1}, t$ such that $x_0 \in I$, $d(x_{i+1}, \varphi(t_i, x_i)) < 1/k$ for $i = 0, \ldots, m-1$; $t_i \geq 1$ for $i = 0, \ldots, m-1$; $y = \varphi(t, x_m)$, $t \geq 0$; $\xi = \varphi(\tau, y)$. Then any point $x \in N_{1/k}(\xi)$ can be taken as x_{m+1}, if we put $t_m = t + \tau$.

Let A_k denote the set
$$A_K = \bigcap_{t \geq 0} \varphi(t, Y_k). \tag{4.3}$$
We show that each of the sets A_k is an attractor. We first prove that A_k is invariant. Fix $z \in A_k$. For any $t \geq 0$ there exists $y \in Y_k$ such that $z = \varphi(t, y)$. Therefore $\varphi(t, z) \in Y_k$ for $t \leq 0$. Since Y_k is positively invariant, $\varphi(t, z) \in Y_k$ for $t \geq 0$. Now for any $\tau \in \mathbb{R}$, $t \geq 0$, we can write $\varphi(\tau, z) = \varphi(t, \varphi(\tau - t, z))$. Since $\varphi(\tau - t, z) \in Y_k$, we obtain $\varphi(\tau, z) \in A_k$, that is, A_k is invariant.

We now prove that A_k is closed. Fix a sequence $z_m \in A_k$, $z_m \xrightarrow[m \to \infty]{} z$. Let $t \geq 0$ be an arbitrary number. Since A_k is invariant, there are points $y_m \in A_k$ (and, consequently, $y_m \in Y_k$) such that $z_m = \varphi(t+1, y_m)$. Denote $\tilde{y}_m = \varphi(1, y_m)$. Let $\tilde{y}$ be a limit point of the sequence $\tilde{y}_m$. We may assume that $\tilde{y}_m \to \tilde{y}$ as $m \to \infty$. Since $\tilde{y}_m \in \varphi(1, Y_k)$, it follows that $\tilde{y} \in \overline{\varphi(1, Y_k)}$. For sufficiently large m we have $d(\tilde{y}_m, \tilde{y}) < 1/k$, and so $\tilde{y} \in N_{1/k}(\varphi(1, Y_k))$; but then it follows from (4.2) that $\tilde{y} \in Y_k$. Letting $m \to \infty$ in the equality $z_m = \varphi(t, \tilde{y}_m)$, we get $z = \varphi(t, \tilde{y})$. Since t is arbitrary, $z \in A_k$.

Thus, A_k is an invariant compact subset of the open set Y_k, and also (4.3) holds. We know that in that case A_k is an attractor [3].

Finally, we prove that the equality $\theta(I, \varphi) = 0$ implies $I = \bigcap_k A_k$.

If $x \in I$, then x, together with any point $\varphi(t, x)$, belongs to each Y_k. Therefore x belongs to each A_k. Hence $I \subset \bigcap_k A_k$.

To prove the inverse inclusion, suppose it is not true. Fix $y \in \bigcap_k A_k \setminus I$. Let $d(y, I) = a > 0$. Since $A_k \subset Y_k$, it follows that for any $k > 0$ there exists a $(1/k, 1)$-chain beginning at $\xi_k \in I$ and passing through y. Using the technique of Lemma 2.4, we construct a sequence of flows φ_k such that $\rho_0(\varphi_k, \varphi) \to 0$ as $k \to \infty$ and the trajectory of φ_k starting at ξ_k passes through y. If $y = \varphi_k(t_k, \xi_k)$ then clearly $t_k \xrightarrow[k \to \infty]{} +\infty$ and therefore $\theta(I, \varphi) \geq a > 0$. This contradiction proves the lemma.

Theorem 3 is a corollary of Lemmas 4.1 and 4.3.

REMARK. We have proved Lemma 4.3 for arbitrary flows, whereas Lemma 4.1 is valid for generic flows only (an example of a flow and a stable set for which Lemma 4.1 is not true is a stationary point that is a center in the plane).

The proof of Theorem 4 is indirect. Let I be a Lyapunov stable set for a flow φ generated by a system of differential equations of class C^1 that satisfies Axiom A and the strict tranversality condition. Suppose that I is not an attractor. Then in any neighborhood of I there is a closed invariant set disjoint from I [15]. Fix a constant $\delta > 0$ with the property that if ξ, η are points of the nonwandering set Ω of φ such that $d(z, y) < \delta$ for some points $z \in W^s(\varphi(t, \xi))$, $y \in W^u(\varphi(t, \eta))$, then

$$W^s(\varphi(t, \xi)) \cap W^u(\varphi(t, \eta)) \neq \varnothing.$$

The existence of such a constant is proved in [11].

Choose an invariant set K disjoint from I, such that $\overline{K} \subset N_\delta(I)$. Fix a point $\eta \in \Omega \cap \overline{K}$ and a neighborhood V of η contained in $N_\delta(I)$ such that $\overline{V} \cap I = \varnothing$. The stationary points and the closed trajectories are dense in Ω. Let η_0 be either a stationary point or a point on a closed trajectory in V. By the choice of V, the trajectory $\varphi(t, \eta_0)$ does not belong to I. Let $\xi \in I$ such that $d(\eta_0, \xi) < \delta$. There exists $z \in \Omega$ such that $\zeta \in W^u(\varphi(t, z))$. Since

$$d(\varphi(t, z), \varphi(t, \zeta)) \xrightarrow[t \to \infty]{} 0,$$

$\varphi(t, \zeta) \in I$, and I is stable, it follows that $z \in I$. Hence $W^u(\varphi(t, z)) \subset I$. If Ω_i is a basis set containing z, then $\Omega_i \subset I$, since $W^u(\varphi(t, z))$ is dense in Ω_i.

Choose a point $\zeta_0 \in \Omega_i$ that is either stationary or lies on a closed trajectory, such that

$$d(\eta_0, W^u(\zeta_0)) < \delta.$$

By the choice of δ,

$$W^u(\varphi(t, \eta_0)) \cap W^u(\zeta_0) \neq \varnothing.$$

Then, by the λ-lemma,

$$W^u(\varphi(t, \eta_0)) \subset \overline{W^u(\zeta_0)} \subset I,$$

but then also $\eta_0 \in I$. This contradiction proves the theorem.

§5. Proofs of Theorems 5, 6

Let G be an arbitrary set in a manifold M. Fix a flow φ and define the set

$$\omega^*(G) = \overline{\bigcup_{x \in \overline{G}} P_x^\omega}.$$

If follows from the definition that $\omega^*(G)$ is invariant and compact.

LEMMA 5.1. *For any flow φ and any set G, the set $\omega^*(G)$ is Lyapunov stable.*

PROOF. Suppose the contrary. Then there exist sequences of points ξ_k, y_k, a sequence of numbers θ_k, and points $\xi \in \omega^*(G)$, $y \notin \omega^*(G)$ such that

$$y_k = \varphi(\theta_k, \xi_k), \quad \xi_k \to \xi, \quad y_k \to y, \quad \theta_k \to +\infty,$$

as $k \to \infty$. By the definition of $\omega^*(G)$, there exist points $x_k \in \overline{G}$ and $\eta_k \in P_{x_k}^\omega$ such that $\eta_k \to \xi$ as $k \to \infty$.

For each η_k there exist a sequence of numbers $t_m^k \xrightarrow[m \to \infty]{} +\infty$ and a sequence of flows $\varphi_m^k \xrightarrow[m \to \infty]{} \varphi$ such that

$$\eta_k = \lim_{m \to \infty} \varphi_m^k(t_m^k, x_k).$$

Choose a sequence $m(k) \xrightarrow[k \to \infty]{} \infty$ such that

$$d(\eta_k, \varphi_{m(k)}^k(t_{m(k)}^k, x_k)) < 1/k.$$

Then, since $\eta_k \xrightarrow[k \to \infty]{} \xi$,

$$\xi = \lim_{k \to \infty} \varphi_{m(k)}^k(t_{m(k)}^k, x_k).$$

Let $x \in \overline{G}$ be a limit point of the sequence x_k. Fix an arbitrary $\eta > 0$. Then there exists $k(\eta)$ such that for any $k \geq k(\eta)$ there is a flow $\psi \in N_\eta(\varphi)$ with the properties

$$\psi(t, x) = \varphi_{m(k)}^k(t, x_k), \quad t \in [1, t_{m(k)}^k - 1],$$
$$\psi(t_{m(k)}^k, x) = \xi,$$
$$\psi(t, \xi) = \varphi(t, \xi_k), \quad t \in [1, \theta_k].$$

(The existence of ψ follows from Lemma 2.4.) It is clear that $y \in P_x^\omega$, but this contradicts our assumption.

LEMMA 5.2. *For any $\varepsilon > 0$ there exists a neighborhood $U(\varepsilon)$ of $\omega(G)$ such that, for any neighborhood U of $\omega(G)$ with $\omega(G) \subset U \subset U(\varepsilon)$,*

$$r(\omega(U), \omega^*(G)) < \varepsilon$$

(recall that the deviation of A from B was defined as $r(A, B) = \sup_{x \in A} d(x, B)$).

PROOF. Fix $\varepsilon > 0$. By Lemma 5.1, there exists $\delta > 0$ such that for $x \in N_\delta(\omega^*(G))$
$$d(\varphi(t, x), \omega^*(G)) < \varepsilon/2, \quad t \geq 0.$$
The inclusion $\omega(G) \subset \omega^*(G)$ is always true. Indeed, let $y \in \omega(G)$. Then there exist points $x_k \in G$ and numbers $t_k \to +\infty$ such that $y = \lim_{k \to \infty} \varphi(t_k, x_k)$. Let $x \in \overline{G}$ be a limit point of the sequence x_k. Then $y \in Q_x^\omega$, but since always $Q_x^\omega \subset P_x^\omega$, it follows that $y \in \omega^*(G)$.

There clearly exists a neighborhood of $\omega(G)$ contained entirely in $N_\delta(\omega^*(G))$; call it $U(\varepsilon)$. This is the desired neighborhood. Indeed, if $U \subset U(\varepsilon)$ and $z \in \omega(U)$, there exist points $\xi_k \in U$ and numbers $\tau_k \to +\infty$ such that $\varphi(\tau_k, \xi_k) \xrightarrow[k \to \infty]{} z$. But by the choice of δ,
$$d(\varphi(\tau_k, \xi_k), \omega^*(G)) < \varepsilon/2.$$
Consequently,
$$d(z, \omega^*(G)) \leq \varepsilon/2 < \varepsilon.$$

LEMMA 5.3. *Let $\varphi \in F_1$. Then for any set G and any neighborhood U of $\omega(G)$,*
$$\omega^*(G) \subset \omega(U).$$

PROOF. Fix $z \in \omega^*(G)$ and a sequence $z_k \in \bigcup_{x \in \overline{G}} P_x^\omega$ that converges to z as $k \to \infty$. There exist points $x_k \in \overline{G}$, flows $\varphi_m^k \xrightarrow[m \to \infty]{} \varphi$, and numbers $t_m^k \xrightarrow[m \to \infty]{} +\infty$ and $m(k)$ such that
$$z_k = \varphi_{m(k)}^k(t_{m(k)}^k, x_k)$$
(see the proof of Lemma 5.1). Let x_0 be a limit point of x_k. As already noted in the remark following Theorem 1 in §2, the ω-limit set ω_{x_0} of the trajectory $\varphi(t, x_0)$ lies in $\omega(G)$.

Fix a neighborhood U of $\omega(G)$. Since $\omega_{x_0} \subset \omega(G)$, there exists $T > 0$ such that the point $y_0 = \varphi(T, x_0)$ belongs to U. For any $\eta > 0$, using Lemma 2.4, we choose $k(\eta)$ such that, for $k \geq k(\eta)$, there exists a flow $\psi \in N_\eta(\varphi)$ with the property
$$\psi(t, y_0) = \varphi_{m(k)}^k(T + t, x_k), \quad t \geq 1.$$
For sufficiently large k,
$$z_k = \psi(t_{m(k)}^k - T, y_0);$$
therefore, $z \in P_{y_0}^\omega$. However, $P_{y_0}^\omega = Q_{y_0}^\omega$ by the choice of φ. Therefore, $z \in Q_{y_0}^\omega$. Since y_0 is an interior point of U, it follows that $z \in \omega(U)$.

REMARK. Lemmas 5.1, 5.2 are true for arbitrary flows φ. The following example shows that Lemma 5.3 is not true for an arbitrary flow.

Let M be the unit circle in the plane. The polar angle θ is the coordinate on M. Let φ be a flow with four stationary points on M: $\theta = 0, \pi/2, \pi, 3\pi/2$; on the arcs complementary to these points, the direction of the motion is the direction of increasing θ. Let G be the point $\{\theta = \pi/4\}$. Then $\omega(G)$ is clearly the stationary point $\{\pi/2\}$, and $\omega(U)$, for a small neighborhood U of $\omega(G)$, is the closed arc $\{\pi/2 \leq \theta \leq \pi\}$. But $\omega^*(G) = M$.

We now prove Theorem 5. Let G be an arbitrary set in M, $\varphi \in F_1$. By Lemma 5.1, $\omega^*(G)$ is a stable set. Then, by Theorem 3, it is a quasiattractor, and
$$\omega^*(G) = \bigcap_{m>0} I_m,$$
where each I_m is an attractor for φ. Fix an arbitrary $\varepsilon > 0$. There exists an attractor I (one of the sets I_m) such that
$$\omega^*(G) \subset I \subset N_\varepsilon(\omega^*(G)).$$
Choose a compact set K such that
$$I \subset \operatorname{Int} K \subset K \subset D(I), \quad K \subset N_\varepsilon(\omega^*(G))$$
(recall that $D(I)$ is the domain of attraction for I). There exists $\Delta > 0$ such that
$$\overline{N_\Delta(I)} \subset \operatorname{Int} K. \tag{5.1}$$
To show that Δ is the desired number, we will prove by induction that for $k \geq 0$ and $\delta \in (0, \Delta)$
$$\omega_\delta^k(G) \subset I. \tag{5.2}$$
For $k = 0$ this follows from the inclusions
$$\omega(G) = \omega_\delta^0(G) \subset \omega^*(G) \subset I$$
(see the proof of Lemma 5.2). If it is already known that (5.2) is true for some k, then
$$\omega_\delta^{k+1}(G) = \omega(N_\delta(\omega_\delta^k(G))) \subset \omega(N_\delta(I)),$$
because $\omega(H)$ depends monotonically on H. By Lemma 4.2 and (5.1),
$$\omega(N_\delta(I)) \subset \omega(\overline{N_\delta(I)}) \subset I,$$
proving (5.2) for all k.

Let us consider two indices $k, \ell \geq 1$, say $k \geq \ell$. It follows from Lemma 5.3 and the previous considerations that
$$\omega^*(G) \subset \omega_\delta^\ell(G) \subset \omega_\delta^k(G) \subset N_\varepsilon(\omega^*(G)),$$
and the desired inequality
$$R(\omega_\delta^\ell(G), \omega_\delta^k(G)) < \varepsilon$$

follows. This completes the proof of Theorem 5.

We now prove Theorem 6. If $\omega(G)$ is stable, then by Theorem 4, it is an attractor. Then, for sufficiently small $\delta > 0$,
$$\omega_\delta^k(G) = \omega(G) \quad \text{for } k \geq 1.$$
But if $\omega(G)$ is not stable, we proceed as follows. Fix a basis set Ω_i. It is clear that there is at most one number $\delta > 0$ such that
$$\overline{N_\delta(\omega(G))} \cap W^s(\Omega_i) \neq \varnothing,$$
$$N_\delta(\omega(G)) \cap W^s(\Omega_i) = \varnothing.$$
Denote this number by δ_i. Choose $\Delta_1 \in (0, \min \delta_i)$. As in Theorem 2, one shows that there is a collection of basis sets $\Omega_{i_1}, \ldots, \Omega_{i_k}$ such that, for any $\delta \in (0, \Delta_1)$, the set $\omega_\delta(G)$ is an attractor and may be expressed as
$$\overline{W^u(\Omega_{i_1})} \cup \cdots \cup \overline{W^u(\Omega_{i_k})}.$$
Therefore, there exists $\Delta_2 > 0$ such that
$$\overline{N_\delta(\omega_\delta^1(G))} \subset D(\omega_\delta^1(G)) \quad \text{for } \delta \in (0, \Delta_2).$$
Clearly, for $\delta \in (0, \min(\Delta_1, \Delta_2))$
$$\omega_\delta^k(G) = \omega_\delta^1(G), \quad k = 2, 3, \ldots.$$
This completes the proof.

References

1. S. Yu. Pilyugin, *Introduction to structurally stable systems of differential equations*, Leningrad Univ., Leningrad, 1988.
2. M. De Oliveira, C^0-*density of structurally stable vector fields*, Bull. Amer. Math. Soc. **82** (1976), 786.
3. M. Harley, *Attractors: Persistence and density of their basins*, Trans. Amer. Math. Soc. **269** (1982), 247–271.
4. S. Yu. Pilyugin, *Limit sets of trajectories of domains in dynamical systems*, Funktsional. Analiz i Prilozhen. **23** (1989), no. 3, 82–83; English transl. in Functional Anal. Appl. **23** (1989).
5. V. A. Dobrynskiĭ, *Genericity of dynamical systems with stable prolongation*, Dynamical systems and questions of stability of the solutions of differential equations, Inst. Mat. Akad. Nauk Ukr. SSR, Kiev, 1973, pp. 43–53. (Russian)
6. N. V. Shcherbina, *On the continuity of one-parameter families of sets*, Dokl. Akad. Nauk SSSR **234** (1977), no. 2, 327–329; English transl. in Soviet Math. Dokl. **18** (1977), no. 3, 688–690.
7. K. Kuratowski, *Topology*, vol. 1,2, Academic Press, New York, 1966, 1968.
8. S. Yu. Pilyugin, C^0-*perturbations of attractors and boundary stability*, Differentsial′nye Uravneniya **22** (1986), no. 10, 1712–1719; English transl. in Differential Equations **22** (1986).
9. V. A. Dobrynskiĭ and A. N. Sharkovskiĭ, *Genericity of dynamical systems, almost all trajectories of which are stable under permanently acting perturbations*, Dokl. Akad. Nauk SSSR **211** (1973), no. 2, 273–276; English transl. in Soviet Math. Dokl. **14** (1973), no. 4, 997–1000.
10. S. Smale, *Differentiable dynamical systems*, Bull. Amer. Math. Soc. **73** (1967), 747–817.

11. V. A. Pliss, *The location of stable and unstable manifolds of hyperbolic systems*, Differentsial'nye Uravneniya **20** (1984), no. 5, 779–785; English transl. in Differential Equations **20** (1984).
12. M. W. Hirsch, J. Palis, G. Pugh, and M. Shub, *Neighbourhoods of hyperbolic sets*, Invent. Math. **2** (1970), no. 2, 121–134.
13. S. Yu. Pilyugin, *Attractors and systems without uniqueness*, Mat. Zametki **42** (1987), no. 5, 703–711; English transl. in Math. Notes **42** (1987).
14. C. Conley, *Isolated invariant sets and the Morse index*, CBMS Regional Conf. Ser. in Math., vol. 38, Amer. Math. Soc., Providence, R.I., 1978.
15. N. I. Zubov, *Stability of motion (Lyapunov methods and their application)*, "Vyssh. Shkola", Moscow, 1984. (Russian)

Translated by V. OPERSTEIN

Recent Titles in This Series

(Continued from the front of this publication)

116 **A. G. Kušnirenko, A. B. Katok, and V. M. Alekseev,** Three Papers on Dynamical Systems
115 **I. S. Belov, et al.,** Twelve Papers in Analysis
114 **M. Š. Birman and M. Z. Solomjak,** Quantitative Analysis in Sobolev Imbedding Theorems and Applications to Spectral Theory
113 **A. F. Lavrik,** Twelve Papers in Logic and Algebra
112 **D. A. Gudkov and G. A. Utkin,** Nine Papers on Hilbert's 16th Problem
111 **V. M. Adamjan, et al.,** Nine Papers on Analysis
110 **M. S. Budjanu, et al.,** Nine Papers on Analysis
109 **D. V. Anosov, et al.,** Twenty Lectures Delivered at the International Congress of Mathematicians in Vancouver, 1974
108 **Ja. L. Geronimus and Gábor Szegő,** Two Papers on Special Functions
107 **A. P. Mišina and L. A. Skornjakov,** Abelian Groups and Modules
106 **M. Ja. Antonovskiĭ, V. G. Boltjanskiĭ, and T. A. Sarymsakov,** Topological Semifields and Their Applications to General Topology
105 **R. A. Aleksandrjan, et al.,** Partial Differential Equations, Proceedings of a Symposium Dedicated to Academician S. L. Sobolev
104 **L. V. Ahlfors, et al.,** Some Problems on Mathematics and Mechanics, On the Occasion of the Seventieth Birthday of Academician M. A. Lavrent′ev
103 **M. S. Brodskiĭ, et al.,** Nine Papers in Analysis
102 **M. S. Budjanu, et al.,** Ten Papers in Analysis
101 **B. M. Levitan, V. A. Marčenko, and B. L. Roždestvenskiĭ,** Six Papers in Analysis
100 **G. S. Ceĭtin, et al.,** Fourteen Papers on Logic, Geometry, Topology and Algebra
99 **G. S. Ceĭtin, et al.,** Five Papers on Logic and Foundations
98 **G. S. Ceĭtin, et al.,** Five Papers on Logic and Foundations
97 **B. M. Budak, et al.,** Eleven Papers on Logic, Algebra, Analysis and Topology
96 **N. D. Filippov, et al.,** Ten Papers on Algebra and Functional Analysis
95 **V. M. Adamjan, et al.,** Eleven Papers in Analysis
94 **V. A. Baranskiĭ, et al.,** Sixteen Papers on Logic and Algebra
93 **Ju. M. Berezanskiĭ, et al.,** Nine Papers on Functional Analysis
92 **A. M. Ančikov, et al.,** Seventeen Papers on Topology and Differential Geometry
91 **L. I. Barklon, et al.,** Eighteen Papers on Analysis and Quantum Mechanics
90 **Z. S. Agranovič, et al.,** Thirteen Papers on Functional Analysis
89 **V. M. Alekseev, et al.,** Thirteen Papers on Differential Equations
88 **I. I. Eremin, et al.,** Twelve Papers on Real and Complex Function Theory
87 **M. A. Aĭzerman, et al.,** Sixteen Papers on Differential and Difference Equations, Functional Analysis, Games and Control
86 **N. I. Ahiezer, et al.,** Fifteen Papers on Real and Complex Functions, Series, Differential and Integral Equations
85 **V. T. Fomenko, et al.,** Twelve Papers on Functional Analysis and Geometry
84 **S. N. Černikov, et al.,** Twelve Papers on Algebra, Algebraic Geometry and Topology
83 **I. S. Aršon, et al.,** Eighteen Papers on Logic and Theory of Functions
82 **A. P. Birjukov, et al.,** Sixteen Papers on Number Theory and Algebra
81 **K. K. Golovkin, V. P. Il′in, and V. A. Solonnikov,** Four Papers on Functions of Real Variables
80 **V. S. Azarin, et al.,** Thirteen Papers on Functions of Real and Complex Variables

(See the AMS catalog for earlier titles)